Altering Nature

Philosophy and Medicine

VOLUME 97

Founding Co-Editor
Stuart F. Spicker

Senior Editor

H. Tristram Engelhardt, Jr., *Department of Philosophy, Rice University, and Baylor College of Medicine, Houston, Texas*

Associate Editor

Lisa M. Rasmussen, *Department of Philosophy, University of North Carolina at Charlotte, Charlotte*

Editorial Board

George J. Agich, *Department of Philosophy, Bowling Green State University, Bowling Green, Ohio*
Nicholas Capaldi, *College of Business Administration, Loyola University New Orleans, New Orleans, Louisiana*
Edmund Erde, *University of Medicine and Dentistry of New Jersey, Stratford, New Jersey*
Christopher Tollefsen, *Department of Philosophy, University of South Carolina, Columbia, South Carolina*
Kevin Wm. Wildes, S.J., *President Loyola University, New Orleans, New Orleans, Louisiana*

For other titles published in this series, go to
www.springer.com/series/6414

B. Andrew Lustig • Baruch A. Brody
Gerald P. McKenny
Editors

Altering Nature

Volume One: Concepts of 'Nature'
and 'The Natural' in Biotechnology Debates

Editors
B. Andrew Lustig
Davidson College

Baruch A. Brody
Baylor College of Medicine

Gerald P. McKenny
University of Notre Dame

ISBN 978-1-4020-6920-8   e-ISBN 978-1-4020-6921-5
DOI: 10.1007/978-1-4020-6921-5

Library of Congress Control Number: 2008925365

© 2008 Springer Science + Business Media B.V.
No part of this work may be reproduced, stored in a retrieval system, or transmitted in any form or by any means, electronic, mechanical, photocopying, microfilming, recording or otherwise, without written permission from the Publisher, with the exception of any material supplied specifically for the purpose of being entered and executed on a computer system, for exclusive use by the purchaser of the work.

Printed on acid-free paper

9 8 7 6 5 4 3 2 1

springer.com

Contents

Contributors vii

Introduction 1
B. Andrew Lustig, Baruch A. Brody, and Gerald P. McKenny

1 Spiritual and Religious Concepts of Nature 13
Aaron L. Mackler, Ebrahim Moosa, Allen Verhey,
Anne Carolyn Klein, and Kurt Peters

2 Philosophical Approaches to Nature 63
John H. Zammito, Philip J. Ivanhoe,
Helen Longino, and Phillip R. Sloan

**3 Scientific and Medical Concepts of Nature
   in the Modern Period in Europe and North America** 137
Laurence B. McCullough, John Caskey,
Thomas R. Cole, and Andrew Wear

**4 Ethical Challenges of Patenting "Nature":
   Legal and Economic Accounts of Altered
   Nature as Property** 199
Mary Anderlik Majumder, Margaret M. Byrne,
Elias Bongmba, Leslie S. Rothenberg,
and Nancy Neveloff Dubler

**5 Technogenesis: Aesthetic Dimensions
   of Art and Biotechnology** 275
Suzanne Anker, Susan Lindee, Edward A. Shanken,
and Dorothy Nelkin

Index 323

Contributors

Suzanne Anker, MFA, is Chair of the Fine Arts Department at the School of Visual Arts in New York City.

Elias Bongmba, Ph.D., is a Professor in the Religious Studies Department at Rice University.

Baruch A. Brody, Ph.D., is Leon Jaworski Professor of Biomedical Ethics and Director of the Center for Medical Ethics and Health Policy at Baylor College of Medicine. He is also the Andrew Mellow Professor of Humanities in the Department of Philosophy at Rice University.

Margaret M. Byrne, Ph.D., is a Research Associate Professor in the Department of Epidemiology and Public Health, Miller School of Medicine, University of Miami.

John Caskey, M.D., Ph.D., is currently in residency training in Pediatrics, Psychiatry, and Child Psychiatry at Brown University's School of Medicine.

Thomas R. Cole, Ph.D., is the McGovern Chair in Medical Humanities and Director of the McGovern Center for Health, Humanities, and the Human Spirit at the University of Texas Health Sciences Center in Houston. He is also a Professor of Humanities in the Department of Religious Studies at Rice University.

Nancy Neveloff Dubler, LL.B., is the Director of the Division of Bioethics and the Center for Ethics and Law in Medicine, Department of Family and Social Medicine, Montefiore Medical Center, and a Professor of Bioethics at The Albert Einstein College of Medicine.

Philip J. Ivanhoe, Ph.D., is Reader-Professor of Philosophy at City University of Hong Kong and an Adjunct Professor of Philosophy at Boston University.

Anne Carolyn Klein, Ph.D., is a Professor of Religious Studies at Rice University and a Founding Director of Dawn Mountain Tibetan Temple, Community Center, and Research Institute.

Susan Lindee, Ph.D., is a Professor in the Department of the History and Sociology of Science at the University of Pennsylvania.

Helen Longino, Ph.D., is a Professor of Philosophy at Stanford University.

B. Andrew Lustig, Ph.D., is the Holmes Rolston III Professor of Religion and Science at Davidson College.

Aaron L. Mackler, Ph.D., is an Associate Professor in the Department of Theology at Duquesne University.

Mary Anderlik Majumder, J.D., Ph.D., is an Assistant Professor of Medicine in the Center for Medical Ethics and Health Policy at Baylor College of Medicine.

Laurence B. McCullough, Ph.D., is a Professor of Medicine and Medical Ethics in the Center for Medical Ethics and Health Policy at Baylor College of Medicine.

Gerald P. McKenny, Ph.D., is an Associate Professor in the Department of Theology at the University of Notre Dame.

Ebrahim Moosa, Ph.D., is an Associate Professor of Islamic Studies in the Department of Religion and Associate Director of Research at the Duke Islamic Studies Center

Dorothy Nelkin, B.A., was a University Professor in the Department of Sociology at New York University until her death in 2003.

Kurt Peters, Ph.D., is an Associate Professor of Ethnic Studies and Director of the Native American Collaborative Institute at Oregon State University.

Leslie S. Rothenberg, J.D., is Professor Emeritus of Clinical Medicine at the David Geffen School of Medicine at UCLA and a clinical ethicist at Kaiser Permanente, Los Angeles.

Edward A. Shanken, Ph.D., is a Visiting Scholar at the California NanoSystems Institute and a Senior Researcher in the Art/Sci Lab at the University of California, Los Angeles.

Phillip R. Sloan, Ph.D., is Professor and former Chair of the Program of Liberal Studies and a Professor in the Graduate Program in History and Philosophy of Science at the University of Notre Dame.

Allen Verhey, Ph.D., is Professor of Christian Ethics at the Duke University Divinity School.

Andrew Wear, Ph.D., is Emeritus Reader in Medicine at the Wellcome Trust Centre for the History of Medicine at University College London.

John H. Zammito, Ph.D., is the John Antony Weir Professor in the Department of History at Rice University.

Introduction

B. Andrew Lustig, Baruch A. Brody, and Gerald P. McKenny

Nearly every week the general public is treated to an announcement of another actual or potential "breakthrough" in biotechnology. Headlines trumpet advances in assisted reproduction, current or prospective experiments in cloning, and developments in regenerative medicine, stem cell technologies, and tissue engineering. Scientific and popular accounts explore the perils and the possibilities of enhancing human capacities by computer-based, biomolecular, or mechanical means through advances in artificial intelligence, genetics, and nanotechnology. Reports abound concerning ever more sophisticated genetic techniques being introduced into agriculture and animal husbandry, as well as efforts to enhance and protect biodiversity. Given the pace of such developments, many insightful commentators have proclaimed the 21st century as the "biotechnology century."

Despite a significant literature on the morality of these particular advances in biotechnology, deeper ethical analysis has often been lacking. Our preliminary review of that literature suggested that current discussions of normative issues in biotechnology have suffered from two major deficiencies. First, the discussions have been too often piecemeal in character, limited to after-the-fact analyses of particular issues that provoked the debate, and unconnected to larger concepts and themes. Second, a crucial missing element of those discussions has been the failure to reflect explicitly on the diverse disciplinary conceptions of nature and the natural that shape moral judgments about the legitimacy of specific forms of research and their applications. Often, only minimal connections are drawn between judgments about particular issues and the deeper conceptual discussions of nature and the natural. These minimal connections tend to reflect simplistic readings rather than richer understandings informed by the diversity of perspectives and interpretations actually found within particular disciplines and modes of discourse. This crucial point deserves further elaboration.

Both current and future developments in biotechnology pose normative questions of profound importance. A careful reading of normative debates on specific areas of biotechnology – on assisted reproduction, genetics and transgenics, human-machine incorporation technologies, and others – reveals that contrasting normative responses to biotechnology – positive or negative – are informed by different conceptual interpretations of nature and the natural, with differing implications for judgments about the rightness or wrongness of particular interventions or "alterations" of nature. These different interpretations often emerge from different disciplinary perspectives and modes of discourse. Nevertheless, despite the fundamental role played by these different interpretations of nature, there has been insufficient discussion or analysis of the various ways that nature and the "natural" are described in different modes of discourse and the relevance of those different descriptive accounts for invocations of nature as a normative appeal, as well as the relevance of such understandings to judgments about particular biotechnological applications. The result is that, while various concepts of nature and the natural underlie debates about biotechnology, key questions remain either unanswered or insufficiently addressed. First, precisely how are nature and the natural understood from various disciplinary perspectives? Second, in what ways, if at all, do particular interpretations of nature have prescriptive implications? Third, what moral insights may be gained not only from specific disciplinary vantages but also from cross-disciplinary comparisons of the ways that concepts of nature function, either in contrasting or in complementary ways? Finally, what implications for public policy discussions and choices emerge as most supportable in light of that rich diversity of perspectives on concepts of nature and the natural?

The research represented in this volume, and a second volume that accompanies it, emerges from a multi-year research project funded by the Ford Foundation and co-sponsored by Rice University, Baylor College of Medicine's Center for Medical Ethics and Health Policy, and Davidson College. The first volume, introduced here, assembles the research from five scholarly groups who explored the ways that concepts of nature and "the natural" frame debates about biotechnology from particular disciplinary perspectives and modes of discourse. The second volume, which will be separately introduced, studies four broad areas of biotechnology – assisted reproduction, genetic therapy, human-machine incorporation, and biodiversity – in light of the research on various concepts of nature in Volume One, and considers the possible implications of cross-disciplinary understandings of those concepts for public policy discussions and choices.

In the present volume, as noted, five groups of scholars explore characteristic meanings of nature and the natural from different disciplinary perspectives. We will review each group's research in more detail below, but a few preliminary remarks are in order here:

1. *Conceptual Framework for the Analyses*. Each of the groups was provided with four broad categories of questions about concepts of nature and the natural: ontological questions, epistemological issues, moral questions, and aesthetic and representational concerns. Ontological questions included concerns about how nature is perceived and described. To what extent is nature seen as fundamentally ordered, broken/fallen, or random? To what extent is nature, considered as a whole or in its parts, viewed in dynamic or static terms? To what extent is nature, taken as whole or in its parts, seen to exhibit *telos*? In light of such considerations, then, what are the relations – if any – between natural teleology and human desires and aims? Epistemological questions focused on the extent to which nature is seen as accessible to human reason and to what degree nature is

deemed a trustworthy source of insight. Moral questions included the following: What is the relationship between humankind and nature (including animals, plants, and environmental niches)? What is the appropriate scope of human intervention into nature? Should such interventions be seen as instances of progress? What is the relationship between nature and understandings of human nature? Are questions of the scale and size of the natural order relevant to interpretations of that relationship? Economic and legal questions focused on such issues as: What is the relationship of nature to human society, politics, and economics? Is nature construed as property? As common heritage? As a commodity? Can it be patented? Finally, aesthetic issues raised the following sorts of questions: To what extent do the arts represent nature as a "given" and what are the moral implications, if any, of such accounts? To what extent do the arts offer accounts of nature as self-consciously constructed, idealized, or abstracted, and what are the moral implications of those accounts? Is nature represented primarily as a source of beauty and goodness, with moral implications? Or is nature represented primarily as a source of chaos and entropy that must be corrected or restored through human intervention?

2. *Complexity in the Concept of Nature*. It was understood by all of the groups that the connections between discussions of specific topics in biotechnology and broader understandings of nature and the natural are beset by complexity precisely because there is such a variety of understandings about that to which those concepts refer. In one useful classificatory scheme, nature is defined according to six categories: (1) everything that exists, including humans; (2) the non-human part of the cosmos; (3) that which occurs without human interference (4) that which is beyond the reach of human agents (i.e., nature in contrast to history and culture); (5) that which is beyond the reach of ordinary human powers (for example, theologically nature stands in contrast to the supernatural or to grace); and (6) the character or structure of a thing, the kind of thing it is (Gregorios, 1978). There are still other senses of "nature." But even a brief listing of the variety of usages warrants the conclusion that there are significant possibilities of confusion about what is being discussed. 'Nature' carries no univocal descriptive meaning.

3. *Historical and Cross-Cultural Considerations*. We charged our groups of scholars to engage in broadly cross-traditional and cross-temporal analyses. Because our research focused on recent developments in biotechnology, the five groups generally limited their historical and cross-cultural inquiry to constructs of nature and the natural during the last 200 or 300 years. However, to avoid being arbitrarily selective, groups also focused on earlier understandings of nature insofar as they continue to influence current discussions. Obviously, as each group worked out its own research agenda, some questions and issues were deemed more relevant to its specific focus. For example, the scientific/medical group and the philosophical group focused greater attention on the ways that particular accounts of nature in their disciplines reflect diverse historical understandings. The religious-spiritual group, while not indifferent to historical variability, also sought to identify patterns of reasoning about nature that serve to

distinguish particular religious and spiritual traditions in characteristic ways across time. Moreover, for traditions that lack explicit concepts of nature and the natural, the religious-spiritual group considered other understandings or families of ideas that provide the closest functional equivalent to ideas of nature and the natural. The aesthetic group, by contrast, chose to focus its research on recent developments – ones that incorporate biomaterials – and, through that prism, to reflect in ways that were more thematic than historical or cross-traditional. The legal-economic group provided both sorts of emphasis by combining a richly historical overview of interpretations of nature as property with a careful analysis of genetic patenting in different traditions.

In short, situating issue-oriented discussions of biotechnology in broader conceptual and normative contexts emerged as a quite complex task. It required both breadth and depth of scholarly attention across a range of epistemological, ontological, moral, and aesthetic issues. It required attention to the many different intuitive conceptions of nature. And it required attention to historical and cross-tradition complexities, especially when a particular interpretation of nature assumes priority with insufficient justification. Keeping these complexities in mind, we turn now to a more detailed discussion of each of the five chapters of this volume of "Altering Nature."

Chapter 1: Religious and Spiritual Perspectives on Nature as a Norm

The first chapter considers concepts of nature as they are understood and invoked in a range of religious and spiritual traditions. It does so by highlighting the conceptual, historical, and methodological touchstones in those traditions that shape their perspectives on concepts of nature and the natural. It is especially concerned with the ways that such concepts function both to authorize and to constrain judgments about divine purposes for the natural order of creation (in theistic traditions) and for judgments about appropriate human responsibility vis-à-vis that order (in all spiritual traditions). The chapter first surveys indigenous religions and spiritualities, worldviews that are often judged to be incompatible with modern, primarily Western, scientific perspectives. The authors identify the conceptual difficulties in analyzing "religion" as a separable dimension of indigenous experience, which is generally understood by indigenous groups in holistic terms. They draw on Native American and African indigenous perspectives as relevant examples of such holism, and identify several characteristic themes of indigenous life that generate a broadly critical resistance to modern, primarily Western, approaches to the "mastery" of nature as a general attitude and to biotechnological applications in particular. Indigenous perspectives view the universe as a living harmonious unity rather than merely an object of mechanical manipulation; indigenous perspectives also underscore a tradition-centered sense of "place" and an intimacy with all of life that resists alterations of what is naturally given.

The chapter then surveys religions of Asian origin, including Hinduism and Buddhism. In each instance, the authors sketch general characteristics of the traditions that frame specific interpretations of nature. In major Hindu texts, nature is often construed as the "body of God," with the implications that all of nature, including human nature, is inseparably related to God, and that all efforts at materialistic reductionism are fundamentally misguided. Nonetheless, the central ethical commitment to the relief of suffering (*ahimsa*) also appears to generate a responsible use of biotechnologies, if that use emerges as an expression of religious duty rather than of reductionistic mastery. As a result of the possible tensions between Hinduism's emphasis on the spiritual unity of God, nature, and human nature and its ethical imperative to relieve suffering, the authors conclude that the complexity of Hindu thought and practice does not allow either blanket affirmations or condemnations of developments in biotechnology.

Buddhism, while emerging from the context of Hinduism, is distinguished by its emphasis on the fundamental reality of impermanence rather than of an enduring spiritual self, and the "inherent mutuality" of all apparently separate substances. For Buddhism, as with Hinduism, the underlying assumptions at work in judgments about "altering nature" must be reframed, since that phrase assumes an act of imposition upon nature as an object of manipulation, rather than nature as a term that describes all of existence. At the deepest level, "nature" cannot be fundamentally altered, in the sense of undoing impermanence. Hence, Buddhist judgments about particular biotechnologies will be made on the grounds of what the virtues of compassion and right-mindedness require in particular circumstances.

The authors then survey the Abrahamic traditions of Judaism, Christianity, and Islam. The concepts of nature and the natural function more centrally in these traditions, in part because the Abrahamic faiths are more closely linked historically with the rise of science and technology. Moreover, these traditions are richly casuistic, with significant resources that are both directed toward, and developed further in response to, modern technologies. At the same time, the Abrahamic traditions generate a broad range of complex and sometimes conflicting judgments about the trustworthiness of nature as a source of moral insight. These general interpretations serve to shape specific responses – both positive and negative – to developments in biotechnology. Such broad perspectives on nature as a theological/moral category are themselves best understood as reflecting distinctive understandings of community authority, appeals to canonical scriptures, and methods of moral reasoning and argument. As with the Asian traditions, our authors identify a wide range of perspectives on nature, and on the alteration of nature, that also can be found in the Abrahamic traditions.

While our authors analyze the concepts of nature and the natural as these are understood in indigenous religious traditions, the Asian traditions, and the Abrahamic faiths, the chapter also emphasizes that the central normative issues are not simply whether it is appropriate to alter nature or to what extent, but also more complex questions of how nature is being altered and toward what end. The authors therefore conclude that religiously informed decisions about altering nature are invariably linked to other broad understandings, including conceptions of what constitutes ultimate reality, the relations between non-human and human nature, virtues such as compassion and right-mindedness, and the scope of appropriate human responsibility, often discussed under the rubric of stewardship.

Chapter 2: Philosophical Perspectives on Nature as a Norm

The second group of scholars considered philosophical discussions of nature and the natural. They focus particularly on developments since the Scientific Revolution of the 17th century, and discuss the alleged "death of nature" or "the world we have lost" associated with the mechanization of the world picture. The authors also discuss a number of alternatives to reductionistic attitudes toward nature that have persisted across the modern era, including organicism in the life sciences, the emphasis on functional explanations, and normative naturalism in such recent developments as deep ecology.

It is a commonplace in histories of Western thought that the Scientific Revolution of the 16th and 17th centuries produced a mechanization of nature, a conception of nature which left no room for the derivation of values from an understanding of nature. This positivistic conception of science was summarized in David Hume's famous observation of the gap between the "is" and the "ought," the claim that moral *oughts* cannot be derived from descriptions of how nature *is*. This mechanistic transformation of the concept of nature to a set of objects to be controlled and manipulated left wide open the question of the origin of values. Some thinkers sought to find that source in human nature, emphasizing that the distinctiveness of human nature might offer a foundation for normative reasoning, but others thought that such a source was as vulnerable as nature in general. As the authors of this chapter point out, there was no uncontroversial source of human values to be found within this new conception of nature.

A major contribution of this chapter is to point out that this mechanization of nature was not unchallenged from the 18th century until now. Instead its adequacy has been challenged during the last three centuries, and this challenge continues in our time in a number of influential modes of thought.

The origins of this challenge are to be found in the work of late 18th century French thinkers reflecting on the life sciences, and in German *Naturphilosophie* of the same period. For these thinkers, nature is both a product of the mechanistic forces discovered in the Scientific Revolution and an intrinsically creative force with great moral value. This duality of conceptions was adopted by American thinkers such as Emerson and Thoreau. Most crucially, as the authors point out, it comes to be adopted by Darwin in his original conception of natural selection. Parenthetically, the authors realize that a more thoroughly mechanistic account of natural selection occurs in the later editions, and they claim that this provides the background to Huxley's and Moore's attacks on the naturalistic fallacy.

The authors then examine how this debate continued in more recent philosophy of science. One area is in the analysis of biological explanations. One type of these explanations, functional explanations, requires the defenders of the mechanization of nature to provide an account of such explanations that can be incorporated into a mechanistic picture of nature. Alternatively, explanations involving molecular genetics (which have proliferated in response to advances in biochemistry, molecular biology and genetics) offer additional support for the mechanistic conception of the natural. The authors also point out the implications of a variety of current modes of thought (especially, post-positivist and feminist philosophies of science) to the issues we have been discussing

Another crucial feature of the authors' analysis is the attention they devote to the impact of certain recent developments in moral theory on the discussion of the normativity of the natural. While agreeing that both utilitarianism and Kantianism devote little attention to the concept of the natural, they call attention to ecological thinkers (such as defenders of Land Ethics and Deep Ecology) who devote considerable attention to the normative implications of the natural order and to deviations from it.

Where do these reflections leave us? In part, the authors argue, they leave us open to a plurality of understandings of the biological enterprise and of the normativity of the natural. But more importantly, the chapter concludes, such reflections lead us to recognize that the choice of one of these understandings is bound up with the acceptance of philosophical, scientific, and social views, and that our views on the normativity of nature and the natural are not dictated by the acceptance of any scientific results.

Chapter 3: Scientific and Medical Interpretations of Nature

The third group of scholars explored scientific and medical interpretations of nature and the natural through the prism of three case studies. The first case study reviews discussions in the early modern period about food, animals, and plants as "pure" or "impure" and considers the relevance of those historical precedents to recent debates about genetically modified foods and organisms. The second case study focuses on medical interventions to correct "deficiencies" in nature, draws on the writings of David Hume and John Gregory to develop an account of the "observably normative" in nature, and considers the relevance of such secular accounts of the normativity of nature to current discussions of biotechnology. The third case study traces the shifts in historical conceptions of aging as "natural" or as "pathological," with implications for interventions to forestall or reverse aging, depending on which conception is invoked.

The discussions in the 16th and 17th centuries about food reveal some interesting tensions. On the one hand, there is a glorification of the time-honored ways of producing food. A good example of this is the farmer's calendar which sets out, often in considerable detail, the work of the farmers in each month, reflecting immutable laws of seasonality. On the other hand, there is a Baconian tradition that approves of modifying the food-producing environment, of introducing new species of plants found in other countries, and even of creating new food species. While the issues are very different, the tensions in those discussions resemble in many ways current tensions surrounding people's attitudes towards genetically modified food.

The discussion of the normativity of the natural in 18th century medical and medico-philosophical thought reveals the complexity of attempting to use a normative account of nature in thinking about disease. It was recognized then, as it is now, that pathologies must be more than departures from a statistical norm; there must be a normative component. This gives rise to the difficult question of how to define that normative component. For Gregory and Hume, 18th century Scottish thinkers, that normativity arises from the sympathetic response to statistical abnormalities. It is this natural response that constitutes the normativity of judgments that abnormalities are pathologies requiring intervention when not self-corrected by nature. This may be very helpful as we think about the question of the use of new technologies to treat diseases versus their use to produce enhancements.

Many of these issues reappear in the third case study, which addresses the issue of how we should think about aging. In the ancient and early Christian world, the emphasis was on aging as a natural process to be accepted rather than resisted. Roger Bacon and some of his contemporaries, by contrast, saw aging as something that could be meliorated, leading to a significant expansion of the normal life span. This attitude was reinforced by Descartes and Francis Bacon, and expanded by some Enlightenment figures who were willing to consider the possibility of immortality through "right thinking and right living." All of this leads to the contemporary anti-aging dispute. By this point, the discussion of whether aging is normal or pathological is obviously no longer a purely scientific or philosophical one.

The chapter ends with a discussion of the lessons that might be learned from these case examples for our contemporary evaluations of technological change. One important lesson relates to the positive and negative valorization of nature. Negative valorizations, supporting improvements on nature, require the identification and justification of the values being employed in the negative valorization. But this is equally true of positive valorizations, since warrants must be provided for moving from a descriptive judgment that something is normal to a positive prescriptive evaluation.

Chapter 4: Legal and Economic Understandings of Nature

The fourth group of scholars researched legal and economic understandings of nature. They focus their discussion on the legal and economic structures that govern the appropriation of nature for private ends. In the biotechnology arena, issues concerning the patenting of nature have emerged as a central locus of conflict over the appropriation of nature for private ends. The authors discuss DNA patenting as a primary case study of issues that arise at the intersection of biotechnology, legal and economic concepts of nature as property, and religion. They supplement their discussion of DNA patenting with a review of other patent controversies (involving the Onco Mouse, the gene for Canavan's disease, the gene mutations responsible for a greater predisposition to develop breast cancer and the price of patented drugs for treating HIV/AIDs) that help to illuminate core concerns.

The authors identify two core issues in this current debate: concerns about the commodification of nature and concerns about injustices in the distribution of the benefits of the patented inventions. The concerns about commodification are concerns about treating things as purely commercial goods even when they are valuable for other reasons as well. Some critics think that any patenting of what is living is inherently wrong because it involves commodification. This is the basis of much of the opposition to DNA patenting and to the patenting of life forms. The concerns about access are concerns that the patents in question (particularly, the drug patents for HIV treatments) prevent large numbers of people (especially those in the third world) from being able to receive the relevant benefits. But these concerns, even if legitimate, must be thoughtfully balanced against the benefits – such as they are – of having a system of patents to encourage innovative development.

The authors then turn to two explicitly theological understandings of these issues: Protestant perspectives and the perspectives of indigenous peoples. They sketch the history of Christian thought about economic issues, showing how earlier views about just prices declined in favor of a more laissez-faire approach, rooted in a strand of individualism in Protestant thought. They also point out that as the newer issues of patenting biotechnology have emerged, recent Protestant thought has become more concerned about a range of additional issues. For some, the issues arise because of a new global and ecological consciousness. For others, the major concerns involve issues of justice. And for still others, the issues are clearly about commodification and about God's sovereignty over the natural order.

The authors point out, using an extensive array of African examples, that indigenous world views often involve radically different approaches to the natural environment, the social community, and the use of nature. Many of these views are challenged by the adoption of a patent system giving individuals control over parts of nature and by the resulting commercial exploitation of those parts of nature. The chapter then sketches the extensive developments in recent international law designed to deal with the resulting tensions by protecting the rights of indigenous people, either by insuring that they receive an equitable share of the profits from the new technology or by protecting and preserving their modes of thought. The latter approach is supported by a desire on the part of many to live in what they feel is a harmonious relation to nature and a feeling that patenting and related commercial developments threaten that harmony.

Chapter 5: Aesthetic and Representational Perspectives on Nature as a Norm

The fifth group of scholars examined how different sensibilities toward nature and "the natural" are reflected in (a) recent visual works that incorporate DNA and other biomaterials and (b) images of the body and of body parts that are employed in popular culture and advertising. In the course of that examination, they explore parallels between religion and art as two ways of dealing with boundaries between the sacred and the secular or between nature and culture.

The authors situate their analysis by initially discussing historic shifts in the practice and ideal of art as an activity. Since the mid-19th century, artistic methods have consistently shifted their focus away from beauty as their primary attribute or mimesis as their primary end. In that shift, especially in practices such as collage and montage, extant materials have been incorporated into artistic media and thereby recontextualized.

The authors then explore the impact on larger cultural perceptions of recent visual works and artistic images that incorporate biomaterials. They propose that representative recent visual media provide occasions to appreciate the "co-evolution" of technical knowledge and animate matter in a process they label *technogenesis*. In that co-evolution, the interactions between technology and biology affect our understandings of nature as a category – as given, as artistically envisioned, and as reconfigured in novel ways. The incorporation of DNA and other biomaterials into artistic objects extends such efforts, and in reflexive ways, provides a new sort of iconography for challenging the meanings of nature and the natural. The authors analyze a number of recent artistic examples that, by incorporating "living" materials into their expression, challenge other perspectives which assume that clear boundaries can be maintained between nature as in some sense a pre-discursive "given" and nature as an imaginative and cultural construct.

In their discussion, the authors review the status of DNA as a cultural icon in recent discussions. Public perceptions of the special status of DNA, while tending at times toward an unwarranted reductionism, provide suggestive evidence for quasi-religious interpretations of genomic information as the key to personal identity. The chapter also offers a series of broader typological comparisons among science, technology, and religion, which in turn provide the context for comparing and contrasting such narrative interpretations with the distinctive meanings of visual objects that incorporate DNA, transgenic life forms, and other biomaterials. In that discussion, the authors analyze specific recent works by artists Tony Cragg, Kevin Clarke, Marc Quinn, and Julia Reodica. The authors also explore the use of bodily images in advertising and popular culture, arguing that images of bodies and body parts often serve as "surrogates" or stand-ins for particular technologies being sold as enhancements to health.

The chapter concludes by emphasizing that images of cells, DNA structures, and embodied transgenic life forms in art, advertising, and popular culture raise provocative ethical questions concerning alterations of nature, because visual art provides a domain where contrasting perspectives on the body can be more freely and imaginatively explored than in scientific discourse. In light of such novel artistic expressions, the authors conclude that biotechnology no longer resides exclusively in the domains of science or science fiction, but has been incorporated into the broader cultural imagination, where it influences both popular and scholarly perspectives on nature and the natural.

Conclusion

As reflected by the chapters in this first volume, the effort to situate issue-oriented discussions of biotechnology in their broader context emerges as a complex and somewhat daunting scholarly task. It requires careful attention to the distinctive understandings of the concepts of nature and the natural at work both within and across different modes of discourse. It requires engagement with fundamental questions of ontology, epistemology, morality, and aesthetics in discussions of nature. In addition, it requires wide-ranging cross-traditional and cross-temporal analyses of the ways that contrasting descriptions of nature lead to different prescriptions about the legitimacy and scope of human decisions to alter nature, including human nature.

Central to the planning and execution of the multidisciplinary research assembled in this volume has been our expectation, richly realized, that judgments about whether and how nature is deemed normative depend, in large measure, on the disciplinary perspectives within which that concept is framed and articulated, as well as on a wide range of historical and cultural variables. In this volume, we have identified the arguments articulated (both traditionally and critically) in several disciplines in order to understand the possibilities for both cooperation and conflict between religious and other perspectives in theoretical and policy debates about biotechnology. In light of that complex discussion, we will turn, in the second volume of *Altering Nature*, to a consideration of the ethical issues raised by four broad areas of biotechnology – assisted reproduction, genetic therapy and enhancement, human-machine incorporation technologies, and biodiversity. Each area will provide the focus for two groups of scholars. A first group will analyze recent humanistic discussions in one of the four areas in light of the broad conceptual research assembled in this first volume. For the same area of biotechnology, a second group will then consider the possible implications of recent humanistic discussions for public policy deliberations and choices.

As we close this volume, several specific acknowledgements are in order. First, we wish to express our gratitude to the Ford Foundation for their generous funding of our research, with a special word of thanks to Constance Buchanan, for many years the Senior Program Officer for Religion, Society, and Culture in Ford's "Knowledge, Creativity and Freedom Program." Ms. Buchanan has been a generous and constructive partner to the project from its inception, and we are deeply indebted to her. Second, we wish to thank the many scholars who participated in the project during the last five-and-a-half years and who have been colleagues in a fascinating multidisciplinary conversation. We trust that our common efforts will make a useful contribution to ongoing debates about the conceptual and policy dimensions of issues posed by new developments in biotechnology. Finally, we wish to thank Dr. Lisa Rasmussen, who has served as managing editor for both volumes of *Altering Nature*. With characteristic diligence and grace, she has proved invaluable in helping us to ready the project for publication.

Reference

Gregorios, Paulos (1978). *The Human Presence: An Orthodox View of Nature*. Geneva, Switzerland: World Council of Churches.

Chapter 1
Spiritual and Religious Concepts of Nature

**Aaron L. Mackler, Ebrahim Moosa, Allen Verhey,
Anne Carolyn Klein, and Kurt Peters**

The Introduction to this volume addressed the difficulties involved in speaking of "religion" in connection with biotechnology. That discussion concluded by noting both the substantive and methodological limitations of this volume with its focus on theological, ethical, and legal discourses largely within the traditions or meta-traditions commonly characterized as "world religions." These limitations should be kept in mind in what follows, which is a selective overview of the religious traditions most discussed throughout the volume along with a treatment of indigenous religions, which are often invoked in debates over biotechnology. This overview has two aims in mind. First, it is meant to provide, in introductory fashion, certain conceptual, historical, and methodological "touchstones" by which to identify and situate these particular religious traditions. This summary is directed toward readers who may desire basic information about religious traditions with which they may be unfamiliar. It also provides a more general background for many of the more nuanced analyses found in particular chapters. Second, while our summary depicts those characteristics that are distinctive within traditions, in the interest of our larger conceptual and policy research agenda, it introduces themes that may suggest areas of common understanding, if not convergence among traditions. The hope is that attention to these themes will stimulate conversations across traditions. While the modes of presentation differ somewhat among the various sections below, each discussion provides some overview of basic features of the tradition itself (including its ethical aspects), treats the meaning of "nature" or "natural" in that tradition, and offers some indications of how the alteration of nature is or might be regarded.

1.1 Indigenous Religions and Spiritualities[1]

We begin our survey by attending to ways of life whose attitudes and practices regarding nature are widely considered to be both incompatible with those of modern biotechnology and uniquely threatened by the expansion of the latter. We immediately face two problems of definition. First, what counts as "indigenous"? According to John Grim, "[t]he term 'indigenous' is a generalized reference to the thousands of small-scale societies who have distinct languages, kinship systems, mythologies, ancestral memories, and homelands. These different societies comprise more than 200 million people throughout the planet today" (Grim, 1998). Second, can we use the term "religion" or even "spirituality" to refer to the attitudes and practices of these societies? If we decide for practical reasons to retain these terms we should do so with Grim's warning that "to analyze religion as a separate system of beliefs and ritual practices apart from subsistence, kinship, language, governance, and landscape is to misunderstand indigenous religion" (Grim, 1998).

Elaborating on the first problem of definition, we note several efforts to give substance to the term "indigenous." The Independent Commission on International Humanitarian Issues issued a report in 1987 which tried to spell out the term:

> There are four major elements in the definition of indigenous peoples: pre-existence (i.e. the population is descendant of those inhabiting an area prior to the arrival of another population); non-dominance; cultural difference; and self-identification as indigenous. Other terms are often used to refer to indigenous peoples: autochthonous, ethnic minorities, tribal people, first nations, fourth world (Independent Commission on International Humanitarian Issues, 1987, 6).

Weighing in somewhat on the other side of the discussion, Stan Stevens has added that other terms used are "native peoples," "primary peoples," "ecosystem peoples," and "aboriginal peoples." "The term, indigenous peoples," he says, "is not in universal use, however, and there are disagreements over how the term should be defined and applied" (Stevens, 1997, 299, fn. 1).

There have also been attempts at a more legally-recognized definition. The Special Rapporteur on the Problem of Discrimination against Indigenous Populations for the United Nations Sub-Commission on Prevention of Discrimination and Protection of Minorities produced this definition in 1986:

> Indigenous communities, peoples and nations are those which, having a historical continuity with pre-invasion and pre-colonial societies that developed on their territories, consider themselves distinct from other sectors of the societies now prevailing in those territories, or parts of them. They form at present non-dominant sectors of society and are determined to preserve, develop and transmit to future generations their ancestral territories, and their ethnic identity, as the basis of their continued existence as peoples, in accordance with their own cultural patterns, social institutions and legal systems.[2]

---

[1] This section includes, in edited form, material originally written by Elias Bongmba for the chapter on "Patenting Nature" in Volume Two, material written by Kurt Peters specifically for this chapter, along with supplemental material supplied by the editors.

[2] United Nations/Document No. E/CN.4/Sub.2/1986/7/Add. 4, para. 379, quoted in Independent Commission on International Humanitarian Issues, 1987, 8, fn. 3.

1 Spiritual and Religious Concepts of Nature

The International Labor Organization Convention 169 defines indigenous and tribal peoples as follows:

> a) [T]ribal peoples in dependent countries whose social, cultural, and economic conditions distinguish them from other sections of the national community, and whose status is regulated wholly or partially by their own customs or traditions or by special laws and regulations; and b) peoples in independent countries who are regarded as indigenous on account of their descent from the populations which inhabited the country, or a geographical region to which the country belongs, at the time of conquest or colonisation or the establishment of present state boundaries and who, irrespective of their legal status, retain some or all of their own social, economic, cultural and political institutions (Beltran, 2000, Annex 3, p. 17).

Of course, it must ultimately be recognized that all of the above definitions are erroneous, at least in the view of the peoples so defined. None have the words of definition in their specific languages, nor did they historically rely on those outside their communities to define them as peoples; such definitions are reflections of the needs and languages of the colonizers of the indigenous communities. All of the individual indigenous societies had particularized names for themselves, and many of the surviving groups – too many to recount here - are globally demanding a return to recognition by those original self-definitions.

With these efforts in mind we may now turn, in the second place, to the question of "religion" and "indigenous peoples." The Native American/American Indian scholar, Vine Deloria, Jr., has written about the difficulty of drawing comparisons between the Western concepts of "religion" and the belief systems of what other scholars have labeled "primitive" peoples. While his categories of "primitive" and "world religions" are problematic, his position is worth quoting in detail. "Primitive peoples," he says, "do not differentiate their world of experience into two realms that oppose or complement each other. They seem to maintain a consistent understanding of the unity of all experiences" (Deloria et al., 1999, 354). They "maintain a sense of mystery through their bond with nature; the world religions sever the relationship and attempt to establish a new, more comprehensible one" (Deloria et al., 1999, 360). According to Deloria, two perceptions are the key to conceptualization of religion and religious experience:

> [P]rimitive peoples' perceptions of reality, particularly their religious experiences and awareness of divinity, occupy a far different place in their lives than do the conceptions of the world religions, their experiences, and their theologies, philosophies, doctrines, dogmas, and creeds. Primitive peoples preserve their experiences fairly intact, understand them as a manifestation of the unity of the natural world, and are content to recognize these experiences as the baseline of reality. World religions take the raw data of religious experience and systemize elements of it, using the temporal or the spatial dimension as a framework, and attempt to project meaning into the unexamined remainder of human experience. Ethics becomes an abstract set of propositions attempting to relate individuals to one another in the world religions while kinship duties, customs, and responsibilities, often patterned after relationships in the natural world, parallel the ethical considerations of religion in the primitive peoples. Primitive peoples always have a concrete reference—the natural world—and the adherents of the world religions continually deal with abstract and ideal situations on an intellectual plane. 'Who is my neighbor?' becomes a question of great debate in the tradition of the world religions, and the face of the neighbor changes continually as new data about people becomes available. Such a question is not even within the worldview of primitives. They know precisely who their relatives are and what their responsibilities toward them entail (Deloria et al., 1999, 364–365).

It is in part for this reason that many scholars of Native American religion are uncomfortable with the term "religion," which in modern usage, as Deloria points out, is more suited to late Western developments than to indigenous ways of life (Deloria, 2002, 120).

Any attempt to represent a Native American perspective must take care to acknowledge the differences among tribal groups. Yet, while acknowledging those tribal differences, we can draw on the genesis story of the Jicarilla Apache, as reported by Erdoes and Ortiz, in order to sketch a characteristic Native American conception of nature:

> In the beginning the earth was covered with water, and all living things were below in the underworld. Then people could talk, the animals could talk, the trees could talk, and the rocks could talk. [When] the earth was all dry, except for the four oceans and the lake in the center [the Great Lakes]...All the people came up. They traveled east until they arrived at the ocean; then they turned south until they again came to the ocean; then they went west to the ocean, and then they turned north. And as they went, each tribe stopped where it wanted to (Erdoes and Ortiz, 1984, 83, 85).

At creation of the Jicarilla Apache world, the humans and the other beings of nature were as one and could speak with each other and the inanimate, the rocks, could speak as well. All had lived below together while the earth was covered by water, and no lesson is taught in this story that any single specie in nature held dominion or stewardship over another; all were equal. Once dry land appeared, all the underworld beings surfaced to live and the "tribes" scattered in various directions and settled in places to their own liking.

While every tribal community today has stories of origins of a similar kind, with appropriate nuances for geographical and cultural differences, there are no extant major deviations from the themes of rootedness in natural cycles as presented in the Jicarilla Apache example. This rootedness has profound ethical significance. What is startling in Native American origin stories is the differentiation between groups based on their locale, declared histories, and so forth, and yet that a single concept binds all of them together: humans in the stories are not above nature or natural beings either as stewards, proprietors, or as being more knowledgeable about the functioning of the world. Humans are merely an equal part of the flow of the natural process and nothing more, and it is not the being human that is integral to the stories, but rather recognizing that "One of the most important roles of the old stories is to remind people of the right way to behave" (Deloria et al., 1999, 357; Bruchac, 1996, 198).

As Marsha Bol writes in *American Indians and the Natural World*, "tribes share in the perspective that abilities and powers are distributed among the many species in the natural world, not restricted to humans" (Bol, 1998, 111). Not only do the animals possess a spiritual power, but are considered in many ways wiser than humans. This wisdom sustains the animals and is often neglected by humans in their quest for status. For example, says Bruchac, "Unlike humans, [animals] do not forget the right way to behave. A bear never forgets that it is a bear, yet human beings often forget what a human must do.... They become confused because of material possessions and power" (Bruchac, 1996, 198). Since creation of the physical earth has occurred prior to human creation in most indigenous origin stories, and humankind is usually an "afterthought" of the creative force, human status in nature's array of beings has already been cast. The Modoc world creation story, "When Grizzlies Walked Upright," outlines exactly the relationship between beings on earth and the innate ability of communication between them. At the time of Modoc creation the animal-human relationship was so inextricable, one from another, that each could assume the other's shape and form as well as converse.

> Before there were people on the earth, the Great Chief of the Sky Spirits grew tired of his home in the Above World, because the air was always brittle with an icy cold. So he carved a hole in the sky with a stone and pushed all the snow and ice down below until he made a great mound that reached from the earth almost to the sky. Today it is known as Mount Shasta [Northern California]...[After the Great Chief's daughter mated with a grizzly]... Those strange creatures, his grandchildren, scattered and wandered over the earth. They were the first Indians, the ancestors of all the Indian tribes (Erdoes and Ortiz, 1984, 85–87).

And as well, the White River Sioux creation story about Rabbit Boy relates a very similar genesis, despite the Modocs (California) being from a vastly dissimilar geographic locale from the White River Sioux (South Dakota). "In the old days," says the White River Sioux creation story, "many people could understand the animal languages; they could talk to a bird, gossip with a butterfly. Animals could change themselves into people, and people into animals. It was a time when the earth was not quite finished, when many kinds of mountains and streams, animals and plants came into being according to nature's plan" (Erdoes and Ortiz, 1984, 5). It is only with the aid of other natural beings that spiritual creation brings humans into physical existence, and therefore, immediately establishes kinship patterns with and between these natural beings, a concept common among indigenous peoples throughout North America. Luther Standing Bear wrote of a "kinship" among all creatures, including the earth, sky, and water, as a "real and active principle" in Indian life, indicating that recognition of origins was not only a common factor in Native American cultures, but that they also acted as a guide for human relations with nature on a day-to-day basis. Stories become the tutor, the mentor, the conscience throughout Native lifetimes (Erdoes and Ortiz, 1984, 85–87; Stone, 1993, 112).

We may sum up these observations by returning to Deloria:

> The task of the tribal religion, if such a religion can be said to have a task, is to determine the proper relationship that the people of the tribe must have with other living things and to develop the self-discipline within the tribal community so that man acts harmoniously with other creatures. The world that he experiences is dominated by the presence of power, the manifestation of life energies, the whole life-flow of a creation.... Tribal religions find a great affinity among species of living creatures, and it is at this point that the fellowship of life is a strong part of the Indian way.... Very important in some of the tribal religions is the idea that humans can change into animals and birds and that other species can change into human beings. In this ways species can communicate and learn from each other.... But many tribal religions go even farther. The manifestation of power is simply not limited to mobile life forms. For some tribes the idea extends to plants, rocks, and natural features that Westerners consider inanimate.... [Finally] [t]he Indian tribal religions would probably suggest that the unity of life is manifested in the existence of the tribal community, for it is only in the tribal community that any Indian religions have relevance (Deloria, 1994, 78–71, 85, 88–90, 93).

Similar observations have been made by scholars of African indigenous or traditional religions, to which we now turn. Jacob Olupona has noted that hermeneutical and phenomenological studies of African religions demonstrate that religious life in Africa is expressed through cultural idioms such as music and arts, and involves ecology and one cannot study religion in isolation from its socio-cultural context (Olupona, 1991, 30). Laurenti Magesa has argued that African traditional religion is communal and *"embraces the whole life*. People do not distinguish between a religious and secular leader. Religious leaders maintain a close relationship with departed ancestors so that things will go well in the visible and invisible world" (Magesa, 1997, 71). Writing about the Akan of Ghana, Elizabeth Amoah writes that there is

> also a demonstrable cultural disposition that emphasizes the spiritual dimension of total human life. This finds expression in a certain specific conception of nature, of humanity and *Nana Onyame*'s (God's) relations with creatures of various types. It also finds expression in the relationships between persons and their physical environment and in the ways in which explanations are sought for the major problems of life, the problem of the meaning of life, the problem of suffering and the problem of evil (Amoah, 1998).

The natural world, human community, and health are important issues in African religious traditions. The universe inhabited by God, spirits, living, and the non-living things all interact with each other. Some African myths state that God created the natural world and human community (see Magesa, 1997, 39). The Yorubas believe that *Olodumare*, the Supreme God, created the natural world and human community through the two divinities, *Obatala* and *Oduduwa*, and established the primordial home of the Yoruba people at *Ile Ife*. The created order consists of *aye*, the visible concrete world that is habitat for humans, wildlife and all things in nature. The invisible world is called *orun*, and is the home of God, the spirits, and the departed ancestors (Drewal and Pemberton, 1989, 14). The Sotho-Tswana believe that God maintains a close relationship with the universe and uses natural forces like thunderbolt, to penetrate into the universe (Setiloane, 1976, 82). The Akans of Ghana believe that humanity has kinship relationship that involves mutual respect with the natural order (Appiah-Kubi, 1981, 8–9). Thus, humans are part of the rhythms of nature. Life is a journey to the other world and a successful journey requires a right relationship with nature, divinities, and other inhabitants of the universe. In order to live a long life and be successful, the Yoruba believe that a person must seek "*ogbon* (wisdom), *imo* (knowledge), and *oye* (understanding)" (Appiah-Kubi, 1981, 16). Divinity has also given humans *ase*, interpreted as power and the ability to "make things happen." *Olodumare*, the creator, has also given *ase* to lesser divinities, spirits, ancestors, plants, rocks, rivers, words, etc. In the social world of human relations, *ase* also means power and authority. Humans are expected to use *ase* to ensure the well-being of nature and other human beings. In order to do this they must employ wisdom, knowledge, and understanding. "Theoretically, every individual possesses a unique blend of performative power and knowledge – the potential for certain achievements. Yet because no one can know with certainty the potential of others, *eso* (caution), *ifarabale* (composure), *owo* (respect), and *suuru* (patience) are highly valued in Yoruba society and shape all social interactions and

1 Spiritual and Religious Concepts of Nature

organization" (Appiah-Kubi, 1981, 16). The Yoruba have a holistic view of power which works at the level of *orun* (heaven), the abode of deity and the *orisas*, and at the level of *aye* (earth) (Lawson, 1985, 66). Living on earth and exploring the natural world calls for a balance that recognizes the realm and power of nature as well as the unique gifts that each individual has received from the creator. One maintains right relationships through mediation. God has created the universe and humans must practice certain graces and act with loving care as they seek to manage and alter nature.[3]

In addition to the natural environment, religious thought in Africa emphasizes the social community. One lives, exercises power, controls, and manages nature in solidarity with other people. Community is the center for fecundity, friendships, hospitality, and healing (Nyamiti, 1973, 9–11). One example where communalism is expressed in relationship to nature is joint ownership of land, forests, waterways, etc. Ideally, these things were considered the property of the community and they shared those resources together, and within that setting, people were given part of the resources to develop for their use and the benefit of other people. We cannot generalize this point because some African states have set up legal systems, which include property rights that are grounded on a secular legal system. This calls for a different attitude towards the acquisition and use of natural resources, especially in sub-Saharan Africa where deforestation and depletion of resources continues at a very fast pace.[4]

We alluded at the outset to the widespread sense that indigenous societies and their ways of life are both incompatible with biotechnology and threatened by the latter. This context strongly shapes the question of the normative status of nature and the ethics of the alteration of nature in relation to indigenous societies. We begin with the sense of incompatibility. Much of the literature regarding indigenous people and ecology incorporates understandings of attitudes of those people toward nature and their relationship to it. Grim has written that "[t]hemes which provide orientation for understanding the relations between indigenous religion and ecology are kinship, spatial and biographical relations with place, traditional environmental knowledge, and cosmology" (Grim, 1998). He offers as examples the Lakota [Native American] "memory of rocks and stones as persons;" the belief of the Temiar people of Malaysia that a cool healing liquid called khayek (the "form taken by the upper soul of a spiritual being from the local Malayan rainforest") "can be imparted to human beings through dreams," specifically the songs in dreams; and the cosmologies of the Dogon peoples of Mali "which elaborate the close relationships which living Dogon share

---

[3] Appiah-Kubi argues that among the Akan: "Proper land use is evidence of faithfulness to God and Mother earth and it is reflected in the health of the soil and nature and economic health or prosperity of the society. Unjust land use on the other hand spells social, economic, spiritual disaster, crop failure, and epidemics" (Appiah-Kubi, 1981, 9).

[4] In a country like Cameroon which is richly endowed with natural resources and an extensive biodiversity, poor management is contributing to the decline of these resources (Blaike, 1996; Atyi, 1988).

with their ancestors, their land, and their animals among which the soon to be living reside" (Grim, 1998). Other examples are those of the Salish speaking Colville peoples on the Columbia River of eastern Washington who "weave ritual forms of knowing and proper approach into the taxonomic discussions of plants," and among the Yekuana people of Venezuela the "ethics of limits with regard to nature consumption" based on the pragmatic use of plants and roots (Grim, 1998).

Deloria sees the problem this way:

> We have reduced our knowledge of the world and the possibility of understanding and relating environment to a wholly mechanical process. We have become dependent, ultimately, on this one quarter of human experience, which is to reduce all human experience to a cause-and-effect situation. When we look at nature and environment through Western European eyes, that is really what we are looking at. That is not the "nature" Indians understand. Indians never made any of those divisions.
>
> In the Indian tradition we find continuous generations of people living in specific lands, or migrating to new lands, and having an extremely intimate relationship with lands, animals, vegetables, and all of life. As Indians look out at the environment and as Indians experience a living universe, relationships become the dominating theme of life and the dominating motif for whatever technological or quasi-scientific approach Indian people have to the land (Deloria et al., 1999, 225–226).

When this environment is threatened, the indigenous people potentially are threatened not only physically but culturally and "religiously." An example can be seen in the description by Julio César Centano and Christopher Elliott of the Yanomami peoples, described as "the largest group of indigenous people still living their traditional life-style in the Americas" and who live on either side of the border between Brazil and Venezuela. They are said "to exist in an interdependent relationship, living in harmony with the environment, practicing low-intensity shifting cultivation, fishing, hunting, and gathering. *Urihi* and *uli* are Yanomami words for forest. Translated from the Amerindian language, they mean 'home' or 'the place where one belongs.'" As that forest area becomes threatened by outsiders who want to take their land, by scientific expeditions who ignore the values of the Yanomami to follow their own agendas, by the Brazilian and Venezuelan armies that are arguably trying to "protect" the Yanomami in what is now a legally protected area, and by ecotourists seeking a "unique jungle experience," the survival of the Yanomami peoples is threatened (Kemf, 1993, 95–103).

Another example can be seen in the beliefs of the Navajo for whom "the complex network of effect makes the human individual part of and dependent on the kinship group and the community of life, including plants, animals, and aspects of the cosmos that share substance and structure" (Schwartz, 1997, 12). For example, the Navajo "have a sentimental attitude toward plants, which they treat with incredible respect" (Reichard, 1950, 22).

In another part of the world:

> [t]he Maori, native peoples of New Zealand, speak of themselves as *tangata whenua*, "people of the land".... Land is the connection both to larger mythologized cosmic forces as well as to the source of personal life.... Interactions with the environment bring humans in contact with the *mana*, or inherent understanding of all reality.... All creatures possess *mana* suggesting that all reality has intrinsic worth, as well as a personal life force, *maori* (Grim, 2001, 41–42).

We have already begun to discuss the threat biotechnology may pose to indigenous ways of life. Debates over the implications of biotechnology for indigenous peoples are inseparable from the much broader histories of conquest and encroachment to which these peoples have been subject, especially during the past five centuries. Ethical and policy discourse in biotechnology tends to perceive the actual and potential threats of biotechnology on indigenous ways of life as problems capable of management by careful legal protections and policy prescriptions guided by the methods and procedures of standard technology assessment. Thus one response has been to grant official recognition to the attitudes and relationships regarding nature sketched above. For example, the World Conservation Union has adopted the principle that "[i]ndigenous and other traditional peoples have long associations with nature and a deep understanding of it. Often they have made significant contributions to the maintenance of many of the earth's most fragile ecosystems, through their traditional sustainable resource practices and culture-based respect for nature" (Kemf, 1993).

Of course, difficult questions arise when actual or proposed biotechnological projects and interventions interfere with indigenous communities and ways of life, and this is the point at which legal recognition is supplemented by technology assessment. However, for many indigenous peoples and others who advocate their causes, such efforts to "manage" the issue are yet one more chapter in the dismal history in which outsiders with little knowledge of or sympathy for indigenous peoples and their ways of life have intentionally or accidentally harmed or destroyed those ways of life by subjecting them to "objective" standards. While today's proponents of biotechnology are often much better disposed than their predecessors, their ignorance of the history of treatment of indigenous peoples in the name of scientific progress and development and of the history of effects of "objective knowledge" are both astounding and morally disturbing. The chapter on biodiversity and biotechnology in Volume Two of this project addresses some of the problems posed by this conflict of perceptions in relation to the effects of genetically modified organisms on indigenous attitudes and practices, and it is clear that more attention to these problems will be needed as agricultural biotechnology expands further into areas where indigenous peoples live.

1.2 Religions of Asian Origin

We turn now to major religious traditions of Asian origin. We immediately face once again the problems noted in the Introduction: that there is something highly artificial about treating these traditions as unified bodies of thought and practice (i.e., as "-isms"), something highly abstract about making their ideas and texts our central focus, and something highly arbitrary about calling them "religions" as distinct from "philosophies" or simply "ways of life." The arbitrariness is especially apparent in the case of Asian religions since this volume treats Chinese conceptions in the chapter on philosophical concepts of nature while the two most prominent traditions of South Asian origin are treated here.

1.2.1 Hinduism[5]

"Hinduism" immediately presents problems of definition since the term itself has a complex history, and only in recent centuries has it been used by adherents of this religion to describe themselves. The definition is further complicated by the sheer variety of practices and textual traditions that can be said to comprise Hinduism, including diverse schools of thought such as *Samkhya, Yoga, Nyaya, Vaisesika*, and *Vedanta*. Present in one way or another in all of these traditions is the tradition of revealed texts known as the *Vedas*, of which the *Rig Veda* is the most important, and the concept of *dharma*, which consolidates and distills a range of Indian cultural and religious influences and which is especially interesting for the purposes of this volume since it can be thought of as the power that governs the social and cosmic orders, somewhat like natural law in Western philosophical and theological traditions. Indeed, it is common for Hindus to refer to their religion as the "Everlasting Law" (*sanatana dharma*). Following the *Vedas*, which are compilations of an oral tradition written in four groups between 1500 and 900 BCE, and the *Brahmanas* which followed them, the *Upanishads* were written around 700 BCE. These texts are central to Hindu philosophical thought and continue to be important for educated Hindus today. Finally, in addition to a wide range of epic, legal, and other texts, a distinctive system of medical practice called *Ayurveda* emerged within the broader context of Hindu beliefs and practices. It was written in the first century CE as the *Carak Samhita*, attributed to a physician by that name.

The *Vedas* contain speculations about the world system, which is called *brahmananda* ("the egg of Brahama"). Brahma is the personification of the divine creative energy from which the universe evolves. More philosophically, Brahma is conceived as the origin and end of the universe. His lower nature is differentiated into various forms, with the evolution of all life, including humanity, emerging from combinations of five elementary principles (*pancamahabhutika*). With this perspective, Hinduism affirms a moral structure to a divinely immanent universe that is permeated by moral values that flow from the power of God (*sakti*). Its literature also depicts a vast array of other supernatural beings, forces of light and darkness, good and evil, which have appeared as central figures in popular beliefs, especially in earlier eras.

Hindu understanding of human nature is dualistic, with the noumenal (essential) self distinguished from the empirical self. The world of the senses, as experienced by the empirical self, is ultimately illusory, the product of an original ignorance (*avidya*). Hinduism describes the empirical self in tripartite fashion – as a complex of three bodies – the physical, subtle, and causal bodies. The physical body is the

---

[5] This section incorporates, in shortened and edited fashion and with supplemental material supplied by the editors, significant portions of two unpublished working drafts by authors in Volume Two of this project: "Hindu Perspectives on Genetic Enhancement," by Cromwell Crawford; and "Asian Perspectives on Incorporation," by David Loy. It also contains material by Cromwell Crawford written specifically for this chapter.

locus of conscious experience. The subtle body – so-called because it is composed of 17 elements more rarefied than those of the physical body – is the basis of dream consciousness, with the presence of the noumenal self inferred from it. Moreover, the subtle body is the instrument for the operation of the law of karma, whereby moral consequences are transmitted in the transmigration of the soul. The just desserts of an individual's life proceed from one birth to the next through the continuity of the subtle body. The third body of the empirical self is the causal body, whose reality is inferred from the state of deep sleep, during which the physical and subtle bodies are in suspension.

The core emphasis in Hinduism is the identity of *Atman* (Self) and *Brahman* (Being), with a consequent deemphasis on the physical body. Yet Hindu dualism is not analogous to a Cartesian cleavage of mind and body. The important contrast is not that between mind and body, but between falsely identifying with one's "mind-body" as opposed to the impersonal and transcendental *Atman* – one's true self.[6] Spiritual practice does not involve transforming the Self – which cannot be improved or damaged – but strengthening and controlling the psycho-physical integration of our mental and physical bodies, an integration that can enable us to disidentify with those bodies and realize the true self. Enlightenment (*moksha*) is the act of becoming fully aware of the real self. The essential self transcends the mind-body complex and is free from all limitations and change. It is eternal and immutable existence (*sat*), pure consciousness (*chit*), and pure bliss (*ananda*). As such, one's essential self (*atman*) is none other than Cosmic Reality (*Brahman*) (Weiss, 1995, 1132).

Three other central features of Hindu thought are its understanding of the law of *karma* ("action"), its theory of rebirth, and its depiction of a social system of castes. *Karma* is a universal law of causation as applied to the rational and moral aspects of human existence. Good and bad actions bear within themselves their own consequences. By linking the present with the past, the law of *karma* provides a "cosmic" context within which to situate the empirical realities of inequalities and suffering. Traditionally, two types of karma have been distinguished. *Anarabdha-karma* refers to *karma* that has not begun to hear fruit; *prarabdha-karma* refers to *karma* that has already come to fruition. The latent karma of the first type can refer either to karma accumulated from past lives (*sanctita*) or to current karma (*vartamana-karma*). Only action arising from selfish motives produces karma. Disinterested actions (*vartamana-karma*) are not only free of binding consequences but can serve to dissipate karma that has not yet begun to bear fruit.

The Hindu understanding of the cycle of rebirth (*samsara*) is deduced from the law of karma. Since the universe is morally structured so that good and evil acts generate appropriate consequences, it must be assumed than an action without current

---

[6] "What the West considers mind is in Hindu thought a more subtle version of the material body, which includes all of one's karma and mental tendencies. At death the physical body dies but these 'mental bodies' transmigrate to another physical body" (Loy, p. 2 of unpublished working draft; see note 5 above).

effects will have consequences in a future life. Thus, Hindu belief posits an immortal spirit that reincarnates itself in a better or worse form, according to the necessity of rewards and punishments for one's actions. The individual continues incarnations until he or she gains enlightenment (*moksha*) and is freed from the cycle of rebirth. The notion of transmigration, therefore, "links all living beings in a single system," without the sharp distinctions between humans and animals one finds in Judaism, Christianity, and Islam (Weiss, 1995, 1132). Moreover, notions of karma and rebirth work together to engender the central value of nonviolence (*ahimsa*) toward all sentient beings.

In Hinduism, there are four basic social classes, which were typically explained to have "emerged at the beginning of time from the body of the Creator as the fundamental basis of society": the *Brahmins*, or priestly caste; the *Ksatriyas*, warriors and rulers; the *Vaisya*, or merchant class; and the *Sudras*, or workers. Beneath those classes were the so-called "untouchables" (Weiss, 1995, 1133). In the principal Upanishads, one finds elaborated a general theory on the basic ends of human life – wealth (*artha*), desire (*karma*), and duty (*dharma*). One's duties were historically understood in light of one's caste status, with different obligations attending various states. However, since *moksha* was often discussed in terms of meditative practice (*yoga*) and transcendence of mundane concerns, the theory of the ends of human life also include a recognition of the duties appropriate to various stages of life (*asramas*). Wealth, desire, and duty are the ends for the family man or householder, a position one assumes after time as celibate student. When one's children mature, the householder pursues other ends, with the highest and ideal stage that of the wandering recluse (*arama*) or *sannyasin* (one focused exclusively on spiritual knowledge).

Having sketched these general characteristics of Hinduism we now turn to the more specific question of the meaning and status of nature in Hindu traditions. Throughout its long history, Hinduism has understood and interpreted Nature in relation to two other basic concepts: Ultimate Reality, and the Individual Self. The fruit of this search has been the development of a "spiritual science," which is relevant to some of the moral problems connected with the subject of "altering nature." Hinduism eminently lends itself to this topic because it is rich in mythology that is ethically nuanced. Among its diverse nature myths is one genre which views *nature as the body of God*. This motif spans the strata of the Hindu textual traditions, appearing in the Vedas, the Brahmanas, the Upanishads, the Bhagavad Gita, and in Vishishtadvaita. Moreover, apart from these sources all schools of Hindu thought agree that animate and inanimate beings do not exist independently of God. There is a pervasive conviction in Hinduism that God and humanity are not segregated in different realms, nor does it separate human beings and the cosmos. Instead, by envisioning God as the source and support of both humanity and the universe, it keeps God in intimate relationship with both of them, allowing no gap between God and humanity, between God and the universe, and between humanity and the universe.

With this clarification of the general position of Hinduism, we now follow the theme of nature as the body of God in diverse Hindu texts. Among the Vedic hymns on cosmic order, one particularly notable one describes the origin of the universe from a primeval sacrifice, in which the male Purusha, a cosmic being, makes a sacrificial offering of himself. By highlighting the universality of Purusha and his role as cosmic sacrifice, the "ritual sacrifice performed on earth by a priestly class eventually was translated into terms of cosmological significance by a process identifying microcosmic with macrocosmic elements" (deBary, 1958, 13). The poet states:

> Thousand-headed Purusha, thousand-eyed, thousand-footed – he, having pervaded the earth on all sides, still extends ten fingers beyond it.
> Purusha alone is all this – whatever has been and whatever is going to be. Further, he is the lord of immortality and also of what grows on account of food.
> Such is his greatness; greater, indeed, than this is Purusha. All creatures constitute but one quarter of him, his three qauarters are the immortal in the heaven.
> With his three quarters did Purusha rise up; one quarter of him again remains here. With it did he variously spread out on all sides over what eats and what eats not.
> When they divided Purusha, in how many different portions did they arrange him? What became of his mouth, what of his two arms? What were his two thighs and his two feet called?
> His mouth became the Brahman; his two arms were made into rajanya; his two thighs the vaishyas; from his two feet the shudra was born.
> The moon was born from the mind, from the eye the sun was born; from the mouth Indra and Agni; from the breath (prana) the wind was born
> From the navel was the atmosphere created, from the head the heaven issued forth; from the two feet was born the earth and the quarters (the cardinal directions) from the ear. Thus did they fashion the worlds (Rig Veda X, 10–13).

In the Brahmanas the motif of nature as the body of God is extended, with ritual sacrifice made to symbolize the creation of the world. The Upanishads affirm the unity between God and man, between God and the universe, and between humans and the universe. The Taittiriya Upanishad establishes Brahman as the source, sustenance, and goal of all beings, animate and inanimate:

> That, verily, from which all these beings are born,
> That by which they live, after being born,
> That into which, when departing, they enter,
> That seek to know. That is Brahman (Taitiriya Upanishad, 3.1.1).

The meaning is that Brahman is the source, sustenance and goal of all beings, animate and inanimate.

The Bhagavadgita presents a dramatic vision of the world as God's body, which has been likened to the dazzling radiance and brightness of a nuclear explosion. Chapter XI begins with Arjuna facing up to the reality that apart from God, no things in the world exist in themselves and maintain themselves. But he is not content to know this as an abstract metaphysical truth; he must see the visible embodiment of the Unseen Divine. This vision of all in the One cannot come empirically, by the human eye, because it is limited to outward forms. Arjuna therefore needs the divine bestowal of the supernatural eye. Once endowed with the eye of the spirit, Arjuna witnesses Krishna's transfiguration, and sees all heavenly and earthly creatures in the Divine Form: "Of many mouths and eyes, of many visions of marvels" (The Bhagavad Gita, 11.10). It is a stupendous, hair-raising vision of the One in the many and the many in the One.

Finally, we turn to the imagery of nature as the body of God in the Vishistadvaita. Ramanuja theorized that though God is one, he is qualified (visista) by humans and the world, both of which constitute his body. The soul-body analogy underscores the belief that though God, humans and nature are different, they are inseparably related like soul and body. The relationship is more than unity; it is an organismic union in which God does not exist in isolation from people and nature. God is the soul (saririn) of humans and nature, and the world is the body (sarira) of God.[7]

First we examine the notion of the world as the body of God. Taking his cue from the Upanishad, which states that prior to creation, there was only Being, the Vishistadvaitin argues that Ultimate Reality must be both the efficient and material cause of the world. He illustrates his point by referring to the spider which constructs its web from the material it produces, and is therefore the material cause. At the same time the spider is the efficient cause of the web, having brought it into being by its own volition. Similarly, Brahman, the One without a second, is simultaneously the material and efficient cause of the world. The world is real, in addition to Brahman, but insentient objects are so integrated with Brahman as to constitute a soul-body relationship. Thus the world is eternal and inseparably related to Brahman, both prior to creation, when it exists in a subtle state, and after creation, when it manifests in a gross state. Given the soul-body relationship between God and the world, the following insights emerge (Balasubramanian, undated, p. 8 ff.):

1. The world, though real and eternal, does not exist independently of God.
2. The world is supported and controlled by God.
3. The world is created and sustained for the purpose of providing objects of experience for souls caught in the wheel of rebirth, with a view to their liberation.

Next, we examine the relation between God and humans in Vishistadvaita. This relationship is characterized in a twofold way. First is the soul-body analogy. Man is conceived as the body of God. On the one hand, God and humans are different; on the other, they are inseparably related as soul and body. Second, a human is also conceived as an "attribute" of God. Being eternal and sentient, he is like God; but because he is finite and dependent, he is different. Though God and humans are different, they are also inseparable, as in a substance-attribute relationship. As an attribute, man derives his substantiality from God, is inseparably related to him, and is dependent on him for his own being. Thus, humans are envisaged in Vishistadvaita in the twofold form of the body of God, and as attributes of God. The meaning of these analogies is that:

1. Divinity and humanity do not occupy water-tight compartments.
2. Divinity is immanent within humanity.
3. Human life is contingent upon divine life.

---

[7] In his unpublished paper, "The Hindu Perspective of Man and the Cosmos," R. Balasubramanian has some valuable suggestions to which we are indebted.

4. The aim of life is the dispelling of ignorance (avidya), which obscures the true nature of the Self and God. When that is achieved, life is all bliss, although it is strictly proportionate to the intrinsic worth of each Self.

This brief survey of the Hindu myth of Nature as Body of God has yielded lineaments of a "spiritual science" that attempts to explain the hidden significance of the world that surrounds us, with hints at how we might interact with nature. The perspective delineated here is relevant to two of the assumptions that accompany efforts to alter nature. One of these assumptions is scientific reductionism. Hindu myth grants that science correctly focuses on parts, but it is wrong when it makes the part into a whole. Over against this reductionist approach, Hindu *mythos* tries to do justice to the interconnectedness and interdependence so evident in nature by adopting a wholistic approach, which then puts questions of altering nature in an entirely different light. In this new frame of reference, ethical directions begin to follow, *naturally*.

The Vishistadvaita of Ramanuja is a strong bulwark against all reductionism. It clearly states that the world and the soul are to Brahman what the body is to the soul. Neither can exist nor be conceived without Him. All three are distinct and eternal, though of unequal status, but they are inseparably associated. This brand of spiritual science is not opposed to science, but to its penchant for reductionism. Through its spiritual lens, it supplies significance to isolated parts, integrating and enhancing them within a larger whole.

The second assumption involves what we may identify as the Baconian impulse in modern technology. Hinduism resonates with the orientation of science to the relief of human suffering; but it cannot support the assumptions that the natural order, in which humans find themselves caught, has no dignity of its own, no internal value, and that the purpose of knowledge is to exercise power over nature.

In the philosophy of Advaita Vedanta, Baconian confidence in the power of technology to solve human ills through a mastery of nature is ill-founded. According to Samkara, an understanding of Brahman provides the philosophic basis for a proper understanding of nature. The natural world is not a commodity which humans may possess, but a community to which they belong. The universe appears to be material, but it is the universal consciousness or Brahman. Since all is one, the control or conquest of nature cannot be true to reality, and our sense of separateness, isolation, and egotism are all products of ignorance.

We referred above to the distinctively Hindu system of medical practice, *Ayurveda*, and it is appropriate to close the treatment of Hindu understanding and practices regarding nature by discussing this tradition. *Ayurveda* draws its understandings of the body and of therapeutics from the conceptions of cosmology and ontology outlined above. It elaborates a humoral theory of physiology "based on the balance of three substances (*dosas*): wind (*vata*), bile (*pitta*), and phlegm (*kapha*)" (Weiss, 1995, 1135). *Ayurveda* gives prominence to the notion of balance. It promotes an ethic of moderation and emphasizes disease prevention and health promotion through a sophisticated knowledge of pharmacology. As Crawford observes, "Health and healing are regarded as acts of nature. In medico-ethical terms: the natural is the good" (Crawford, 2003, 6).

The implications of Hindu beliefs for assessments of altering nature and biotechnological developments are mixed. With regard to therapeutic possibilities, the evolutionary orientation of Hindu ethics allies it with scientific progress. On the one hand, the complete transcendence and indifference of *Atman*, one's true self – which cannot be hurt or otherwise affected – suggests that there is no principled reason to oppose interventions that alter phenomenal nature or the physical body. On the other hand, the spiritual thrust of Hindu beliefs prompts caution in assessing alterations of nature and human nature. The traditional Hindu attitude toward death and disability is quite different from that of modern Western medicine. If each of us is reincarnated many times, according to our karma, it may be deemed spiritually immature to concern ourselves with any particular physical body, even if disabled, all the more so if disability is the karmic effect of one's past actions. To be sure, in a situation where life is threatened by disease or disability, active compassion to alleviate suffering may, as an unselfish act, be viewed as a moral duty that dissipates lingering negative karmic effects and produces positive ones.

In summary, the evolutionary orientation of Hindu thought does not permit us to offer blanket affirmations or condemnations of developments in biotechnology. Hindu understandings do not proceed from judgments against "meddling with nature" or "playing God." Rather, they frame normative assessments within a holistic account of nature and human nature. Thus, all efforts to alter nature or human nature must be viewed in light of the core Hindu principle of *ahimsa*, which dictates non-violence and the avoidance of harm.

1.2.2 Buddhism[8]

Whether Buddhism is thought of as a "religion" or a "philosophy" it is at its core a practice-based quest for liberation. In its classical forms, it does not affirm a "Creator" deity; and although it embraces core texts, they are not viewed seen as foundationally canonical in the same sense one finds, for example, in Judaism, Islam, or Christianity. Buddhism is best viewed as a practice-based way of life, based on practice-oriented teachings (*Dharma*) that seeks to identify those existential factors that promote or hinder transcendental enlightenment (*Nirvana*), and to inculcate habits and virtues that lead to that state.

The language of "Buddhist ethics" is seen as problematic, according to some scholars of Buddhism, if that rubric is meant to suggest that Buddhism is reducible to extrinsic systematic or philosophical categories (Keown, 1992). Nonetheless, Buddhism does exhibit central features that perform most, if not all, of the functions

---

[8] This section incorporates, in edited form, material from drafts by Cromwell Crawford and David Loy (see note 5 above).

that "religions" perform in the West; i.e., concern with an ultimate good and certain core practices as the bases for ethics as well as for devotion. Thus, other scholars provide an account of Buddhist practice that sees ethics as primary. Kathleen Nolan, for example, observes that "[e]arly Buddhist texts plac[ed] morality first among the three central doctrines of the Buddha's teachings – *sila* (morality), *samadhi* (concentration), and *prajna* (wisdom).[9] Moreover, Buddhist ethics, despite its distinctiveness, may well share affinities with other recognizable versions of "ethics." For example, Ninian Smart observes that all Buddhist practice should be undergirded by "a general attitude of compassion (*karuna*) which holds a place in Buddhism comparable to that of *agape* in Christianity."[10]

Buddhism arose in the fifth century BCE in India, and is identified, in its origins, with the example and teaching of Sakyamuni Buddha (563–483 BCE). In his first discourse, delivered shortly after his enlightenment at Sarnath, Sakyamuni emphasized the so-called "Four Noble Truths" that are the touchstones of the quest for liberation and enlightenment. The First Noble Truth is the fact of the omnipresence of suffering (*dukkha*). The Second Noble Truth identifies the cause of suffering to be attachment and desire. The Third Noble Truth proclaims that there is cessation (*nirodha*) to suffering through the "vanishing of afflictions so that no passion remains" (Nakamura, 1978, 135). The Fourth Noble Truth sets out the way to eliminate suffering and achieve enlightenment – through specific, practice-based strategies for overcoming suffering and achieving enlightenment called the "Eight-Fold Path." A word about each of the Four Noble Truths is in order here.

With reference to the first truth about suffering, while earlier Hindu teaching had emphasized that ultimate reality is Brahman, that Brahman is Atman (the true reality of the individual), and that the empirical world is illusion (*Maya*), Buddha's teaching in the First Noble Truth is that of the non-substantiality of the self (*anatma*) rather than the subsistence of some "true" or enduring self. All is change (*anitya*), including the ego; there is a ceaseless "co-arising" of patterns and energies that cannot be reified as individuated natures or "selves." All existence, and all existents, are impermanent and interpenetrating, with an "inherent mutuality" (*pratityasamutpada*) of all conditions, energies, and aggregates.

In reference to the Second Noble Truth, on the origin of suffering, Buddha identifies the cause of suffering as desire, craving, or thirst (*taha*) in all its forms – thirst for sensory and mental experiences, as well as attachment to various beliefs about

---

[9] Nolan expands upon this point: "Despite its mystical character, Buddhism is … an extremely practical religion, a religion of the present life and of the present moment. In fact, enlightenment itself can be understood as wisdom (awakening to the true nature of reality) expressed in the activity of compassion. Moral conduct thus forms the foundation for all spiritual attainment, and enlightenment without morality is not true enlightenment" (Nolan, 1993, 185).

[10] However, as Anne Klein observes, in qualifying Smart's comparison, it is important to determine whether agape is something that Christians cultivate or a distinguishing feature of God. If the latter, its place is not equivalent to compassion in Buddhism, which everyone cultivates to the point that theirs is identical with the compassion of the Buddha (private correspondence).

the seeming solidity of endlessly transitory phenomena. Such thirst, according to Sakyamuni, is the primary cause of the suffering of sentient beings.

The Third Noble Truth, that there is a way for suffering to cease, points to the "vanishing of afflictions so that no passion remains" (Nakamura, 1978, 135). Etymologically, "cessation" (*nirodha*) connoted "control." The linkage here between the "vanishing of afflictions" and the "control" of suffering may have positive implications for judgments concerning the use of biotechnology, although there is little on that point in the literature to date.

The Fourth Noble Truth sets forth the prescription for eliminating suffering and achieving *Nirvana*, the "Noble Eight-Fold Path." The eight steps of that Path are right views, right aspiration, right speech, right conduct, right mode of livelihood, right effort, right mindfulness, and right concentration. The "right," as a designation for each aspect of the Path, reveals an approach that characterizes the *Dharma* of the Buddha; *viz.*, that the way to enlightenment entails a "middle way," a right path between extremes of undue mortification or undue indulgence. Buddha therefore enjoined wisdom (*prajna*) and virtuous conduct (*sila*) as the avenues to enlightenment.

The moral precepts of Buddhism are expressed in terms of what is to be avoided in the so-called "Five Precepts" (*pancasila*), which are directed to both monks and laypersons. One should refrain from: taking life, from taking what is not given, from sexual misconduct, from wrong speech, and from intoxicants. As Ninian Smart observes, the first precept, to refrain from deliberately taking life, is framed within two larger truths: "the sense of kinship with other living beings implicit in the doctrine of rebirth" and the nonviolent attitude (*ahimsa*) central to Buddhist understanding" (Smart, 1986, 66).

After the death of Sakyamuni, Buddhism developed into three major branches – Theravada ("the way of the elders"),[11] Mahayana ("The Greater Vehicle"),[12] and Vajrayana, as reflected in some forms of Tantric and Tibetan Buddhism. Theravada Buddhism developed within 100 years of Buddha's death. In the Pali Canon, a collection of the oral tradition written down about 200 years after Buddha's death, the "three jewels" of practice were stressed: the Buddha, the *Dharma*, and the *sangha* (monastic community). The textual tradition of Theravada Buddhism centered on monastic life and practice. The *Vinaya Pitaka*, along with later commentaries, provides detailed rules for monastic life. In the first century BCE, Mahayana Buddhism arose, with a greater emphasis on practice extending to laypersons as well as monks. Mahayana introduced as a central theme the ideal of the *bodhisattva*, a "being intent on enlightenment" by enlarging the Theravadin quest for enlightenment of self to the quest for the enlightenment of all beings. That same emphasis on the need to end the universal suffering of all beings is also found in the various cultural expressions of the Vajrayana school.

---

[11] Theravada is actually the only extant school of 18 early Buddhist schools. These 18 are referred to as "Hinayana" (the "Lesser Vehicle"), a term adopted by the Mahayana (or "Great Vehicle"). Jon Strong has coined the term "Nikaya" Buddhism, which, unlike "Hinayana," is non-pejorative (Anne Klein, private correspondence).

[12] Mahayana is itself divided into sutra and tantra.

There has been a tendency in Buddhist scholarship to distinguish various schools of Buddhism by their different ways of interpreting core Buddhist precepts. Thus Theravada (dominant today in Sri Lanka, Thailand, Cambodia, and Laos) "takes a more literalistic and prescriptive stance" while in the Mahayana and Vajrayana schools, "the vow to end the suffering of all beings takes precedence over any dogmatic adherence to rules, and the precepts must function in accord with particular circumstances" (Nolan, 1993, 188). This is a useful way to characterize the presentation of concrete moral practice and instruction in the textual traditions of these schools, though it is not clear whether these differences also apply to actual moral practice and instruction as it is carried out in everyday life in these schools.

With reference to the issues of altering nature through biotechnology that are the focus of this volume, we can think of Buddhist perspectives in two ways. One way is to relate key Buddhist themes directly to issues involved in biotechnology. For example, the centrality of compassion, as evidenced by the *bodhisattva* ideal, would seem to have positive implications for interventions that are clearly therapeutic in their intent. At the same time, compassion has also been invoked to support the loving acceptance of disability, especially within the context of belief in the patient acceptance of particular karmic effects as an opportunity for growth toward enlightenment. This in turn is connected with the insight, endemic to the First Noble Truth, regarding the fundamental unsatisfactory character of existence. Moreover, all forms of intervention must be assessed in light of the central focus of Buddhist understanding; viz., "whether such interventions would render the recipient unable (or less likely to pursue) the Buddhist path of self-transformation" (Loy, 2003, 6–7). Finally, the central Buddhist notion of the mutual interdependence of all beings has, to date, been applied extensively to issues of environmental ethics, and is likely to be an important value in developing consensus approaches to topics such as global warming and the protection of biodiversity.

These core insights have provided an important context for Buddhists to participate in contemporary debates. However, we should be careful not to assume that there is "a Buddhist perspective" on these issues as they are formulated in contemporary debates. For one thing, the principles of compassion, interdependence, and a view of the world as by nature unsatisfactory, as presented in the previous paragraph, are central in Buddhist thought, but of course Buddhist traditions elaborate the significance of these principles in a wide variety of ways. Moreover, the implications Buddhists are likely to draw from these principles may involve significant departures from assumptions imbedded in the issues raised throughout this book regarding the altering of nature. For example, to raise the issue of "altering nature" assumes that there is some kind of leverage on nature from something outside it. Discussions in the West also tend to presuppose a virtually "natural" tension between science and religion. Both these orientations are deeply challenged by many Buddhist traditions. Finally, while Buddhists certainly recognize the distinction between common-sense meanings of "natural" and "civilized", a more fundamental distinction for them, philosophically speaking, is between the conditioned and the unconditioned (about which more is said below).

With this in mind we can think of a second way in which Buddhist perspectives might engage issues of biotechnology. This alternative begins by saying more about both nature and the question of its alteration from a Buddhist perspective. In the Buddhist culture of India and Tibet the strongest associations with the term "nature" have to do with the nature that is the ultimate status quo of all things. It is the nature of all phenomena, natural and unnatural, conditioned and unconditioned. The Sanskrit term *svabh)va* and its Tibetan equivalent *rang bzhin* mean, most literally, "own nature." Every existent phenomenon has its own nature, and the ultimate own-nature of all things is, in many Buddhist philosophical systems, identical. This is not a term, as in common Western usage, by which certain categories of phenomena are distinguished from others. "Own-nature" is all pervasive. It characterizes all phenomena: both the caused or conditioned as well as the uncaused or unconditioned. The distinction between conditioned and unconditioned is not a distinction that maps easily onto common Western distinctions such as noumena and phenomena, external and internal, or, as we have said, natural and unnatural. It is the nature of conditioned things to fall apart. Metaphysically, the arising of the world, and all the other arisings and ceasings of life in general, are possible precisely because everything is a dependent arising. To be a dependent arising means to depend on causes or conditions, or on parts (as the space of a room depends on the space of the front and back of the room, though space itself does not arise from causes), or on a mind which impute it. To be a dependent arising means to be empty of existing self-sufficiently, or inherently. This character of conditioned things bequeaths alertness to the inevitability of age, pain, and mortality, and it is in recognition of this that the entire round of birth and death known as *samsara* is declared fundamentally unsatisfactory. These dissatisfactions are not alien to life; they are intrinsic to it.

How might this general understanding of nature impact our taking stock of the place of nature in biotechnology, and most especially of the significance of the body as subject of biotechnological interventions? It must be admitted that at this time, when these perspectives are only beginning to become known to each other, we cannot yet predict what that impact will be. However, to indicate one way in which a Buddhist perspective illuminates the issue we will close by considering a poem by Longchen Rabjam (1308–1363), a great Tibetan philosopher and one of the ultimate authorities of the Tibetan Buddhist tradition. From Longchenpa's perspective, the natural world is an ideal place from which to seek liberation into the ultimate nature of everything. At the same time, the natural world, like all the rest of samsara, is fundamentally unsatisfactory and deceptive. Yet, it offers enough of a reprieve from distraction to allow one to become aware of this:

> In a forest, naturally there are few distractions and entertainments
> One is far from all the suffering of danger and violence…
> Blossoming flowers and leaves emit sweet odors.
> Fragrant scents fill the air….
> In forests, emotions decline naturally.
> There no one speaks unharmonious words.

1 Spiritual and Religious Concepts of Nature

> As it is far from the distractions of entertainment in towns.
> In forests the peace of absorption grows naturally....
> And (one) achieves the happiness of ultimate peace.[13]

It is worth noting that Longchenpa does not here have a generic term for the first kind of "nature." He refers always to some specific natural element such as flowers, scents, forests, These fit deceptively easily into a modern Western construct of "natural world," a concept which has no real equivalent in Sanskrit or Tibetan perhaps for the very reason noted earlier, that so called "nature" is in fact just another expression of the ultimate nature, and not "apart" in any meaningful ontological way. At the same time, obviously, Longchenpa draws a contrast between the forest and the town, but he does not categorize one as more "natural" than the other, only as having different properties pertinent to human spiritual development. There is no appeal to nature as such, only a recognition of an environment supportive the goal of liberation. Suffering, or at the very least a fundamentally unsatisfying uncertainty, is the hallmark of being in the world, and "nature" offers endless examples that can awaken one to this reality:

> Life is impermanent like the clouds of autumn
> Youth is impermanent like the flowers of spring.
> The body is impermanent like borrowed property...[14]

The principles of nature, then, are not something against which to pit the autonomy of human ingenuity, but rather are indicative of a fundamentally problematic situation which Buddhist practice, much like science, seeks to address. There is really no question of "altering" nature in the sense either of undoing impermanence – cosmetic cover-ups are no threat whatever to the inexorable existential truth of mortality – there is only an impetus to understand how best to live in the face of it. Translated into contemporary terms, this might mean that quality of life issues and other questions of value are crucial elements to be integrated into any kind of technological achievement. In any case, this offers one example of the kind of conversation we may hope will occur between Buddhism and biotechnology.

1.3 The "Abrahamic" Traditions

Treatment of the meaning of nature and its alteration in Judaism, Christianity, and Islam has the advantage that these traditions are closely intertwined with the history of Western science and technology. The advantage is even greater in the cases of Judaism and Christianity, which both have traditions of casuistical and other forms

---

[13] Quoted and translated by Tulku Thondup in *Buddha Mind*, 168–169 from NKT *Nags-tshal Kuntu dGa'I ba'I gTam* (ff. 66–72) Vol. *I gSung Thorbu*.

[14] *Ibid.*, p. 177. *Extract from TRL gTer-Byung Rin-po'che'I Lo* rgyus (f. 53) (Vol Om mkna'-gro Yang Tig, Ya-bzhi) by Dri-Med A'od Zer, published by Sherab Gyaltsen Lama 50b.3

of moral discourse whose modern iterations developed in close connection with modern technology. However, these advantages are offset by the difficulty of characterizing in a few paragraphs the complexity and the sheer volume of material relevant to our topic in these traditions, especially in the cases of Judaism and Christianity, but increasingly in the case of Islam as well.

1.3.1 Judaism

Ethical values lie at the heart of Judaism and are central to all aspects of Jewish life. The core concepts of Judaism are "God; Torah, or 'Teaching'; and the community of Israel, the Jewish people" (Mackler, 2003, 44). The origin of Judaism is God's covenant (*berit*) with the Jewish people, and the giving of Torah as the pattern for community. Torah consists of Scripture (written Torah) and Tradition (oral Torah). The written Torah consists of the first five books of the Hebrew Bible, while the oral Torah includes "all Jewish traditional teaching – in fact, all authentic Jewish thought and practice" (Mackler, 2003, 44). The bulk of the oral Torah compiles rabbinical teachings from the so-called Talmudic period (first century BCE to the sixth century CE).

Halakhah ("path" or "way") designates the Jewish legal system, with its source in God's revelation. There are several sources of normative authority for *halakah*: 613 positive and negative commands, drawn from the written Torah; and the full compendium of rabbinical opinions on passages of the written Torah, as well as "rabbinical decrees, customs, positive enactments, and negative enactments" (Steinberg, 1991, 179).

In addition to the *halakhah* is a second source of Jewish tradition, the *aggadah* ("narrative" or "lore"). *Aggadah* refers, in general, to "Jewish theological reflection, lore, articulation of values, expressions of meaning, and cultivation of virtues" (Mackler, 2003, 45). In the written and oral Torah, one finds, therefore, two levels of authority. *Halakhah*, deemed to have legal authority, was historically enforceable in Jewish courts. *Aggadah*, while seen as an important part of the oral Torah, historically generated greater diversity of opinion in its interpretation.

Traditional authority has been vested in a decentralized rabbinic leadership in local communities. At the same time, certain rabbis, as well as "leaders of various academies" (Mackler, 2003, 51) have been acknowledged as major authorities in light of the expertise of their judgments and scholarship. While Scriptural law is open to interpretation, "it cannot be amended or repealed … because it is taken to be the direct word of God" (Novak, 1995, 1302). By contrast, rabbinic law "at least in theory admits of amendment and repeal," although "reinterpretation of already existing norms" is the general method of amending rabbinical interpretations (Novak, 1995, 1302).

Among the milestones in the history of rabbinical opinion, several are especially important (Steinberg, 1991, 179–180). Rabbi Judah Hanasi, in the early third century CE, compiled the *mishnah*, which collected scholarly discussion to that point. A second set of writings was the *midrash*, with commentaries on the written Torah.

During the Amoraic period (220–470 CE), rabbinic scholars assembled the *gemara*, a collection of further interpretations. The latter, in combination with the *mishnah*, was called the *Talmud*. In the sixth and seventh centuries CE, Palestinian authorities assembled the so-called Jerusalem Talmud, and Babylonian rabbis the so-called Babylonian Talmud, with the latter subsequently deemed the more authoritative.[15]

Other important sources for Jewish thought are later Talmudic commentaries and a group of writings called *responsa*, which are rabbinic decisions on particular issues or cases that involve *halakhic* matters, and which "collectively ... constitute the case law of Judaism" (Mackler, 2003, 49). Legal codes are a third source. In the twelfth century, Moses Maimonides assembled the *Mishneh Torah*, the first systematic codification of all *halakhic* opinions until that time. In the sixteenth century, Joseph Karo compiled and edited *Shulhan Arukh*, a systematic codification subsequently viewed as especially authoritative.

In its use of *halakhic* materials, Jewish ethical method is characteristically legalistic, pluralistic, casuistic, and open to further interpretation and application (Brody, 1998). It is pluralistic because it draws on a range of sources of authority and interpretation, within the overall context of *halakhic* understanding. According to a recent commentator, "Judaism is opposed to absolutizing any single precept; rather, a golden path, a middle way, is always advocated and encouraged" (Steinberg, 1991, 181). When values come into conflict, Judaism adopts the method of casuistry: the scriptural norm is determined; rabbinic interpretations, especially in the *responsa* literature are studied; precedents are identified and compared; and principles and rules are derived and applied (Novak, 1995, 1302).

However, while "[a]ll Jewish thinkers acknowledge a role for reason and experience in shaping ethics," there are significant differences among the three major movements in Judaism – Orthodox, Conservative, and Reform – in the way that reason and experience are seen to function in relation to Scripture and Tradition (Mackler, 2003, 46). Reform Judaism, arising in the wake of the Enlightenment, tends to emphasize reason and experience and "do[es] not consider Jewish law to be binding" (Dorff, 1997, 76). Orthodox Judaism, in response to the Reform movement, emphasizes revelation and tradition as the decisive guides for Jewish ethics (Mackler, 2003, 47). Conservative Judaism, while viewing *halakhah* as binding, is "more willing than Orthodox [Judaism] to make changes in its content in response to modern needs" (Dorff, 1997, 76).

In theory, Judaism views seven fundamental commandments in the written Torah as binding on Jews and Gentiles alike (the so-called "Noahide Law," based on the covenant made with Noah and his descendants (Gen. 9) that predates the call of Abraham). These commandments are six proscriptions: against blasphemy, idolatry, stealing, murder, sexual offenses, cruelty to living animals; and a positive injunction: to establish courts to administer justice.

---

[15] According to Steinberg, "With the closing of the *Talmud*, this work virtually became the infallible source of the *Halakhah*, and the history of post-Talmudic *Halakhah* is founded on recourse to the *Talmud* as the final and overriding authority" (Steinberg, 1991, 179).

Jewish understandings of nature begin with the teaching of the Hebrew Bible that God created the world. The first chapter of Genesis describes God as creating the world in an orderly process of seven days. The narrative of creation is marked by a repeated refrain, "and God saw that this was good"; the first chapter concludes, "And God saw all that He had made, and found it very good" (Gen. 1:31).[16]

According to the biblical account, humans are part of nature, but also are distinguished from the rest of the natural world. In Genesis 1, God creates humans on the same day that land animals were created; but God creates humans in God's image. God blesses the first humans, "Be fertile and increase," the same blessing given to animals, and continues, "fill the earth and master it; and rule the fish of the sea, the birds of the sky, and all the living things that creep on the earth" (Gen. 1:27–28). In the account of Genesis 2, God forms the first human (*adam*) from the dust of the earth (*adamah*), and then God breathes in the breath of life. God sets the first human in the garden of Eden, to work with it and to preserve it (Gen. 2:7, 15).[17]

References to creation are found throughout the Hebrew Bible, with some variation. The canvas of Genesis 1 is vast, the entire cosmos and all it contains. Humans are created near the end of the process, on the same "day" as beasts and cattle. The account of Genesis 2–3 is more anthropocentric. It begins with the creation of man, after which God plants a garden and places man within it, and then creates animals as potential companions for man, before the first woman was formed. In both of these accounts, God creates calmly and with regal power, without any hint of opposition (until the first humans disobey His command). Other biblical passages reflect a battle from which God emerges victorious. "It was You who drove back the sea with your might, who smashed the heads of the monsters of the waters; it was You who crushed the heads of Leviathan, who left him as food for the denizens of the desert; it was You who released springs and torrents, who made mighty rivers run dry" (Ps. 74:13–15).[18] Many medieval philosophers and mystics believed that our world came about through emanation from the primordial divine source.[19] Some thinkers today reappropriate this image; creation is understood as emanation, "a process, a flowing forth of the divine self, rather than the creation of a wholly other out of nothing" (Green, 1992, 59). Accordingly, "God and the universe are related

---

[16] Biblical sources in this section generally are given in accord with the New Jewish Publication Society translation, *Tanakh* (Philadelphia, PA: Jewish Publication Society, 1985).

[17] An ancient midrash (creative exegesis) reflects appreciation for the benefits that the natural world offers for humans, while emphasizing that the world is God's and humans have responsibility for their stewardship. "In the hour when the Holy One, Blessed be He, created the first person, He showed him all the trees of the garden of Eden". He said to him, "See my works, how pleasant and praiseworthy they are. All of this I created for you. Be careful that you do not ruin and destroy my world, for if you ruin it there is no one who will repair it after you" (*Ecclesiastes Rabbah* 7:13).

[18] Neil Gillman traces four differing accounts of creation in the Hebrew Bible and Jewish tradition, arguing that each of the four provides a partial view that complements the others. Because of the inherent mystery of creation, no single account can convey the whole truth (Gillman, 2000, 127–144).

[19] See *Encyclopaedia Judaica*, s.v. "Emanation," 6: 694–696, by J. Kramer and G. Scholem.

not primarily as Creator and creature, but a deep structure and surface" (Green, 2002, 4). Feminist thinker Judith Plaskow likewise asserts that "God is not a great king who rules the world, but the power that sustains and moves it, ... present in the whole of reality" (Plaskow, 1990, 144).

The Bible and most of Judaism, however, understand God as distinct from (though involved with) nature. Still, God is manifest to humans in and through nature. For the Book of Psalms, "The heavens declare the glory of God, the sky proclaims His handiwork." God causes water to flow, and causes plants to grow to feed animals and humans. "How many are the things You have made, O Lord; You have made them all with wisdom, the earth is full of Your creations" (Pss. 19, 104). The Psalmist poetically calls upon the natural world to praise God. "Let the heavens rejoice and the earth exult; let the sea and all within it thunder, the fields and everything in them exult; then shall all the trees of the forest shout for joy" (Ps. 96:11–12). For the Book of Job, God's power and wisdom is evidenced by nature's terrifying anomalies as well as by its regular patterns (Job 38–41). In the Torah, the rainbow is proclaimed a sign of God's covenant with Noah, and God's revelation at Mount Sinai is accompanied by thunder and lightning (Gen. 9, Ex. 20). The community of Israel would enjoy rainfall and natural fertility if they lived righteously and in accord with their covenant with God, but drought and natural disaster would follow sin (e. g., Lev. 26, Deut. 11, 28). The ultimate redemption will be marked by the flourishing of nature and by the restoration of the harmony that marked the original creation, as "the wolf shall lie down with the lamb" (e.g., Amos 9, Jer. 33, Is. 33, 11:6).

Reflecting on the Bible's understanding of nature, the modern Jewish thinker Abraham Heschel writes that

[T]here are three aspects of nature that command our attention: its power, its beauty, and its grandeur. Accordingly, there are three ways in which we relate ourselves to the world– we may exploit it, we may enjoy it, we may accept it in awe. ... Our age is one in which usefulness is thought to be the chief merit of nature; in which the attainment of power, the utilization of its resources is taken to be the chief purpose of man in God's creation (Heschel, 1955, 33–34).

In contrast, the Bible and rabbinic Judaism call attention to nature's grandeur and to the sublime, which is understood as "a way in which things react to the presence of God" (Heschel, 1955, 40),

Rabbinic Judaism developed responses to God's presence as manifest in nature. God's creation and sustenance of the world is celebrated in the traditional daily liturgy every morning and evening. A central morning prayer declares: "Praised are You, Lord our God, King of the universe, who forms light and creates darkness, makes peace and creates everything. He illumines the earth and those who dwell upon it with mercy, and in His goodness He renews the works of creation each day, continually." (Some feminists and others reject the traditional image of God as king. New blessings have been suggested that instead praise "the source of life" [Plaskow, 1990, 129, 142, citing Falk, 1987].) Shorter blessings acknowledge God's presence as one encounters it in nature. Upon hearing thunder one traditionally recites, "Praised are You, Lord our God, King of the universe, whose power and might fill the world." On seeing high mountains, one recites, "Praised are You...

who makes the works of creation." Blessings are recited on other occasions, such as seeing a rainbow, smelling fragrant plants, and seeing fruit trees in bloom. Traditionally, a blessing is recited upon relieving oneself: "Praised are you ... who with wisdom fashioned the human body, creating openings, arteries, glands and organs, marvelous in structure, intricate in design.... Praise are You, Lord, healer of all flesh who sustains our bodies in wondrous ways" (Harlow, 1985, 7).[20] One also recites various blessings before (as well as after) enjoying food or drink. For example, before eating fruit, one recites, "Praised are You ... who creates fruit of the tree." Before consuming many types of food and drink, including water, one recites, "Praised are You, Lord our God, King of the universe, by whose word all things come into being." As Heschel observes, "Each time we are about to drink a glass of water, we remind ourselves of the eternal mystery of creation, 'Blessed be Thou ... by Whose word all things came into being.' A trivial act and a reference to the supreme miracle" (Heschel, 1955, 49).

Humans are called to respect and preserve the natural world, responding to God's presence as manifest in creation. At the same time, Judaism understands humans to have an active role as partners with God, helping to improve the world and bring it closer to perfection. Creation is understood at least in some ways to be an ongoing process (as reflected in the liturgy's praise of God's daily renewing of creation). The Talmud and later tradition speak of humans as partners with God (Talmud, *Shabbat* 10a, 119b). A traditional reading finds warrant for human activity in Genesis 2:3, which tells that God on the seventh day of creation ceased from "all the work of creation that He had done" – a phrase that could be read, "all His work that God created to do," suggesting a mandate for humans "to do" further work.[21] Contemporary rabbi Irving Greenberg writes: "There is a covenant between God and humanity .... The human role in this covenant is to perfect the world which was brought into being by God .... To perfect the world means that the human is called upon not to accept the world as it is, but to improve it and to complete it. ... The key to constructive use of power is partnership" (Greenberg, 1986, 128–134).[22]

---

[20] Liberal thinker Eugene B. Borowitz writes:

> I was once in the hospital for a torn kidney and was very grateful when it began to heal. As a sophisticated modern who felt refined by high culture, I had thought it rather crude, even vulgar, for the rabbis to have prescribed a *berakhah* [blessing] after urination and defecation. In the hospital I discovered that urination was an extraordinary gift. In deep appreciation of the wonder and grace of my body's functioning, I taught myself to say that *berakhah*. Saying it in innumerable repetitions over the years has taught me a good deal about the physicality of my spirituality and of the intimate depths of my dependence on God (Borowitz, 1991, 112).

[21] Genesis 2:3 and commentaries of Abraham ibn Ezra and David Kimchi.

[22] According to some mystics, the original creation involved God "contracting" from some sphere to make space for the world, followed by divine emanation; some of the "vessels" intended to hold the divine light shattered in a cosmic catastrophe. Humans have a task of *tikkun*, repair and restoration of the world. Many who do not accept the full mystical cosmology value its dramatic imagery of the vital role of humans in improving the world. See Scholem (1974, 128–144).

1 Spiritual and Religious Concepts of Nature

Some have seen in the commandment of circumcision a warrant for humans changing, rather than simply accepting, the natural condition.[23] Healing represents another example of interference in the natural order that is not only accepted, but mandated. One classical narrative depicts two rabbis as offering medical advice to an ill individual, only to have the patient accuse them of interfering with God's will of making the person ill. The rabbis asked the man, a farmer, why he interfered with the God-given state of his vineyard by fertilizing and weeding. When he responded that crops would not grow well without human care, the rabbis replied: "Just as plants, if not weeded, fertilized, and plowed, will not grow and bring forth fruits, so with the human body. The fertilizer is the medicine and the means of healing, and the tiller of the earth is the physician."[24]

The Bible and later Judaism mandate *mitzvot* (singular *mitzvah*, commandment) specifying responsibilities to treat the works of nature with appropriate respect. Deuteronomy commands that soldiers besieging a city may not cut down fruit trees (Deut. 20:19–20). This became the basis for a general prohibition of wanton destruction, *bal tashchit* (Maimonides, *Mishneh Torah*, Kings 6:10; *Sefer Hachinukh*, no. 530). Animals should be properly treated, and cruelty to animals is forbidden by the tradition.[25]

Traditional observance of the weekly Sabbath reflects the tradition's value of nature as God's creation. Six days humans should work creatively, but on the Sabbath acts of physical creation or manufacture are forbidden. On this day, one should accept and appreciate the world as it is, respecting God's sovereignty. The Bible also mandates a sabbatical year in which agriculture is not practiced. On the seventh year, "the land shall observe a sabbath of the Lord." The sabbatical and jubilee year reflect the belief that ultimately the land belongs to God. God allows humans to enjoy the world, and humans appropriately exercise stewardship actively in preserving and improving the world, but this should be done with a sense of responsibility and respect for nature as God's creation.

In the context of God as Creator and Author of the Covenant, certain core ethical values shape a characteristically Jewish approach to the ethics of altering nature and the assessment of biotechnologies. Three aspects of Jewish thought are especially salient.

---

[23] The medieval *Midrash Tanhuma* (*Tazria* 7) presents a debate between Rabbi Akiba and a pagan, Turnus Rufus. The pagan asked which works were preferable, those of God or those of humans. Akiba answered that human works could be preferable, giving the example of stalks of grain and loaves of bread. Turnus Rufus argued that if the circumcised state were preferable, God should have made boys be born circumcised. Akiba responded that there is value in humans being given an opportunity for significant action. Circumcision, like other commandments, can serve to purify the character of those who perform it.

[24] *Midrash Temurah*, translation based on that of David M. Feldman (1986, 166). Rabbi Joseph Soloveitchik writes, "Only the man who builds hospitals, discovers therapeutic techniques and saves lives is blessed with dignity .... Man reaching for the distant stars is acting in harmony with his nature which was created, willed, and directed by his Maker. It is a manifestation of obedience to rather than rebellion against God" (Soloveitchik, 1965, 14, 16).

[25] As well, religious holidays are linked to the natural cycle of the agricultural year. Passover is observed in the spring, and Shavuot is celebrated with first fruits seven weeks later. The harvest festival of Sukkot is celebrated with lulav and etrog, tree branches and a "beautiful fruit," or citron (Lev. 23).

First, when life is at risk, Jewish ethics requires that all laws of Torah that would appear to prevent action must be set aside, except for the prohibitions against idolatry, murder, and sexual sin. The basic value expressed here is captured by a passage from the *mishnah*: "Whoever saves even one human life, it is as if he saved an entire world" (Novak, 1995, 1303). God as Creator sets the context for that ethical norm (*pikkuah nefesh*). We are not owners of our own lives, but stewards of the life that God has given: "Therefore, [one] is positively obliged to seek medical help, and to do all that is necessary to preserve … life and health" (Steinberg, 1991, 187). The implications of this value for beginning- and end-of-life decisions have been the focus of significant recent rabbinical discussion and debate, especially in relation to issues of treatment withdrawal and the definition of death. More pointedly, for biotechnologies that are accurately described as therapeutic in their intent and effects, that value might also be determinative, though much would depend on an assessment of risks and the specific arguments yet to be offered in the *responsa* literature.

Second, Jewish thought does not elaborate a doctrine of "natural law" as such. Thus, absent specific *halakhic* prohibition, "man is free to utilize scientific knowledge in order to overcome impediments on nature" (Steinberg, 1991, 183), under the rubric of "repairing the world" (*tikkun olam*). Moreover, Judaism accepts both natural and artificial means in responding to illness; thus, "[t]he mere fact that human beings created a specific therapy rather than finding it in nature does not impugn its legitimacy" (Dorff, 2000, C-3).

Third, there is a "duty to care for the sick, and to heal them whenever possible (*biqur holim*, literally, 'visitation of the sick')" (Novak, 1995, 1303–1304), which is based on a range of rabbinical sources and arguments. Most prominent among these is the interpretation of Moses Nahmanides (1194–1270), who describes healing as a participation in divine activity, a form of "following after God's attributes" (*middotav*) (Novak, 1995, 1303–1304).

Finally, in the context of justice in health care, questions about the distribution of benefits (and burdens) of new forms of biotechnology provide a broader context for the assessment of particular developments. Judaism understands the provision of health care as a responsibility of the community, a commitment that "flows from traditional values and practices of *tzedakah* – a word meaning justice and referring to support for poor people" (Mackler, 2003, 194). In the context of justice, the tension to be negotiated is the need to "balance applications of [new technologies] with the legitimate right of a private company to make a profit on its efforts to develop and market these applications" (Dorff, 2000, C-4).

1.3.2 Christianity

1.3.2.1 Roman Catholicism

Roman Catholic moral theology arose as a distinct discipline in the sixteenth century. At its core, in contrast to Protestant theological ethics, Roman Catholicism embodies a relatively positive appraisal of natural moral insight available to persons as a function of God's grace in creation. In Aquinas's formulation, human beings share in God's Eternal Law by virtue of their rational nature (*Summa Theologica* (*ST*) I–II, 79–85).

From the time of High Scholasticism and until the Second Vatican Council (1961–1965), Roman Catholic moral method has emphasized natural law as a source of moral knowledge in principle available to all persons. The primary natural law directive in Catholic thought, available to moral reason in a process called "synderesis," is to "do good and avoid evil." In concretizing particular moral demands in light of that general directive, human reason, according to Aquinas and later Scholastics, reflects on certain natural inclinations as "givens." In light of that reflection, certain basic goods of human flourishing become obvious to practical reasoning. In turn, that "natural law" reasoning enables human beings to determine appropriate action guides (natural law directives) in light of the goods so identified.

Both nature and human nature are teleologically suffused in Catholic thought. While original sin has distorted human capacity to know and to do the good, Catholic moral method is distinguished from Protestantism by its greater confidence in our natural moral capacity to discern the good and our natural moral will to do the good (at least partially). At the same time, though often underemphasized in Catholic moral theology, Special Revelation (Scripture and Church Tradition) functions crucially in two respects: first, to illuminate and empower human beings to know and to realize their supernatural ends (*beatitudo*, or final union with God) and to bring greater clarity to the conclusions of "natural" moral knowledge that, although in principle available to moral reason, may be distorted by sin.

Drawing on Aristotelian categories, Aquinas depicts nature as Creation. The human capacity to reason about nature and human nature allows men and women, upon reflection, to understand the ends and purposes of the created order, including the ends and purposes of human life. For Thomas, the basic goods of human flourishing transparent to reflection (*per se nota* truths) are self-preservation, the begetting and education of children, the good of community, and knowledge of and communion with God.[26]

As a central tenet of Catholic moral method, a morally licit choice requires that one never choose *directly* against any of the basic goods available to practical reason.

---

[26] *ST*, I–II, q. 94, art. 2: "There is an order of precepts of natural law corresponding to the order of natural inclinations. First, there is an inclination in man towards the good corresponding to what he has in common with all individual beings, the desire to continue in existence in accordance with their nature. In accordance with this inclination, those matters which conserve man's life or are contrary to it are governed by natural law. Secondly, there is in man an inclination to some more specific objects in accordance with the nature which he has in common with other animals. According to this, those matters are said to be of natural law 'which nature has taught all animals,' such as the union of male and female, the bringing up of children, and the like. Thirdly, there is in man an inclination to do good according to the nature of reason which is peculiar to him. He has a natural inclination to know the truth about God and to live in society. Accordingly, those matters which concern this inclination are matter for natural law, such as that a man avoid ignorance, that he not offend others with whom he should have converse, and other matters relating to this."

Given that stricture, the discussion of legitimate exceptions to moral norms, so construed, has been a significant focus of Catholic moral theology, especially during recent decades. The language of "directness" is crucial to that formulation, and the principle of double effect emerges as an effort to maintain moral consistency by distinguishing, in complex cases, between the good effects one directly intends and evil effects that, while foreseen, are unintended and therefore "indirect."[27]

The foregoing example of double effect reveals a central characteristic of Catholic moral method, *viz.*, its employment of casuistry (Jonsen and Toulmin, 1988; Keenan and Shannon, 1995). Casuistry is the attempt to "formulate expert opinion about the existence and stringency of moral obligations in situations when some general precept would seem to require interpretation due to circumstances" (Jonsen, 1986, 78). For example, the meaning of the precept's terms may be unclear, or in conflict with those of another equally binding precept. In other cases, a situation may pose novel features that require assessing the application of earlier moral formulations. While casuistry is often viewed negatively (especially by much of the Protestant tradition), it is an important feature of Judaism, Islam, and Catholicism.

Especially since the time of Vatican II, natural law thought has been subject to critical scrutiny and major efforts at reformulation, with three new emphases emerging. First, a stress on historicism has challenged an earlier largely static model of individuals and society. While the primary directive of natural law – to do good and avoid evil – retains its general, largely formal, force, many recent discussions point to the variability of cultural and historical values in the specification and application of "secondary" natural law directives in particular circumstances (e.g., McCormick, 1989). To be sure, natural law method remains a bulwark against moral relativism – since the basic goods Thomas identified are thought to endure as identifiable *teloi* across cultures and eras – but the possibilities for clear development in moral doctrine are now emphasized.

In that light, critical studies, especially feminist approaches, have explored the ways that past interpretations of nature and human nature have tended to reify constructs of the human good in essentialist terms, especially in relation to issues of sexuality, gender, and parenting, often to the detriment of women in particular cultural settings (Shannon and Cahill, 1988). More broadly, critics have noted a "physicalism" (Curran, 1999) at work in the natural tradition, particularly within the so-called "manualist" tradition developed in seminary settings to train future confessors.[28] Especially on matters of sexual ethics, such physicalism has tended to

---

[27] Recently, so-called "proportionalists" have called into question the status of putatively "exceptionless" moral norms. They argue that the language of "intrinsic evil" (*malum in se*) cannot be coherently applied independently of the intentions of the agent and the fuller context of decision making. See McCormick and Ramsey (1978); Hoose, (1989).

[28] "[T]he single most influential factor in the development of the practice and of the discipline of moral theology is to be found in the growth and spread of confession in the Church" (Mahoney, 1987, 84). With the growth of private confession, so-called 'penitential books' were written and circulated – in Wales and Ireland, then France, England, Italy, and Spain, that provided priests with specific questions to ask of penitents about various sins in order to assign appropriate penances.

emphasize the structure of acts independent of broader concerns about person-centered agency or intentionality.

As a second emphasis since the time of Vatican II, Catholic moral theology has increasingly adopted the approach and language of "personalism." That perspective has significantly broadened the focus of natural law reflection prevalent before the Second Vatican Council. The goods of human embodiment – individual, sexual, and communal – are contextualized in more existential fashion, with the constitutive elements of human flourishing now considered in terms of the overall good of persons "integrally and adequately considered").[29] The extent to which personalism's shift of focus alters moral judgments based on more traditional natural law grounds remains the subject of ongoing debate.

A third emphasis in the recent tradition has been a renewed interest in, and appreciation for, the place of Scripture in moral reflection. Scripture retains its place as a central authority in virtually all forms of Protestant moral reflection (though *that* Scripture is authoritative is clearer in Protestantism than precisely how it functions in that capacity). Nonetheless, while recent Catholic moral discussions have appealed to Scripture as an important context for moral reflection, as Aaron Mackler observes, "detailed analysis of Scripture generally receives relatively little attention in the formulation of specific ethical judgments" (Mackler, 2003, 32). Although John Paul II has emphasized broad Scriptural motifs in recent encyclicals, the influence of Scripture on moral reasoning and judgment remains primarily at the level of motivation and disposition (see John Paul II, 1993). Thus, despite greater overt attention to Scripture, Catholic moral theology retains its characteristic emphases on human reason, natural law, and the pronouncements of the magisterium.

Roman Catholicism is perhaps most distinct from other traditions in its official understanding of the magisterium as its central teaching authority. As defined by the First Vatican Council, the term "magisterium" refers to the centralized teaching authority of the Church's hierarchy, headed by the Pope. To be sure, the charism of teaching is a widely shared function in the church, but "this prerogative resides in a special way with the Pope" (McCormick, 1986, 363). The foci of magisterial teaching are matters of faith and morals. While natural law remains a fundamental source of moral insight, the magisterium as central authority embodies the Church ecclesiological self-understanding; *viz.*, that the Pope claims the legacy of Peter as first pontiff in guarding and handing on the "deposit of faith" (Mackler, 2003, 35). That understanding claims the power of the Holy Spirit in aiding the Pope to provide unique insights concerning the essentials of the Catholic faith and fundamental implications of natural law.

---

[29] "... personalism refers to that modality of application of theological principles whereby an emphasis is placed on the entire personal complex of the act in its human dimensions, circumstances, and consequences ... personalism does not limit its scope to the physical or biological qualities of the action, but rather extends its purview to psychological, social, and spiritual dimensions" (Kelly, 1979, 419, quoted in Mackler, 2003, 40).

Such confidence about the magisterium's role in clarifying moral conclusions that remain in principle derivable from natural law flows from Catholicism's traditional confidence in the mutuality between Revelation and natural moral insights. As James Gustafson observes, "There are no serious cleavages between the revealed moral will of God and the natural moral law, as both have the same ultimate source" (Gustafson, 1978, 26, quoted in Mackler, 2003, 42).

Nonetheless, the role of the magisterium in the Catholic Church has also been subject to significant misinterpretation. While the Church "understands its teaching authority to extend to absolutely definitive and irreversible proclamations" (infallible teachings), such pronouncements are in fact rare. Indeed, most Church teaching, especially on moral matters, assumes the form of "authoritative but noninfallible teaching that does not exclude error in principle" (McCormick, 1986, 363). Thus, while deference is owed to magisterial authority, other sources of moral insight – natural law and human experience broadly construed – are brought to bear on the discussion by Catholic moral theologians on a number of conflicted questions, especially regarding sexual morality. In light of the pluralism of sources available for moral insight in Catholic moral theology – including relevant data from the natural and social sciences – there is clear room in the tradition, at least in principle, for change and development in moral teaching (Noonan, 1993).

The implications of Catholic moral theology for issues of altering nature and the assessment of new biotechnologies remain mixed. Official Roman Catholic teaching on new reproductive technologies is based on what has been called the "inseparability principle;" viz., procreative potential and conjugal intimacy should not be separated in any act of marital intercourse. Hence, "all methods of artificial reproduction which replace sexual intercourse as the cause of human generation are judged immoral" (Boyle, 1991, 8). Official Catholic teaching also rejects all forms of manipulation (including embryonic stem cell research) that involve destruction of embryos as a form of early abortion. But in other areas of biotechnology (including genetics) that do not involve such destruction, particular developments will be assessed in light of the goods of human flourishing that are served or disserved in any given instance.

1.3.2.2 Protestantism

Allen Verhey observes that "Protestant reflection about morality, including medical morality, defies generalization" (Verhey, 1995, 2117). From its historical roots in the Reformation thought of Martin Luther, John Calvin, Ulrich Zwingli, and other figures, "Protestantism" captures a broad range of tendencies and emphases, both theologically and ethically. Protestant thought ranges from the prophetic stance of Reformation piety and polity through the high liberalism of the nineteenth century, from the Social Gospel movement of the early twentieth century to the neo-orthodoxy of Karl Barth, from the existentialism of Rudolf Bultmann and Paul Tillich to the moral realism of Reinhold Niebuhr, and with myriad of other perspectives as well. In bioethics proper, one finds represented in the pioneering work of Joseph Fletcher and Paul Ramsey two quite different emphases native to Protestant thought. Fletcher discusses sacrificial love (*agape*) in situational terms, with an emphasis on human freedom and rationality. By contrast, Paul Ramsey, in his bioethical work, emphasizes covenant love (*hesed*) and deduces fairly stringent principles and rules that govern the concrete dimensions of love and justice.

There are both theological and institutional reasons for Protestant pluralism; *viz.*, "freedom of thought, within quite wide boundaries, is essentially Protestant, and it is impossible to say – as one can say for the Roman Catholic Church – that this or that is 'the Protestant position' at any place and time" (Johnson, 1978, 1364–1365). Nonetheless, in what follows, we will prescind from the wide variety of Protestant thought in order to outline certain shared characteristics of a broadly Protestant perspective, especially by contrasting tendencies in Protestantism with those in Roman Catholicism.

In overview, two features of Martin Luther's perspective remain critical to Protestant understanding: first, an emphasis on the sovereignty of God, and second, God's unmerited grace that enables the Christian to respond in freedom to the concrete requirements of loving one's neighbor. In contrast to the medieval Catholic understanding of vocation in terms of calling to clerical or vowed religious life, Luther democratized the meaning of vocation, with Christian witness now "mediated through the various professions and occupations" of ordinary life. In his or her calling, each person stands radically before God, without the need for clerical intermediaries. In contrast again to Catholic understanding, Luther preached "the priesthood of all believers." Moreover, rejecting the central place of penitential casuistry in Catholic method, Luther saw each individual as responsible for hearing God's word, and therefore distrusted casuistry as an unevangelical legalism. So, too, Catholic sacramentalism was deemed an encroachment on the prerogatives of grace.

Equally important to the thought of Luther and Calvin was their discussion of the nature of, and the relations between, "law" and "gospel" as theological and ethical "categories." For Luther, law is that which God demands, the gospel that which God gives. On this reading, law precedes gospel, or, as often formulated, the imperative (law) precedes the indicative (grace). Central to Protestant understanding here is an emphasis on sin as the general condition of a broken relationship between God and man as the result of Adam's Fall. (That emphasis on the general condition of sin as an existential and moral reality stands in marked contrast to the penitentially based focus on sins as individual choices and acts in Roman Catholicism).

In this context, for Luther, the law has only two functions: an "elenctic" or judging function, which offers a mirror to our depravity in failing to heed God's command; and a civil use, as a dike against the effects of sin through what Luther called the "orders of nature," including the function of government in maintaining public order. For John Calvin, law also has a third or spiritual use, whereby the law can school believers in the path of sanctification and spur them to holiness.

Of general relevance to the focus of our project are Protestant interpretations of the appropriate norms for Christian participation in the structures and processes of a fallen world. Luther developed his doctrine of the "two realms" of creation and redemption. In the realm of redemption, "the Redeemer rules all regenerate believers through Christ and the gospel" (Lazareth, 1986a, 360)." In the realm of creation, even after the Fall, God's love remains for all creatures in the "dependable order which furthers human well being." Although creation remains under the law and not the gospel, "there is good in the world which is not derived from the reconciling act of Calvary but from the fact of Creation" (Lazareth, 1986b). Insofar as redeemed individuals, living out their callings in the world (including the professions of science and medicine), can positively affect the orders of creation, progress in biotechnology for the good of others is a general implication of Luther's understanding. For Calvin, who often describes creation, even though fallen, as the theater of God's glory, creation is seen in even more positive terms: "[i]n covenant with God ... believers can participate in God's creative activity" (Lazareth, 1986b).

Distinctively Protestant ethical reflection proceeds from seeing Scripture as a foundational authority. However, as Allen Verhey observes, "To affirm the authority of Scripture is to invoke the use of the Bible in moral discernment and judgment, but it is not to prescribe how the Bible is to be used" (Verhey, 1986, 57). Scripture functions in Protestant ethical discernment, therefore, in a variety of ways, depending on several factors. First is a judgment about its nature and message. How is Scripture conceived – in literal terms, in broadly thematic fashion, as a specific source of rules and principles applicable to all times and circumstances, or as an influence primarily at the level of character and virtue rather than as a source of specific insight for particular moral issues? If Scripture functions non-fundamentalistically, how are its different voices brought to bear with some sense of moral resolution, if not moral harmony? In the absence of centralized teaching authority (as in Roman Catholicism) or recognized interpretive expertise (as in Jewish rabbinical opinions), what appropriate limits can be placed on the process of individual or communal discernment? And of particular importance to the issues of altering nature and biotechnology we discuss in this volume, what is the relevance, as well as the relevant weight, of non-Scriptural sources of moral wisdom and insight, including insights from the natural and social sciences?

The ways that Scripture functions as an authority vary widely, in keeping, perhaps, with the Protestant pluralism noted above. But in that process of moral discernment, several tensions have featured prominently in recent Protestant bioethics and are directly relevant to issues of altering nature and assessing biotechnologies.[30] On the one hand, there has been, from the Reformation onward, a central Protestant emphasis on the freedom of the Christian to respond to God's call. However, that freedom is constrained by the fundamental distinction to be drawn between the realms of creation and redemption, and much of the variety of Protestant thought is based on the different emphases at work in framing the nature and scope of human responsibility, in light of sin. Some thinkers tend to emphasize the continued goodness

---

[30] These themes are explored in far greater detail in Chapter 3, Volume Two of this project, "Religious Traditions and Genetic Enhancement."

of God's creation, despite the Fall, and see in new possibilities the extension of traditional notions of responsible stewardship (e.g., Cole-Turner, 1993). More recently, some Protestant thinkers have developed an account of positive Christian responsibility under the rubric of "co-creation" in accord with God's good purposes (e.g., Peters, 1997, 1998). Other Protestant thinkers remain more literally "conservative" in their assessment of altering nature and biotechnologies. They therefore tend to emphasize the pervasive effects of sin on human epistemological and moral capacities, and the likelihood of human hubris in usurping God's prerogative (Ramsey, 1972). The linkage between their generally negative anthropology and specific judgments, however, often remains vague or unspecific.

In addition, there are varying judgments about the nature of human nature itself, which in some respects parallel the debates between "historicists" and "classicists" in Roman Catholic debates about "human nature" as the basis of natural law reasoning. The judgment about whether human nature is (relatively) fixed or open-ended will influence the interpretations of, warrants for, and limits on, particular interventions. Moreover, judgments about the nature and scope of health and disease, normalcy and abnormalcy, will tend to frame moral conclusions about what love and justice will require in response to such conditions (Lebacqz, 1998).

In overview, then, the pluralism of Protestantism will exhibit several central themes that are often held in tension – creation and redemption, grace and freedom, law and gospel, and revelation and reason. Protestant theological reflections will therefore exhibit a range of responses, depending on how the above themes are interpreted and the above tensions navigated. In turn, Protestant judgments on particular issues are likely to express and reflect that same variety of theological interpretation.

1.3.2.3 Eastern Orthodoxy

Eastern Orthodox Christianity views itself as one of "life, method, doctrine, ecclesial organization, history, spirituality, sacramental life and worship, canon law, and ethical teaching with the one united church of Jesus Christ of the first eight centuries" (Harakas, 1986, 166). Its major authority is Revelation in a somewhat broader sense than the scope of that term as used in Catholic or Protestant thought. Orthodoxy emphasizes the "living tradition" – liturgy, canonical Scriptures, the judgments of the first seven ecumenical councils, and the monastic tradition of Eastern spirituality. The Orthodox Church is hierarchical in structure, with an ordained clergy, and with national and ethnic churches organized under patriarchs.

According to Stanley Harakas, the "ethical" approach of Orthodoxy is distinguished from Roman Catholic and Protestant Christianity by its central focus on corporate worship more than "formulations of abstract ethical constructs" (Harakas, 1995, 85). All ethical reflection proceeds in light of foundational theological doctrines of the Holy Trinity, of creation, of Christ's incarnation and redemption, and the church. As Trinity, God is a "community of persons in organic relationship" rather than "an abstract impersonal essence." In his inmost nature, God is transcendent and unknowable, but he sustains His creation through what are called his "divine energies." Here Orthodoxy is distinctive in its understanding of the divine energies as "grace" – "not a substance or thing imparted to creation, but rather the very presence of God" (Harakas, 1995, 85).

Humans are made uniquely in the "image" and "likeness" of God. Image refers to capacities that distinguish humans: intelligence, self-determination, moral perceptivity, creativity, and the capacity for interpersonal relations. "Likeness" refers to "the potential open to such a creature to become God-like" (Harakas, 1978, 643–644). This potential for deification (*theosis*) has been lost through sin, and has distorted all of creation, but the possibility of deification has been restored through the saving merits of Christ. The Church, in the sacraments or "mysteries," is the locus for "cooperation between the human and divine" (*synergy*) in the process of theosis (Harakas, 1978, 643–644). As Harakas observes, "The relationship of God with human life as created and redeemed and as growing toward God in *theosis*" provides for an " 'ought' not based on the 'facts' of a fallen creation and distorted humanity, but rather on the *telos*, or goal, toward which human beings are directed by their calling to be fully human" (Harakas, 1986, 168).

Three approaches have dominated Orthodox ethical reflection during recent decades. One perspective, emphasizing creation, has identified "the continuities ... between philosophical understandings ... and foundational Christian views." Here, although the triune God is understood as the source of good, there are "inborn" goods as well; thus other approaches to ethics, in considering such goods, share "in a portion of the truth" (Harakas, 1986, 168). Moreover, unlike classically more pessimistic Protestant understandings, this view recognizes the capacity of even unredeemed human beings to discern and do the good, at least partially.

A second view has stressed redemption as a central category, emphasizing the radical disjunctions between "unredeemed" and "redeemed" human existence. The Bible and Patristic writings from the fourth and fifth centuries are its source texts. A third ethical approach emphasizes the "mystical and ascetic" insights of later Eastern Christian traditions in its discussions. The *Philokalia*, a collection of texts written between the fourth and fifteenth centuries and first published in Greek in 1782, has been the primary text influencing the latter approach.

Orthodoxy remains closer to Catholicism than to Protestant thought in its positive assessment of the continuing presence and exercise of natural moral capacities, despite the effects of sin. Thus, one finds in Patristic writings references to what the Fathers call the "natural moral law." That term refers to basic norms that are shared across societies, with the Second Table of the Decalogue their paradigmatic expression (Harakas, 1986, 169).

Moral issues raised by developments in biotechnology involve novel applications for Orthodox ethical reflection, but a distinguishably Orthodox approach will exhibit several characteristic features. First, the "living consciousness" or "mind" of the church provides the ethos for ethical reflection. Central to that experience, as noted above, is the Tradition writ large – the Bible, the Patristic writings, the decisions of the first seven Church councils, and a rich tradition of mystical and ascetical writings. Under the tutelage of appropriate patriarchal and episcopal authority, and through the workings of the Spirit, the "mind of the Church" is embodied. In Harakas's words, the Orthodox ethos "is a mindset, rather than a set of rules or propositions." Attending to the various sources in the Tradition allows the church to determine "whether those sources speak either directly, or indirectly, or by analogy" to new issues (Harakas, 1986, 169).

Second, Orthodoxy blends elements of both "apophatic" (negative) and "kataphatic" (positive) theology. Since God is, and will always be, transcendent, he is "best described by what He is not" (Harakas, 1999, 90). At the same time, God remains, through his divine energies, "in touch" with the world (Harakas, 1999, 90), and on that basis a positive (though limited) theology of His attributes, as known through His energies, is possible. As noted, the natural moral law provides, at least in principle, a universal basis of conscience, despite the presence of sin. Moreover, from the "paradoxical relationship between God and the world, one that is both continuous and discontinuous," there emerges a "relative independence and autonomy" to the world that allows for its use and development (Harakas, 1999, 90). In that attribution of relative autonomy, Orthodoxy shares affinities with certain Calvinist strands of Protestant thought.

Third, the Orthodox understanding of human persons is that they are made in the image and likeness of God. Human "likeness" to God embodies a dynamism, a movement toward theosis, which frames Orthodox understanding of human nature in more open-ended rather than static terms. In this respect, Orthodox thought is much closer to recent Protestant interpretations of humans as "co-creators" (Peters, 1997) than to traditional, fairly static, Roman Catholic depictions of natural law, and the human nature on which it is based.

Finally, however, Orthodoxy understands personhood in robustly holistic terms: we are ensouled bodies, embodied souls, in psychosomatic unity. In contrast to accounts, both secular and religious, that emphasize an "existentially understood personhood" (Harakas, 1999, 93), with the capacity for autonomous choice, Orthodoxy emphasizes the enduring importance of embodiment to personhood, even when autonomy is absent. As a result, Orthodoxy exhibits both a "progressive" vector in its dynamic reading of human nature, and a "conservative" vector with its emphasis on protecting embodied human life. That seeming paradox suggests that Orthodoxy will likely hold both emphases "in tension" when it assesses new developments in biotechnology.

1.3.3 Islam

This third of the Abrahamic religions follows Judaism and Christianity, both of which Islam recognizes and whose moral code Islam shares. However, Islam has also developed a comprehensive system of religious law, the *shari'a*, that governs all aspects of personal and social life. Islam arose in the seventh century CE, with the call of the prophet Muhammad. Its two central authoritative texts are the

Qur'an, read by Muslims as direct revelations from Allah to Muhammad through the angel Gabriel, and the *Sunna* ("trodden path") and *Hadith*, which are the tradition and sayings of Muhammad developed by Islamic jurists. As Allah's direct revelation, the *Qur'an* is viewed as unchanging and without error, and therefore requiring submission. Islam is literally translated as "surrender" or "submission" to the will of God. That etymology is instructive, for Islam is a religion with foundationally divine-command elements, with a characteristic stress on obeying the divine law, elaborated as *shari'a*. There is therefore no distinction for Muslims, in this legal framework, for distinguishing, as in other traditions, "religious" and "secular" spheres.

The *Hadith* sets forth, as fundamental elements of faith and practice, the so-called "Five Pillars of Islam": (1) the profession of faith in God and in Muhammad as His prophet (*shahada*); (2) daily worship, with five designated times for prayer (*salat*); (3) almsgiving (*zakat*); (4) fasting during the month of Ramadan (*saum*); and (5) a pilgrimage to Mecca (*hajj*) made at least once in a lifetime if possible.

Islam has no organized "church" or ordained clergy. The *Qur'an* and *hadith* function as foundational sources of authority. When an issue is clearly settled by those sources, the verdict is deemed final, for "It is not fitting for a believer, man or woman, when a matter has been decided by Allah and His messenger, to have any option about their decision. Anyone who disobeys Allah and His messenger is clearly on a wrong path" (*Qur'an* 33:36). Much of Islamic jurisprudence, however, is an application of *Qur'an* and *hadith* to situations that have not been directly addressed by those sources. In these cases, Islamic jurisprudence (*Fiqh*) has been developed, applied, and extended by recognized scholars of *shari'a*,[31] the muftis, who issued opinions (*fatwa*) to guide Muslims in civic and economic matters, as well as individual practices in daily life. In their opinions, core principles are interpreted and applied in light of two other key features of Islamic thought: the *ijma* (the consensus opinion of Islamic scholars during the first centuries of Islam) and by the use of *quiyas* ("analogies") in order to offer rulings on novel cases in light of precedents established earlier. In the history of Islamic jurisprudence, especially during its first three centuries, certain figures in the tradition further codified and systematized the *shari'a*, with Abu Hanafi (d. 767), Malik ibn Anas (d. 795), Al-Shafi'i (d. 820), Ahmad ibn Hanbal (d. 855), and Ja'far al-Sadiq (d. 765) among the most prominent. *Fatwas* may result in different judgments on conflicted questions, depending upon the legal school of interpretation of a particular *imam*. Generally, Muslims accept the judgments of "the legal schools prevalent in their region" (Sachedina, 1995, 1293). Therefore, among Sunni Muslims, Hanafi or Shafi'i schools predominate, while Ja'fari interpretations are authoritative among Shiites.

As set forth within the *shari'a*, actions are judged according to five categories: as required (*fard* or *wajib*), recommended but not required (*mandub*, *sunna*), neutral

---

[31] Gamal I. Serour offers several general examples of the role of *Fiqh*: "...harm should be removed, the lesser of two harms should be chosen, and the public interest should take priority over private benefit" (Serour, 1997, 172).

or permitted (*mubah, ja'iz*), disapproved but not forbidden (*makruh*), or forbidden (*haram*). There are five fundamental objectives of the *shari'a*: the protection of faith, of life, of mind, of ownership, and of offspring. In interpreting the tradition in relation to those objectives, a basic premise in Islam is that "everything is lawful unless it is otherwise specified by the Qur'an and the Tradition or unless it conflicts with the objectives of the *shari'a*" (Hathout and Lustig, 1993, 133).

We now turn to the more specific questions regarding the meaning and status of nature in Islam. It is almost trite to say that there is no single and monolithic view of nature in Islam, yet such disclaimers have to be repeated for a simple reason in order to stress the diversity of viewpoints against attempts to present uniformity where it does not exist. Given the diverse cultural moorings of early Islam from Arabia to China eastwards and the Iberian Peninsula westwards, coupled with the absence of a centralized religious authority, there was little that theology could do in order to impose uniformity. The pre-Islamic worldviews of new Muslim societies who joined the early Muslim commonwealth, known as the *umma*, invariably gave local color to iterations of Islam as a religion and the secular manifestation of Muslim practices.

So, despite the bonds of faith that bind Arabs, Indians, Chinese and European Muslims, each would have very different views on how nature functions for reasons related to moral economies and worldviews that refract the Islamic impulse contextually. In fact, Arab science is only one iteration of an Islamic moral economy and should not be conflated with other cultural substrates within which Islam flourished in Africa, Asia and Europe and the scientific achievements of these cultures.

Even though the technical Arabic term for nature, *tabi'a* or *tab'*, is used in several Muslim languages, there are far-reaching differences over the interpretation of nature and how humans relate to nature. In part, and for reasons that cannot be elaborated here, the growing diversity of Muslim viewpoints on nature has grown exponentially. This has in large part to do with the ambivalence towards modernity and modern science, a debate that is possibly unique in its intensity in Muslim societies, although not limited to this faith tradition.

Politics of the past three centuries from colonization to post-colonial blues have as much to do with the shaping of people's moral economies and worldviews as does ethics (see Euben, 2003). Hence, one finds an explicit account of the "coloniality of knowledge" in the calculus of contemporary Muslim theologies and religious understandings of nature: the contentious status of western knowledge systems that forcefully displaced indigenous ones resulted in hybrid epistemic configurations. This hybridity also produced conditions of complex moral and ethical ambiguities in a rapidly globalizing universe. Since there is no centralized authority for Islam or an ecclesiastical society, the locus of authority in Muslim communities have always been the communities of the learned and experts in the discursive tradition. With the imposition of secularism in many Muslim societies these traditional knowledge communities in religion have been displaced, fragmented and marginalized, resulting in a crisis of authority that compounds any semblance of trying to reach a broad consensus and coherence.

This diversity of voice quite unsurprisingly coincides with a diversity of interpretations of nature in Islam. Such interpretations have always been mediated by larger framing discourses, namely, theology (*kalam*) and jurisprudence (*fiqh*), also called juridical-theology, which would form the basis of what we would today call ethics. Even the Qur'an's rhetoric in framing nature as creation (*khalq*) and references to God as the Creator (*fatir*) are concepts that are mediated in a variety of ways in the early Muslim communities. According to Muslim teachings, God had stamped a nature (*fitra*) into all creation, ranging from humans to inanimate objects. This is a God-created nature (*fitrat allah*). There is, of course, an extensive debate as to how one discovers that DNA-like imprint of nature.

Early Muslim understandings easily assimilated with certain modes of Aristotelian thinking where grasping the essence of things was one way to understand the nature of things. It suffices to say that a variety of medieval influences from Aristotelianism to Neoplatonism to notions of Scripturalism colored the historical debates about nature in Islam and they invariably shape contemporary Muslim discussions on nature.

In early and middle Islam the whole idea of "nature" quickly fell into the flux of contentious theological debates. The pro-rationalist or pro-Stoic Muslim theologians influenced by philosophical currents came to be known as the Mu'tazila. They argued that things had an inherent nature that was discernible by reason. A thing has a natural disposition that necessarily determines its particular behavior.[32] Here *tabi'a* is synonymous to *khilqa*, meaning a natural disposition or peculiarity of character. For the Mu'tazilite theologians nature had a natural causal agency that brought about the real phenomena experienced in the world. But the idea of equating nature and natural causation would not remain uncontested for long.

If all physical bodies consist of atoms and accidents, says the Jewish philosopher, Musa ibn Maymun (Maimonides), then everything is created by God's will. The critical issue is that these accidents are not caused by the imperative of necessity, nor do they arise out of a nature (*tab'*) that is inherent in things. Rather, they are the product of God's direct and creative causal action. Maimonides provided an accurate picture of the Ash'ari theologians, who represented an early 'orthodoxy' that combined both scriptural and rational authority in Islam, arguing that God is free to create as he wills and is not limited by the 'nature' of things. In other words, the nature of things is always contingent and in theory open to Divine molestation.

For the Ash'aris what we call nature is actually a repeatedly observed and predictable habit or a custom of a thing or natural phenomena. The foremost exponent of this view was Abu Hamid al-Ghazali (d. 1111) who explains that the reason why fire burns or why a boy never turns into a dog is not due to anything essential in fire or in the essence of "boyness." It is due to a succession of events that God had caused within things by way of a divinely created custom and a divinely created knowledge, that we call nature. Therefore, fire will continue to burn and a child will not turn into a dog due to the divine creation of custom and knowledge; it has nothing to do, says Ghazali, with some autonomous qualities inherent in fire or in a child. Ghazali denied natural causation since he contested the occurrence of an

---

[32] Encyclopedia of Islam 2, s.v. tabi'a.

autonomous and inherent predisposition (*tab'*) in things. However, natural objects in his view do exhibit certain dispositions. So when a two horses mate the disposition is that the offspring will be a horse and not a cow. If anything, then in Ghazali's view nature is a space of divine sovereignty; it does not act on its own but is metaphysically subject to a higher will.

More mystically and metaphysically sensitive viewpoints assert that nature is properly understood and discovered only when human nature complied with its optimum condition; human nature is subdued by restraining the appetite that averts its distortion. Under such ideal conditions the harmony between nature and human nature is self-evident. Either revelation or reason, depending on one's view, provides the proper lens with which one can properly discover nature.

In the earliest Muslim ethical teachings humans are required to balance their own needs with other species that inhabited the earth. Hence, a Muslim is to ensure that he or she shares the environment with plant and animal life. For instance, there are dire spiritual consequences in killing animals for frivolous reasons (Ajmal, 1984, 218). Animals used for transportation and for domestic purposes are to be treated with due deference to their dignity. The Prophet Muhammad once reprimanded a person who branded a donkey in the face. 'Umar, the second caliph, eloquently captures the general sense of responsibility towards nature. "If a she-mule stumbled in Iraq," 'Umar, whose capital was in Madina in Arabia, said, "I would be responsible in the eyes of God for neglecting to pave the roads for her." Similarly, 'Umar restricted the weight of loads packed on camels from a 1,000 to 600 pounds. From the early literature we get the sense that the relationship with nature must reflect compassion and dignity.

We turn now to contemporary Muslim perspectives. All the major debates regarding nature in contemporary Islam share a common vocabulary and rhetoric. For instance, humans in their relationship with nature must act as stewards (*khalifa*), not as sovereigns. Stewardship is part of an ethical narrative of moral responsibility (*amana*) towards nature. Stewardship and more responsibility are at the base of Islam's doctrinal pyramid. Therefore, humans have fiduciary rights over nature. As such it is not an absolute right: there are concomitant duties that are designated by the revealed law (*shari'a*). Similarly, when treating the body, it is often the case that moral reasoning prizes interpersonal justice which take precedence over private rights and autonomy (Sachedina and Ainuddin, 2004, 44).

While the corpora of prophetic traditions carry a good deal of embodied narratives as to how one deals with animals and plant life, it is slightly different in the Qur'an (see Haq, 2001, esp. 162–170; Dallal, 2004). In the Qur'an there are references to the naturalistic phenomena and the cosmos. If anything, the Qur'an stresses that nature is a reflection or representation of transcendence. "Nature," says Nomanul Haq, "is an emblem of God; it is a means through which God communicates with humanity" (Haq, 2001, 146). Because nature does not speak for itself, but is itself a "sign", like all signs, nature requires an interpretation. And like all interpretative traditions, there are variant Muslim interpretations of nature.

These interpretative differences center on certain Qur'anic passages and other canonical traditions that discuss natural phenomena in their relation to humans.

Given the covenantal relationship between God and humans, stewardship is mediated by a sense of proportionality and measurement (*qadr*) as well as a balance (*mizan*) between contending demands. Furthermore, the Qur'an repeatedly states that whatever is in the heavens and the earth or in the cosmos has been placed at the disposal of humans. In fact, it is the utility of nature that comes into play, a view derived from the notion of *taskhir*, which means to make subservient or to make available, a theme that is repeatedly reinforced in the Qur'an over twenty times, some of which include passages like 14:32, 33; 16:12; 22:65; 29:61 and 31: 20.

The language of nature's "subservience" or "availability" to human needs is on face-value disconcerting since it gives the misleading idea that humans are unbridled in their exploitation of nature. To the contrary, Muslim ethical interpretations require such passages to be read in conjunction with the duty of stewardship and moral responsibility, which circumscribes human authority over nature and understands it as limited by responsibility. Even the fourteenth century thinker Ibn Taymiyya, who is sometimes excessively doctrinaire, explains that one should bear in mind that despite the fact that nature is at the service of human needs, humans are only one constituency to whom nature is given, implying that other species also have a similar right.

If in the pre-modern period hermeneutical interpretation of the canonical sources mediated the human-nature relationship, then in the modern period there is the added gaze of modern science that also refracts the way we view nature. Hence, much discussion about nature and its malleability, purpose and ends is found in voluminous Muslim debates over modern science. Given that the modern scientific tradition reaches Muslim societies via a dark history of colonialism that displaced the cultural, political and scientific traditions that preceded European modernity, the contemporary discussions about nature are marked by these discourses of the "coloniality of knowledge." The politics of science as a secular endeavor and its impact on religious communities is a continuous cause of concern among certain Muslim religious and scholarly circles. Even those Muslim communities and individuals who live in North America and Europe can not insulate themselves from questions about the commensurability of modern science and the religio-moral system of Islam.

Of particular concern is the invasive and colonizing character of modern science and technology over nature – the body, the natural habitat and the cosmos. Technology can now deliver science in such discreet ways that it effectively changes practices, ways of life and natural habitats on a global scale, ranging from forms of consumption to modes of dress. Globalizing capitalist economies generate new modes of being that colonize entire cultures on the globe. It is not only a case of benign science harnessing nature in order to produce more efficient and mass-scale medicine which result in cures for life-threatening diseases and interventions in poverty-relief and development. In fact, science-based technologies produce distinct ethical relationships vis-à-vis nature that create tremendous anxieties in many Muslim societies precisely because science is not ethically neutral. The prevailing view, especially among traditional Muslim religious thinkers, is that in the modern West science had catastrophically supplanted religion, a prospect that Muslim societies should avoid at any cost (al-Asad, 1998). These moral and political concerns are at the forefront of discussions about nature and are related to the contentious issues of modernity and development. One of the biggest challenges that Muslim thinkers identify is finding ways in which science can be placed under the moral restraint of religion.

The Iranian emigre scholar Seyyed Hoosein Nasr is one of the more astute critics of modern science from a Muslim metaphysical viewpoint. Nasr, who identifies with the perennialist school of Muslim philosophy as a form of traditionalism, regards modernity as a lamentable condition that has desacralized nature. Nature, he argues, was historically viewed as a sacred manifestation of a larger cosmology. Both humans and nature were governed by cosmological principles. Cosmological principles, he explains, governed the lower levels of existence and also regulated the world of nature (Nasr, 1996, 60). Thus, in pre-modern Muslim worldviews, nature was not an independent domain of reality with an independent order, but its principle resided in a Divine reality. In this description there are echoes of early Muslim theology. Nature in this view is a sacred expression of the Transcendent sphere, which makes for a holistic manifestation of the cosmos. In short, nature is but a specific manifestation of a metaphysical order.

In Nasr's view the disenchantment of nature and the world is a result of modernity and historicist interpretations of history, elements that in his view have alarmingly infected the thinking and practices of all religions. Thus, modern human subjectivities cause humans to abdicate their trust and responsibility in the governance of nature. A similar view is articulated by the Malaysian scholar, Naquib who points out that the most important concern is not so much the control of nature in the scientific sense nor governance in the socio-political sense, but the governance of the human self (al-Attas, 1995, 145). With the advent of modernity the coherence of age-old cosmological principles was disrupted by the secularization and the desacralization of nature. For Nasr the secularization of science had resulted in modern man's obsession with the domination of nature. Therefore, modern science had irreparably harmed the ecological system to the extent that the very existence of the planet is in danger.

Another critic of modern science on sociological grounds is the British writer Ziauddin Sardar who points out that nature is the testing ground for humans. "Man is enjoined to read its 'signs'," Sardar says,

> [W]hich reflect both man's position in creation and the glory of God. As such nature is created orderly and knowable. Were it unruly, capricious and erratic, morality would be impossible. It would be both oppressive and degrading for man who humble himself before its slightest whim...As such, the orderliness of nature and its amenability to rational enquiry are an essential prerequisite for morality (Sardar, 1986, 226).

Both Sardar and Nasr agree that western science is in crisis and that science is not neutral but is deeply attached to cultural practices.

They both argue that an alternative view of science and by implication of nature is possible. However, both critics of modern science have not demonstrated how their views would translate into practice. Furthermore, their discourses of science and nature are detached from the day-to-day reality of Muslim societies. While there is much rhetorical resistance to scientism and the corrosive effects of modernity in these circles there has been a paucity of alternative models to imagine science and nature in practice.

In order to map representative samples of how nature is imagined we need to turn to a select number of juridical decisions and discussions of moral reasoning generated in Muslim societies. Some of them can be deduced from the practices of the 'ulama, the learned scholars, whose orientations differ from one context to another. Therefore, it is difficult to make generalizations. Moral reasoning among the 'ulama is steeped in the discourse of law, *fiqh*. Often the discourses of law and moral reasoning are anchored in a traditional worldview, one that resists the encroachments of modernity, but there are also 'ulama who are prepared to theorize the realities of the modern condition into their craft of moral reasoning and law. There is thus a spectrum of opinions among contemporary Muslim 'ulama on any given issue. It might be helpful to think of contemporary discourses of Muslim law and moral reasoning as translations. Modern realities and practices are translated into the idiom of Muslim moral reasoning in order to generate moral verdicts. It is a discourse that is strong in realism but not very self-reflexive.

The operative keyword in 'ulama moral discourse is to preserve cherished Islamic values and the retention of a thick Muslim subjectivity. The 'ulama exhibit a strong skepticism towards modern modes of cultural thinking but are open to consider the beneficial aspects of technology albeit an instrumentalist appropriation. When the 'ulama endorse or oppose scientifically mediated notions of nature, such moral calculations are measured on a standard of what is morally beneficial or harmful to the community and the individual.

Many 'ulama also admit that the pre-modern language of theology through which nature was viewed had been surpassed by science which provides a totally different picture of nature. In the scientifically mediated view of nature that many 'ulama do not condemn but seek to reconcile with theology, there is an admission that the "concept" of nature as imagined by early Muslim theologians is different from a modern scientific one or that they can co-exist with each other. In the scientific picture of nature, it is malleable and deliverable to consumers in the form of products? Critics like Nasr believe that many 'ulama are either woefully naïve or ignorant of the scientism and secularism of modern science. Their endorsement of scientifically mediated notions of nature is catastrophic to traditional Muslim moral discourses and metaphysics. Many Muslim thinkers who are trained in the modern sciences view the discourses of the 'ulama as uninformed and blunt instruments of tradition that do not provide intellectual comfort.

The nexus of conversation in contemporary Muslim discussions are not so much centered on nature as they are concerned with the technologies of nature, namely science and technology. When contemporary Muslim theologians do deploy traditional Muslim theological arguments, they use these, as a bulwark against a science that they think will be harmful to Islamic conceptions of nature such as their reluctance to embrace reproductive cloning.

In the view of many 'ulama, the emergence of modern science is nothing but another means to discover nature. As one representative of the 'ulama viewpoint,

Nasir Ahmad Uthmani, put it: "despite the progress of science all the laws and patterns of nature are not known." In other words, while there still may be an enchanted and unknowable aspect of nature worthy of reflection and contemplation, it would be equally foolhardy to deny and reject science (Uthmani, 1970, 23).

In an attempt to reconcile modern science with the Muslim moral tradition, the view of internationally renowned traditionalist Indian Muslim thinker, Abul Hasan Ali Nadvi, is illustrative. In discussing Muslim responses to space exploration Nadvi starts with the view of the fourteenth century thinker, Ibn Taymiyya, who argued that there is no conflict between "sound reason" and "sound religion." In Nadvi's translation of this dictum, we should bear in mind, science is an iteration of reason. Hence, if both science and religion are correctly interpreted there should be no conflict: harmony between science and religion is the desideratum. Nadvi is also aware of crude attempts to invoke science in order to bolster claims made by religion, especially references to natural phenomena in the Qur'an that many persons have used to demonstrate that the scripture is a true revelation. He warns against some Muslim thinkers who try to package established scientific discoveries in a religious idiom drawing on religious texts and then giving it a "garb of religious sacrality" (Nadvi, 1970, 13). Such folks, Nadvi says, believe that they do their faith a tremendous service by presenting this scientific discourse in a religious idiom. But he calls such persons the "naïve friends of religion" who are unwittingly greater enemies of religion than many an intelligent opponent of religion. Nadvi is also critical of scientists who make absolute claims about scientific matters that are in their nature a work-in-progress such. Science, he says, should be treated as provisional knowledge and always subject to change and enhancement by later discoveries.

Nadvi is joined by many traditional 'ulama who view modernity and especially modern science as compatible and commensurable with Muslim religious discourse. Thus, the modern iterations of nature subtly seep into Muslim discourse and get embroidered into the moral discourse with varying effects. In India and Pakistan, Muslim religious scholars are opposed to organ transplantation and cadaver transplantations, whereas in Egypt and other Middle Eastern countries religious scholars in that region deem the same practices permissible (see Moazam, 2004, 37; Moosa, 2002). Similarly, Muslim thinkers in different traditional contexts offer variant bioethical decisions about brain death, cloning and abortion because the translation from one ethical register into another varies and is creatively supplemented by local and contextual discursivities and subjectivities. Of course, we recognize that science is also an ethical language that refracts into different moral languages of religion and human experiences.

If some 'ulama view organ transplantation and especially cadaver transplantations to be an act of supreme altruism, which enhances human dignity, then 'ulama holding an opposing view would argue that a donation that involves invasive and disfiguring bodily surgery, is a violation of human dignity, and is corrosive to the integrity of the body. Furthermore, many 'ulama opponents of invasive surgeries for transplantation purposes fear that such practices may result in the commodification of the body (Sambhali, undated; al-Din, undated).

From any sample of Muslim moral reflections on issues related to science and nature one will be able to infer the ambivalence towards modernity, science and contested new understandings of nature. There is no single and monolithic view, nor is there a single authority that decides for all. The understanding of nature is not only caught in a web of moral discourses of differing perceptions and readings of nature but also, questions of the geo-politics of knowledge weigh heavily in such considerations. Thus, attempts to map understandings of nature are ultimately highly contextual and display a remarkable transitional character towards emancipatory knowledge. Given that Muslim knowledge traditions in particular are caught in the crosshairs of the complex binary geo-politics of modernity and tradition it is difficult to make unambiguous determinations about views of nature in the form of a blueprint. Overall, the variety of Muslim responses to understanding nature range from outright resistance of modern interpretations of nature to more critical attempts to find alternative frames and languages of speaking about nature.

1.4 Concluding Thoughts

This chapter has surveyed a number of religious and spiritual traditions. Each presents a unique and nuanced understanding of nature. Each in its particularity provides concepts of nature that illuminate understanding and challenge conventional assumptions, and the traditions elude easy summary.[33] Indeed, the clearest implications of this chapter may be to problematize and question common assumptions. Nonetheless, some general characteristics may be suggested.

The traditions examined generally agree on some basic points. The word "nature" has a range of meanings, and careful attention is required in articulating and working with understandings of nature. Humans in their bodily existence are part of nature. "Altering nature" should not be construed as human subjects doing something to an alienated nature that is simply object. Humans must deal with nature in a respectful and responsible manner. Different traditions may understand this responsibility to be best expressed as stewardship, or may understand humans to be simply one part of the whole of nature. They agree that arrogant anthropocentrism must be avoided.

For all the traditions in different ways, the central question of altering nature is not whether the existing physical world may be altered or should be left unchanged. Rather, the more basic issue is the need for mindfulness of the deeper and spiritual significance reflected in nature as we experience it. All traditions agree in different ways that the existing world includes suffering, leading to the challenge for humans to accept or to find liberation from suffering. Helping persons in need, and generally living in accord with true values and goods, is paramount.

---

[33] Not to mention the differences that may be found among views within the same tradition.

References

Ajmal, Mohammad (1984). "Islam and Ecological Problems," in Rais Ahmad and Syed Naseem Ahmad (eds.), *Quest for New Science*. Aligarh, India: Centre for Studies on Science.
Al-Asad, Nasir al-Din 1998. "al-Islam fi Muwajaha al-hadatha al-shamila," *Majalla Majma' al-Fiqh al-Islami* 11, 3.
Al-Attas, Syed Muhammad N. (1995). *Prolegomena to the Metaphysics of Islam*. Kuala Lumpur: International Institute of Islamic Thought and Civilization.
Amoah, Elizabeth (1998). "African Indigenous Religions and Inter-Religious Relationship": 1998 Autumn Lecture. Oxford International Interfaith Centre. Retrieved December 24, 2002 from the World Wide Web: http://www.interfaith-center.org/lectures/amoah98.htm.
Appiah-Kubi, Kofi (1981). *Man Cures, God Heals: Religion and Medical Practice Among the Akans of Ghana*. Totowa, NJ: Allanheld & Osumn.
Atyi, Eba'a R. (1988). "Cameroon's Logging Industry: Structure, Economic Importance and Effects of Devaluation," Occasional Paper No 14. Jakarta: Center for International Forestry Research.
Balasubramanian, R. (undated). "The Hindu Perspective of Man and the Cosmos," unpublished paper.
Beltran, Javier (ed.) (2000). *Indigenous and Traditional Peoples and Protected Areas: Principles, Guidelines and Case Studies*. Gland, Switzerland: IUCN – The World Conservation Union.
Blaike, Piers (1996). "Environmental Conservation in Cameroon: On Paper and in Practice," Occasional Papers. Copenhagen: Center for African Studies, University of Copenhagen.
Bol, Marsha C. (1998). *American Indians and the Natural World*. Niwot, CO: Roberts Rinehart Publishers.
Borowitz, Eugene B. (1991). *Renewing the Covenant: A Theology for the Postmodern Jew* Philadelphia, PA: Jewish Publication Society.
Boyle, Joseph (1991). "The Roman Catholic Tradition and Bioethics," in Warren Reich (ed.), *Bioethics Yearbook, Volume 1*. New York: MacMillan.
Brody, Baruch (1998). "Themes and Methods in Jewish Ethics," unpublished lecture as part of the "Recovering the Traditions" conference. Texas: Institute of Religion, Texas Medical Center.
Bruchac, Joseph (1996). *Roots of Survival: Native American Storytelling and the Sacred*. Golden, CO: Fulcrum.
Cole-Turner, Ronald (1993). *The New Genesis: Theology and the Genetic Revolution*. Louisville, KY: Westminster.
Crawford, Cromwell (2003). "Hindu Perspectives on Genetic Enhancement." Unpublished manuscript.
Curran, Charles (1999). *The Catholic Moral Tradition Today: A Synthesis*. Washington, DC: Georgetown University Press.
Dallal, Ahmad (2004). "Science and the Qur'an," in Jane McAuliffe (ed.), *Encyclopedia of the Qur'an, Volume 4*. London: Oxford University Press, pp. 540–558.
deBary, William T. (ed.). (1958). *Sources of Indian Tradition, Volume 1*. New York: Columbia University Press.
Deloria, Barbara, Kristen Foehner, and Sam Scinta (eds.) (1999). *Spirit and Reason: The Vine Deloria, Jr. Reader*. Golden, CO: Fulcrum.
Deloria, Vine Jr. (1994). *God is Red: A Native View of Religion*. Golden, CO: Fulcrum.
Deloria, Vine Jr. (2002). *Evolution, Creationism, and Other Modern Myths: A Critical Inquiry*. Golden, CO: Fulcrum.
Dorff, Elliot (1997). "Review of Recent Work in Jewish Bioethics," in B. Andrew Lustig (ed.), *Bioethics Yearbook, Volume 5*. Dordrecht, The Netherlands: Kluwer.
Dorff, Elliot (2000). "Fundamental Theological Convictions" [on stem cell research], in *Ethical Issues in Stem Cell Research, Volume III: Religious Perspectives*. Rockville, MD: National Bioethics Advisory Commission, C-3.
Drewal, Henry J., and John Pemberton III with Roland Abiodun (1989). *Yoruba: Nine Centuries of African Art and Thought*. New York: The Center for African Art.

Erdoes, Richard, and Alfonso Ortiz (eds.) (1984). *American Indian Myths and Legends*. New York: Pantheon Books.
Euben, Roxanne L. (2003). "A Counternarrative of Shared Ambivalence: Some Muslim and Western Perspectives on Science and Reason," *Common Knowledge* 9(1), 50–77.
Falk, Marcia (1987). "Notes on Composing New Blessings: Toward a Feminist-Jewish Reconstruction of Prayer," *Journal of Feminist Studies in Religion* 3 (Spring), 43–53.
Feldman, David M. (1986). *Health and Medicine in the Jewish Tradition*. New York: Crossroad.
Gillman, Neil (2000). *The Way into Encountering God in Judaism*. Woodstock, VT: Jewish Lights Publishing.
Green, Arthur (1992). *Seek My Face, Speak My Name: A Contemporary Jewish Theology*. Northvale, NJ: Jewish Lights Publishing.
Green, Arthur (2002). "A Kabbalah for the Environmental Age," in Hava Tirosh-Samuelson (ed.), *Judaism and Ecology*. Cambridge: Harvard University Press, Center for the Study of World Religions.
Greenberg, Irving (1986). "Toward a Covenantal Ethic of Medicine," in Levi Meier (ed.), *Jewish Values in Bioethics*. New York: Human Sciences Press, pp. 128–134.
Grim, John A. (1998). "Indigenous Traditions and Ecology." Retrieved April 10, 2003, from the World Wide Web: http://environment.harvard.edu/religion/religion/indigenous.
Grim, John A. (2001). "Indigenous Traditions and Deep Ecology," in David Landis Barnhill and Roger S. Gottlieb (eds.), *Deep Ecology and World Religions: New Essays on Sacred Grounds*. Albany, NY: State University of New York Press, pp. 35–57.
Gustafson, James (1978). *Protestant and Roman Catholic Ethics: Prospects for Rapprochement*. Chicago, IL: University of Chicago Press.
Haq, Nomanul (2001). "Islam and Ecology: Toward Retrieval and Reconstruction," *Daedalus* 130(4).
Harakas, Stanley (1978). "Eastern Orthodox Christianity," in Warren Reich (ed.), *Encyclopedia of Bioethics, Volume 1*. New York: Free Press, pp. 643–44.
Harakas, Stanley (1986). "Eastern Orthodox Christian Ethics," in James Childress and John MacQuarrie (eds.), *The Westminster Dictionary of Christian Ethics*. Philadelphia, PA: Westminster.
Harakas, Stanley (1995). "Eastern Orthodoxy," in Warren Reich (ed.), *Encyclopedia of Bioethics, Volume 1*, 2nd ed. New York: MacMillan.
Harakas, Stanley (1999). *Wholeness of Faith and Life: Orthodox Christian Ethics*. Brookline, MA: Holy Cross Orthodox Press.
Harlow, Jules (ed.) (1985). *Siddur Sim Shalom*. New York: Rabbinical Assembly and United Synagogue of America.
Hathout, Hassan, and Andrew Lustig B. (1993). "Bioethical Developments in Islam," in B. Andrew Lustig (ed.), *Bioethics Yearbook, Volume 3*. Dordrecht, The Netherlands: Kluwer, p. 133.
Heschel, Abraham J. (1955). *God in Search of Man: A Philosophy of Judaism*. New York: Farrar, Straus & Cudahy.
Hoose, Bernard (1989). *Proportionalism: The American Debate and Its European Roots*. Washington, DC: Georgetown University Press.
Independent Commission on International Humanitarian Issues (1987). "Indigenous Peoples: A Global Quest for Justice: A Report for the Independent Commission on International Humanitarian Issues." London: Zed Books.
John Paul II (1993). *Veritatis Splendor*, *Origins* 23(18) (October 14).
Johnson, James T. (1978). "History of Protestant Medical Ethics," in Warren Reich (ed.), *Encyclopedia of Bioethics, Volume 3*. New York: Free Press.
Jonsen, Albert (1986). "Casuistry," in James Childress and John MacQuarrie (eds.), *The Westminster Dictionary of Christian Ethics*. Philadelphia, PA: Westminster.
Jonsen, Albert, and Stephen Toulmin (1988). *The Abuse of Casuistry: A History of Moral Reasoning*. Berkeley, CA: University of California Press.

Keenan, James, and Thomas Shannon (eds.) (1995). *The Context of Casuistry*, Washington, DC: Georgetown University Press.
Kelly, David (1979). *The Emergence of Roman Catholic Medical Ethics in North America*. New York: Mellen.
Kemf, Elizabeth (ed.) (1993). *The Law of the Mother: Protecting Indigenous Peoples in Protected Areas*. San Francisco, CA: Sierra Club Books.
Keown, Damien (1992). *The Nature of Buddhist Ethics*. Basingstoke, England: Macmillan.
Lawson, Thomas (1985). *Religions of Africa: Traditions in Transformations*. San Francisco, CA: Harper San Francisco.
Lazareth, William (1986a). "Lutheran Ethics," in James Childress and John MacQuarrie (eds.), *The Westminster Dictionary of Christian Ethics*. Philadelphia, PA: Westminster.
Lazareth, William (1986b). "Orders of Nature," in James Childress and John MacQuarrie (eds.), *The Westminster Dictionary of Christian Ethics*. Philadelphia: The Westminster Press.
Lebacqz, Karen (1998). "Fair Shares: Is the Genome Project Just?," in Ted Peters (ed.), *Genetics: Issues of Social Justice*. Cleveland, OH: Pilgrim Press.
Loy, David (2003). "Asian Perspectives on Incorporation." Unpublished manuscript.
Mackler, Aaron L. (2003). *Introduction to Jewish and Catholic Bioethics*. Washington, DC: Georgetown University Press.
Magesa, Laurenti (1997). *African Religion: The Moral Traditions of Abundant Life*. Nairobi: Paulines Publications Africa.
Mahoney, John (1987). *The Making of Moral Theology: A Study of the Roman Catholic Tradition*. New York: Oxford University Press.
McCormick, Richard (1986). "Magisterium," in James Childress and John MacQuarrie (eds.), *The Westminster Dictionary of Christian Ethics*. Philadelphia, PA: Westminster.
McCormick, Richard (1989). *The Critical Calling*. Washington, DC: Georgetown University Press.
McCormick, Richard, and Paul Ramsey (eds.) (1978). *Doing Evil to Achieve Good*. Chicago, IL: Loyola University Press.
Moazam, Farhat (2004). "'The Sands of the Ocean:' Live, Related, Renal Transplantation in Pakistan: Who Shall Give a Kidney, Who Shall Receive It?," Unpublished PhD dissertation, Department of Religious Studies, University of Virginia, May 2004. Published as *Bioethics and Organ Transplantation in a Muslim Society: A Study in Culture, Ethnography, and Religion* (2006), Indianapolis: Indiana University Press.
Moosa, Ebrahim (2002). "Interface of Science and Jurisprudence: Dissonant Gazes at the Body in Modern Muslim Ethics," in Ted Peters, Muzaffar Iqbal, and Syed Nomanul Haq (eds.), *God, Life and the Cosmos*. Aldershot, England: Ashgate.
Nadvi, Abul Hasan A. (1970). "Foreword," in A. Nadvi Abul Hasan (ed.), *Chand ki Taskhir Qur'an ki Nazar mai*. Bangalore: Furqaniyya Academy
Nadvi, Muhammad Shahabuddin, *Chand ki Taskhir, Qur'an ki Nazar men*: Cand Afaqi Dala'il ka Ja'iza (Bangalore; Furqaniyya Academy, 1970)
Nasr, Seyyed H. (1996). *Religion and the Order of Nature*. London: Oxford University Press.
Nakamura, Hajime (1978). "Buddhism," in Warren Reich (ed.), *Encyclopedia of Bioethics, Volume 1*, New York: Free Press.
Nolan, Kathleen (1993). "Buddhism, Zen, and Bioethics," in B. Andrew Lustig (ed.), *Bioethics Yearbook, Volume 3*. Dordrecht, The Netherlands: Kluwer.
Noonan, John (1993). "Development in Moral Doctrine," *Theological Studies* 54(4), 662–678.
Novak, David (1995). "Judaism," in Warren Reich (ed.), *Encyclopedia of Bioethics, Volume 3*. New York: MacMillan.
Nyamiti, Charles (1973). *The Scope of African Theology*. Kampala: Gaba Publications.
Olupona, Jacob K. (ed.) (1991). *African Traditional Religions in Contemporary Society*. New York: Paragon House.
Peters, Ted (1997). *Playing God: Genetic Determinism and Human Freedom*. New York: Routledge.
Peters, Ted (ed.) (1998). *Genetics: Issues of Social Justice*. Cleveland, OH: Pilgrim Press.

Plaskow, Judith (1990). *Standing Again at Sinai: Judaism from a Feminist Perspective*. New York: HarperCollins.

Ramsey, Paul (1972). "Genetic Therapy: A Theologian's Response," in Michael Hamilton (ed.), *The New Genetics and the Future of Man*. Grand Rapids, MI: William B. Eerdmans.

Reichard, Gladys A. (1950). *Navaho Religion: A Study of Symbolism*. Princeton, NJ: Princeton University Press.

Sachedina, Abdulaziz (1995). "Islam," in Warren Reich (ed.), *Encyclopedia of Bioethics, Volume 3*. New York: MacMillan.

Sachedina, Abdulaziz and Nageen Ainuddin (2004). *Islamic Biomedical Ethics: Issues and Resources*. Islamabad: COMSTECH.

Sambhali, Mawlana Burhan al-Din (undated). "A'da ki paywandkari," in Jadid Fiqhi Mabahith (ed.), *Mawlana Mujahid al-Islam Qasimi, Volume 1*. Karachi, Pakistan: Idarat al-Qur'an wa al-Ulum al-Islamiyya.

Sardar, Ziauddin (1986). *Islamic Futures: The Shape of Ideas to Come*. New York: Mansell.

Scholem, Gershom (1974). *Kabbalah*. New York: New American Library.

Schwartz, Maureen Trudelle (1997). *Molded in the Image of Changing Woman: Navajo Views on the Human Body and Personhood*. Tucson, AZ: University of Arizona Press.

Serour, Gamal (1997). "Islamic Developments in Bioethics," in B. Andrew Lustig (ed.), *Bioethics Yearbook, Volume 5*. Dordrecht, The Netherlands: Kluwer, p. 172.

Setiloane, Gabriel (1976). *The Image of God Among the Sotho-Tswana*. Rotterdam, The Netherlands: A. A. Balkema.

Shannon, Thomas, and Lisa Cahill (1988). *Religion and Artificial Reproduction: An Inquiry into the Vatican Instruction on Respect for Human Life*. New York: Crossroad.

Smart, Ninian (1986). "Buddhist Ethics," in James Childress and John MacQuarrie (eds.), *The Westminster Dictionary of Christian Ethics*. Philadelphia, PA: Westminster.

Soloveitchik, Joseph (1965). "The Lonely Man of Faith," *Tradition* 7, 5–67.

Steinberg, Avraham (1991). "Jewish Medical Ethics," in B. Andrew Lustig (ed.), *Bioethics Yearbook, Volume 1*. Dordrecht, The Netherlands: Kluwer.

Stevens, Stan (ed.) (1997). *Conservation Through Cultural Survival: Indigenous Peoples and Protected Areas*. Washington, DC: Island Press.

Stone, Jana (ed.) (1993). *Every Part of This Earth Is Sacred: Native American Voices in Praise of Nature*. New York: Harper Collins.

Uthmani, Nasir A. (1970). "Foreword (2)", in Muhammad Shihab al-Din Nadvi (ed.), *Chand ki Taskhir Qur'an ki Nazar mai*.

Verhey, Allen (1986). "Bible in Christian Ethics," in James Childress and John MacQuarrie (eds.), *The Westminster Dictionary of Christian Ethics*. Philadelphia, PA: Westminster.

Verhey, Allen (1995). "Protestantism," in Warren Reich (ed.), *Encyclopedia of Bioethics, Volume 4*, 2nd ed. New York: MacMillan.

Weiss, Mitchell (1995). "Hinduism," in Warren Reich (ed.), *Encyclopedia of Bioethics, Volume 2*, revised ed. New York: MacMillan, 1132–1139.

Chapter 2
Philosophical Approaches to Nature

John H. Zammito, Philip J. Ivanhoe, Helen Longino, and Phillip R. Sloan

2.1 Introduction

"Do we have the ethical resources to use our genetic powers wisely and humanely? ... Do existing ethical theories, concepts, and principles provide the materials for constructing more adequate instruments for moral navigation?" These questions posed at the outset of *From Chance to Choice* go to the heart of the whole project upon which we are embarked. Its authors observe that "even if we were more assured than we should be that our technical control will be complete, we would continue to wonder whether we will be able to distinguish between what we can do and what we ought to do." They resolve that "something more is needed. A systematic vision of the moral character of the world we hope to be moving toward is required" (Buchanan et al., 2000, 4).

What *philosophical* resources can be brought to bear upon the choices, social and personal, that innovations in biotechnology will be forcing upon us imminently and with unprecedented gravity? The first place to turn would be explicit moral philosophies currently invoked and asserted both theoretically and in practice. Two types – utilitarian and deontological – dominate current usage. But the tacit premise of this entire research project is that such explicit ethical theorizing, confronted with the challenges of the new biotechnologies, may require substantial supplementation when the public policy controversy reaches full tide. When fundamental issues of this gravity present themselves in the public policy arena, it has been conspicuous that established *religious* commitments have been called into play. This is due not only to the traditional authority religious communities may exercise upon their adherents, but also, and as one basis of that authority, to the *metaphysical grounding* of ethical judgments in a more comprehensive, explicitly theoretical as well as evaluative, conception of humankind's *place* in the world. In that light, what resources – *normative* resources – can the philosophical discourse *on nature* supply? In this chapter we wish to consider what modern philosophical thought might provide in the way of connection or grounding for ethics via conceptions of nature and humankind's participation in that nature – including, but not restricted to, the notion of a "human nature." We do not believe that any *one* definitive philosophical position can be identified that would resolve the issues facing contemporary culture, but rather that a variety of historical and contemporary approaches bring mutually contesting views which will need to be negotiated and elected in the ongoing debates.

It appears that in modern Western thought the philosophical resources available by taking nature as a normative principle have sharply diminished. Not only has the recourse to such metaphysical grounding become vanishingly thin in contemporary ethical thought, but conversely ethical concerns appear quite insubstantial over against the forward lurch of science and technology as a project of instrumental mastery of the world. With regard to ethical connections, our obliviousness rooted in a confining expertise, what Veblen once called our "trained incapacity," may indeed have been the price of our scientific-technical facility. Radical impoverishment of ethical purchase manifests itself even as we come to face some of the most staggering ethical choices in our species and planetary history. Alasdair MacIntyre has drawn the starkest conclusion: "There seems to be no rational way of securing moral agreement in our culture," he wrote in 1981 (MacIntyre, 1984, 6). That is, "rival premises are such that we possess no rational way of weighing the claims of one as against another" (8). Worse, in his view, it has become commonplace to believe that "all moral judgments are *nothing but* expressions of preference," and "agreement in moral judgment is not to be secured by any rational method, for there are none" (12).

In one of the paradigm statements of the modern Western worldview, Max Weber asserted that an "ethic of ultimate ends apparently must go to pieces on the problem of the justification of means by ends," for "the proponent of an ethic of absolute ends cannot stand up under the ethical irrationality of the world" (Weber, 1946, 122). This "ethical irrationality" is the simple consequence of a new science by which "the world is disenchanted" (139). The world of ethics is sundered from the world of knowledge: "increasing intellectualization and rationalization do *not*, therefore, indicate an increased and general knowledge of the conditions under which one lives" (139). If the question is *What shall we do and how shall we live?* then, Weber replies: "That science does not give an answer to this question is indisputable" (143). Instead and exclusively, "science contributes to the technology of controlling life by calculating external objects as well as man's activities" (150). As MacIntyre aptly summarizes, Weber concludes that "questions of ends are questions of values, and on value reason is silent; conflict between rival values cannot be rationally settled. Instead one must simply choose" (MacIntyre, 1984, 26).

An abyss – between *is* and *ought*, between science and ethics – began to open up with and through the emergence of modern Western philosophy in the seventeenth and eighteenth centuries, the epoch of the "Scientific Revolution" and the Enlightenment. The modern West gradually enshrined "a novel conception of science: ... a type of understanding which stands apart from all value-judgment and value-determination" (Leiss, 1972, 109). The way of understanding (and manipulating) nature became disengaged utterly from any normative restraint grounded in such an understanding. That abyss between science and ethics remains at the core of contemporary science and technology, notwithstanding the sea-change that has occurred in the theory of the world, and perhaps even in the idea of "scientific method," in the interval. It behooves us to reckon with the power of the image of a new science which was projected into Western culture from the epoch of the "Scientific Revolution," constituting a pervasive positivism and scientism in modern culture.[1] In a classic history of Western science expressive of this powerful positivist mindset, Charles Gillispie opined: "neither in public nor in private life can science establish an ethics. It tells us what we can do, never what we should. Its absolute incompetence in the realm of values is a necessary consequence of the objective posture" (Gillispie, 1960, 154). Accordingly, "After Galileo, science could no longer be humane in the deep, internal sense of its forerunner in classical antiquity" (81). Objectivity, whose "cruel edge" Gillispie hones and hallows, signifies that "the act of understanding is an act of alienation" (44, 82). Modern science is "impersonal and objective," and it "seeks both to comprehend and control nature" as part of a "thrust toward mastery of the world" (10). Gillispie affirms "western man's most characteristic and successful campaign, which must doom to conquer" (199–200). He dismisses those who "want science to give us a world we can fit, as Greek science did, and not just a world like any external object that we can first measure, and then destroy" (155). This, Gillispie insists, is "the condition that nature sets ... that science communicate in the language of mathematics, the measure of quantity, in which no terms exist for good or bad, kind or cruel" (43). In short, "Newton's world offered virtue no purchase" (182). There is no place in such a world for a "language of will and purpose and hope" (43).

For MacIntyre the inevitable failure of modern ethics cries out for a historical retrieval of the premodern "world we have lost."[2] Such phrases as "the disenchantment of the world," the "death of nature," "the world we have lost," and "after virtue" all betoken a sense of radical severance between the modern from the premodern cultural experiences in the West. Part of the argument in many such texts is that a substantial solidarity can be discerned between the forsaken premodern traditions of the West and the still living, though embattled cultures of the non-West, and that the prospects of global flourishing depend largely upon rallying these dispelled, dispersed and desperate elements together to withstand the scientific-technological juggernaut launched by the modern West (Callicott, 1994).[3] In the global context in which the issues of biotechnology will need to be negotiated, it is dangerously parochial to consider only contemporary Western philosophical notions. These need to be nuanced, indeed, challenged by other ideas. In our view, resources reinforcing some dissenting directions in contemporary Western thought (virtue ethics and environmental ethics, in particular) can be found and should be sought in non-Western

---

[1] Auguste Comte, father of positivism, wrote of the "grand movement impressed upon the human mind ... by the combined action of Bacon's precepts, Descartes' conceptions, and Galileo's discoveries, ... the moment when the spirit of positive philosophy began to pronounce itself in the world..." (Comte, cited in Cohen, 1994, 36).

[2] The *phrase* is not his; it is an ironic *geste* from Peter Laslett (1971), but the *sense*, without irony, is unquestionably MacIntyre's.

[3] For critical reflections on the relation of modern Western thought with non-Western philosophies, see Restivo (1978, 1982).

approaches, as we have tried to suggest by a brief consideration, later in this chapter, of East Asian notions of ethics in relation to nature. It remains to be seen, of course, how effective this incorporation of non-Western and more traditional Western approaches will be. In stressing the break at the inception of the "Scientific Revolution," our sketch problematizes contemporary access to and employment of premodern categories for current ethical dispute. What survives of the older world and what can be resurrected on new foundations will have to be worked out with caution. We recognize that historical retrieval alone can hardly accomplish ethical empowerment of "all those Aristotelian and quasi-Aristotelian views of the world in which a teleological perspective provided a context in which evaluative claims functioned as a particular kind of factual claim" (MacIntyre, 1984, 77).

In what follows, we propose not simply to recount the inaugural rupture or to describe the contemporary situation, but also to discern the persistence of traditional values within science and beyond science over the modern period, to retrieve creative moments of resistance and contestation within the dialectic of modern science and modern ethics, and finally to articulate the plurality of possibilities that philosophy still holds out for a notion of nature with ethical purchase that might be of avail in our impending decisions about biotechnology. In Part I, we examine the seventeenth-century moment of rupture, seeking to assess both what the "mechanization of nature" meant and how its cultural implications emerged then and later. The growing ascendancy of this conception in the modern West never betokened the extinction of challenges, either within or beyond the realm of science. The remaining parts of the chapter trace out these challenges in terms, first (Part II), of the persistence of more traditional ideas about nature and its ethical implications, (especially in the life sciences from the eighteenth to the twentieth centuries, pivoting around Darwin), and then (Part III) of the rise in the twentieth century of new ideas both philosophical (pluralism, naturalism, feminism) and scientific (environmentalism, sociobiology, evolutionary ethics). Thus, we recount the dialectic of challenge and response between the dominant model, logical positivism, and a series of challengers within twentieth-century philosophy of science. Part IV develops a conspectus of current ethical philosophies that incorporates both the dominant and the dissenting viewpoints, drawing parallels with traditional Western and non-Western ethical stances. We also discern *new* impulses. In recent years, starting from very different vantages, two new efforts have come to prominence which seek in contemporary science, especially biological science, a grounding for ethics. One, radical environmentalism, seeks to derive a new "ecocentric environmental ethics" from evolutionary biology and ecology (Callicott, 1994; Warren, 1990). In large measure, this endeavor is driven by its environmentalist *telos*. From a very different vantage, sociobiologists and others have proposed "evolutionary ethics," i.e., prescriptions for conduct encoded in genetics and shaped by evolutionary constraints (Maienschein and Ruse, 1999). In contrast to the first, this endeavor is starkly determinist, seeking in nature causes which propel human conduct (Wilson, 1980; Dawkins, 1989; for a careful rebuttal see Midgley, 1995). In the Conclusion we make the case that the "fact/value dichotomy" should not be taken to preempt inquiry into the normative potential in nature and we highlight views that make such inquiries. We recognize not only the *plurality* of positions that have resulted, but the *fruitfulness* as well of this complex contestation.

2.2 The "Mechanization of the World Picture": The Departure of the Seventeenth Century

"What is really peculiar to the modern world is the belief that scientific knowledge can be used at will by man to master and exploit nature for his own ends" (Dubos, 1961, 16). This crucial observation by biologist René Dubos in 1961 sets the terms of our inquiry. It suggests that there is a connection between the historical uniqueness of the modern scientific worldview and what is assuredly a *moral* claim for human entitlement to dominate nature. The implication of Dubos' statement is that such a belief in entitlement is unique, indeed aberrant. Since in most premodern and non-Western societies the conception of a normativity of nature is quite common, its relative absence in Western thought since the Enlightenment demands explanation. Thus it behooves us to establish when and how this distinctive cultural position arose and what it betokens for our current ethical conversation.[4]

In *After Virtue* MacIntyre makes clear that "the concepts we employ have in at least some cases changed their character in the past 300 years … 'virtue' and 'justice' and 'piety' and 'duty' and even 'ought' have become other than they once were" (MacIntyre, 1984, 10). He associates the historical origin of this mutation of ethical theory with the epoch of "Enlightenment" in an extended sense ("from say 1630 to 1850") (39). This idea of historical rupture has resonated in such theoretical formulations as Thomas Kuhn's notion of paradigms and revolutions, Michel Foucault's notion of incommensurable epistemes, and Reinhard Koselleck's notion of a *Sattelzeit*, a semantic divide in the conceptual history of the West (Kuhn, 1970; Foucault, 1966; Koselleck, 1985). Foucault and Koselleck would set the watershed around 1800, but there are grounds to set it earlier – in the seventeenth century.[5] Whether it makes sense any longer to conceive of *the* "Scientific Revolution," there is no doubt that over the course of the seventeenth century a "mechanization of the world picture" set in to guide philosophical and scientific thinking about the physical world.[6] Concomitant with that *cognitive* revision was a shift in the *purpose* of

---

[4] For other considerations of the idea of nature in early modernism, see Nobis (1967) and Spaemann (1967).

[5] Nicholas Jardine (1999) gives a careful account of the positions of Foucault and Koselleck.

[6] Recent historiographical discussion has made the concept of a singular Scientific Revolution extremely problematic; Cohen describes it aptly as "an ongoing effort to turn the apparent inevitability of the birth and subsequent development of modern science into an array of contingencies" (Cohen, 1994, 16). See Lindberg and Westman (1990), Porter (1986), Cunningham and Williams (1993). On the other hand, there is still widespread agreement regarding the prominence in the seventeenth century of a "mechanization of the world picture" or the ascendancy of a "mechanical philosophy." See Hall (1941), Brown (1941), and Osler (1994).

scientific inquiry: the crucial goal, articulated bluntly by both Bacon and Descartes, was domination of nature.[7]

2.2.1 The Significance of Mechanization

What exactly *was* the "mechanical philosophy" of the seventeenth century? How did the "mechanization of the world picture" work? Did it, from the outset, knowingly embrace the ethical and cultural consequences that interpreters have subsequently ascribed to it, or are we dealing here with a cardinal instance of "unintended consequences" in history? As a preliminary, three maxims of historical method need to be acknowledged. First, we must be clear of *whom* we speak: the *scope* of that mechanical philosophy was limited starkly to a few pioneering, radical thinkers and their immediate circles of followers; wider socio-cultural diffusion of their ideas took considerable time. Second, even among these revolutionaries we cannot presume monolithic unity; what became *the* mechanical philosophy or *the* mechanization of the world picture over the course of a century was in fact a congeries of mechanical philosophies and mechanizations, different in motive, in articulation and in efficacy, though we should not leap to the other extreme and despair of any coalescence. Finally, it behooves us not to project anachronistic categories uncritically upon the seventeenth century: it is not clear that they would recognize our concept of *science* as equivalent to their "natural philosophy" or that they would discern or affirm the *revolution* that has subsequently been attributed to them. Precisely because we have made of them the inaugurators of our worldview, it is crucial that we take seriously that they may well have had other concerns.

"Mechanical philosophy" was at least explicitly avowed by the key figures of the seventeenth century, though perhaps with a register of significations different from that which we have synthesized retrospectively. Similarly, the "mechanization of the world picture" was a deliberate endeavor of seventeenth-century natural philosophers, though, again, what each of them endeavored under this rubric may well have differed both from contemporaries and from our retrospective integral. Because our interest is in the long-term legacy rather than the immediate richness and diversity of their historical project, we can allow ourselves to attempt a retrospective synthesis, having admitted from the outset its element of arbitrariness.

---

[7] "For the end which this science of mine proposes is the invention not of arguments but of arts; not of things in accordance with principles, but of principles themselves; not of probable reasons, but of designations and directions for works. And as the intention is different, so accordingly is the effect; the effect of the one being to overcome an opponent in argument, of the other to command nature in action" (Bacon, 2000, 15–16). "[I]nstead of that speculative philosophy which is taught in the Schools, we may find a practical philosophy by means of which, knowing the force and action of fire, water, air, the stars, heavens and all other bodies that environ us, as distinctly as we know the different crafts of our artisans, we can in the same way employ them in all those uses to which they are adapted, and thus render ourselves the masters and possessors of nature" (Descartes, 1985, vol. I, 142–143).

The question *why* mechanism appeared so indispensable to the seventeenth-century intellectual radicals is too complex to take up here. Rather, it seems expedient to articulate *what* mechanism signified. The ontological foundation is most accessible in Cartesian terms, in the discrimination (dualism) of extension (matter) from thought (mind) as incommensurable substances (Descartes, 1985). Not only did they constitute radically opposed orders of being, but that led to a further discrimination, from the perspective of thought, about the ontological status of qualities present to mind. *Primary* qualities were taken to have to do with the "real" character of the extended, material world (Locke, 1975, II:viii:9–26). These were objective and not coincidentally quantitative attributes – size, shape, location in space and time, mass, motion.[8] *Secondary* qualities were taken to be subjective impressions occasioned in the mind by sensory encounter with these primary qualities.[9] Thus the entire register of qualitative sensibility – color, savor, scent or sound – came to lose ontological status. Across the seventeenth century, increasing precision about the exact meaning of the inertness of matter, especially as corpuscular theory came to prominence, created a further ontological conundrum (Henry, 1986; Hutchison, 1983, 1997). Given that matter by definition was inert, i.e., could not alter its motion without external interference, the questions of how the whole universe got set into motion, how it was preserved in its motions, and how motions (among other things) *changed* became cardinal for natural philosophy (Westfall, 1971). Transcendent creation resolved most such matters for seventeenth-century thinkers, but for many of them, "occult" properties ("etherial principles," "forces") in the extended world seemed both indispensable and unaccountable.[10] "Force" became the name for an ontological element *sine qua non* for mechanistic accounts, yet utterly incongruous with the concept of inert matter.[11]

---

[8] "The geometrical form is homogeneous with matter, and this is why geometrical laws have a real significance, and dominate physics" (Koyré, 1978, 204).

[9] "[T]he *ideas* of sensible qualities, which we have in our minds, can, by us, be no way deduced from bodily causes, nor any correspondence or connection be found between them and those primary qualities which (experience shows us) produced them in us..." (Locke, 1975, IV:iii:28).

[10] See Isaac Newton, "General Scholium" (1713) to *Mathematical Principles of Natural Philosophy*: "And now we might add something concerning a most subtle spirit which pervades and lies hid in all gross bodies, by the force and action of which spirit the particles of bodies attract one another at near distances and cohere, if contiguous; and electric bodies operate to greater distances, as well repelling as attracting the neighboring corpuscles; and light is emitted, reflected, refracted, inflected, and heats bodies; and all sensation is excited, and the members of animal bodies move at the command of the will, namely, by the vibrations of this spirit, mutually propagated along the solid filaments of the nerves, from the outward organs of sense to the brain and from the brain into the muscles. But these are things that cannot be explained in few words; nor are we furnished with that sufficiency of experiments which is required to an accurate determination and demonstration of the laws by which this electric and elastic spirit operates" (Newton, 1953, 45–46).

[11] "The inability of mechanical philosophers to consider any conception of force except the 'force of a moving body' became an obstacle to the development of a mathematical dynamics and tended to confine mechanics within kinematic problems, in which motions were described without reference to the forces that caused them" (Westfall, 1971, 123).

All these ontological elements had epistemological concomitants.[12] It swiftly became clear that if extension and thought were radically disparate ontologically, and thus primary and secondary qualities had such drastically different claims to reality, it would be very difficult to move from subjective judgments to objective validity.[13] This came to formulation as a tension between "real" and "nominal" essences, with philosophers increasingly persuaded that human understanding faced insuperable "limits" in accessing real essences.[14] That, in turn, threw nominal essences into a condition of epistemological insecurity. The most compelling case was to presume a necessary connection between effects in consciousness (albeit colored by secondary qualities) and causes in the extended world of truly real primary qualities.[15] The inference from known effects to hypothetical causes drew strength from the discrimination of *mathematical*

---

[12] "[T]he corpuscularian programme entailed the reduction of visible phenomena to the behaviour of unobservable entities whose properties were obviously conjectural. Once one accepts the corpuscularian theory of matter, and with it the theory of knowledge which makes corpuscles unobservable in principle, then it is altogether natural to adopt a hypothetical methodology. In short, the metaphysics and epistemology of the mechanical philosophy led, by its own inner logic, to the acceptance of a certain methodology" (Laudan, 1966, 96).

[13] "Koyré never tired of emphasizing that naive, daily experience of the Aristotelian kind cannot lead to knowledge of a type that may count as science proper in the new universe of precision" (Cohen, 1994, 81).

[14] Perhaps the classic statement is John Locke, *Essay Concerning Human Understanding*: "as to the *real essences* of substances, we only suppose their being, without precisely knowing what they are…" (III:vi:6). "But *as to the powers of substances* to change the sensible qualities of other bodies, which make a great part of our inquiries about them, and is no inconsiderable branch of our knowledge; I doubt, as to these, whether *our knowledge reaches* much farther than our experience; or whether we can come to the discovery of most of these powers, and be certain that they are in any subject by the connection with any of those *ideas*, which to us make its essence. Because the active and passive powers of bodies, and their ways of operating, consisting in a texture and motion of parts, which we cannot by any means come to discover: 'tis but in very few cases, we can be able to perceive their dependence on, or repugnance to any of those *ideas*, which make our complex one of that sort of things. I have here instanced in the corpuscularian hypothesis, as that which is thought to go farthest in an intelligible explication of the qualities of bodies; and I fear the weakness of human understanding is scarce able to substitute another, which will afford us a fuller and clearer discovery of the necessary connection, and *coexistence*, of the powers, which are to be observed united in several sorts of them… I doubt, whether, with those faculties we have, we shall ever be able to carry our general knowledge (I say not particular experience) in this part much farther…" (IV:iii:16) "And therefore I am apt to doubt, how far soever human industry may advance useful and *experimental* philosophy *in physical things, scientifical* will still be out of our reach… [W]e are not capable of *scientifical knowledge*; nor shall ever be able to discover general, instructive, unquestionable truths concerning them…" (IV:iii:26).

[15] "[M]any Atomists and other Naturalists, presume to know the true and genuine causes of the things they attempt to explicate; yet very often the utmost they can attain to, in their explications, is, that the explicated phenomena may be produced after such a manner, as they deliver, but not that they really are so" (Boyle, 1664). Of course, Hume would pose a monumental challenge to this in *Enquiry Concerning Human Understanding*.

2 Philosophical Approaches to Nature

regularities in these effects.[16] Mathematical form became a crucial warrant for actual knowledge.[17] But clearly the traditional philosophical standard of absolute knowledge – necessary certainty – began to appear hopeless, and the question of probability and its warrant became increasingly pressing (Hacking, 1975; Daston, 1988). Indeed, mathematics itself, in its most creative seventeenth-century articulations concerning both infinity and infinitesimals, posed grave conceptual challenges to a mechanistic conception of space, matter and "force."[18] Similarly, the gap between subjective experience and objective knowledge occasioned a sharper need to discriminate mere *observation*, beset with all the subjective distortions of sensibility, from adequate *experience*. "Experiment" came to mean precisely a deliberate intervention not only into the material world (to discipline its manifestations – Bacon's infamous "putting nature to the wrack") but into subjective experience as well (to minimize the distortions of human understanding – Bacon's famous "idols").[19] Here advances in instrumentation – the telescope and the microscope – had a decisive impact, especially as these accentuated the questions of infinity and the infinitesimal (van Helden, 1997; Wilson, 1995).

What were the ethical and cultural concomitants? One of the most acute commentators of the seventeenth century was Blaise Pascal. Reflecting upon both the ontology and the epistemology of the "mechanical philosophy," he observed: "Man is himself the most wonderful object of nature: for he cannot conceive what the body is, still less what the mind is, and least of all how a body should be united to a mind" (Pascal, 1941, 27). Altogether cognizant of the mathematical intricacies of the new mechanization of the world picture, he noted,

---

[16] "[T]he moderns, laying aside substantial forms and occult qualities, have endeavoured to subject the phenomena of nature to the laws of mathematics..." (Newton, Preface to *Mathematical Principles of Natural Philosophy*, in Newton, 1953. "As in Mathematics, so in Natural Philosophy, the Investigation of difficult Things by the Method of Analysis, ought ever to precede the Method of Composition. This Analysis consists in making Experiments and Observations, and in drawing general Conclusions from them by Induction, and admitting of no Objections against the Conclusions, but such as are taken from Experiments, or other certain Truths. For Hypotheses are not to be regarded in experimental Philosophy. And although the arguing from Experiments and Observations by Induction be no Demonstration of general Conclusions; yet it is the best way of arguing which the Nature of Things admits of, and may be looked upon as so much the stronger, by how much the Induction is more general. And if no Exception occur from Phaenomena, the Conclusion may be pronounced generally" (Newton, 1952, 404–405).

[17] "Our purpose is only to trace out the quantity and properties of this force [attraction] from the phenomena and the apply what we discover in some simple cases, as principles, by which, in a mathematical way, we may estimate the effects thereof in more involved cases. For it would be endless and impossible to bring every particular to direct and immediate observation. We said, *in a mathematical way*, to avoid all questions about the nature or quality of this force (attraction), which we would not be understood to determine by any hypothesis" (Newton, 1953). Immanuel Kant maintained "that in every special doctrine of nature only so much science proper can be found as there is mathematics in it" (Kant, 1786; AA:4:470).

[18] "The still deeper difficulty is that the very notion of infinite number or magnitude, which we know to be truly predicable of reality, at the same time involves our thought in insoluble antinomies" (Lovejoy, 1936, 128). See Mancosu (1996, ch. 5) and Bray (1988).

[19] Bacon, *New Organon*. Cohen (1994, 184) describes these perplexities. Shapin and Schaffer have made them the centerpiece of their provocative study, *Leviathan and the Air Pump* (1985).

> Our intellect holds the same position in the world of thought as our body occupies in the expanse of nature… Our senses perceive no extremes. Too much sound deafens us; too much light dazzles us… This is our true state; this is what makes us incapable of certain knowledge and of absolute ignorance. We sail within a vast sphere, ever drifting in uncertainty, driven from end to end… (Pascal, 1941, 25).

But the epistemological indeterminacy carried with it, for Pascal, an existential *Angst*:

> I see those frightful spaces of the universe which surround me, and I find myself tied to one corner of this vast expanse, without knowing why the short time which is given me to live is assigned to me at this point rather than at another of the whole eternity which was before me and shall come after me. I see nothing but infinites on all sides, which surround me as an atom, and as a shadow which endures only for an instant and returns no more. All I know is that I must soon die, but what I know least is this very death which I cannot escape… (Pascal, 1941, 68).

Not everyone, to be sure, reached this level of anxiety or insight, but "uneasiness," to borrow the more temperate term of John Locke, rode the winds of these changes (Locke, 1975, II:21:§29–40). If the prospect of Providence, which mathematical lawfulness in nature appeared to confirm, was supposed to console and edify, still the inscrutability of that divine nature and will became increasingly obvious (Goldmann, 1964).[20] In a climactic disputation in which the cultural as much as the cognitive aspects of the seventeenth-century adventures with mechanical philosophy figured centrally, the Leibniz-Clarke correspondence, the question was essentially what could be inferred from *natura naturata* to *natura naturans* (Leibniz and Clarke, 2000). Indeed, one fear behind the controversy was that some "atheists" (Hobbes, Spinoza) aimed to dissolve the difference – *Deus sive Natura* (Mason, 1997). What was feared most was that deterministic materialism would transform man's spiritual elevation beyond nature into man's entrapment in a meaningless machine. By 1700 that "cruel edge" Gillispie ascribes to the new science was clearly beginning to lacerate Western cultural sensibilities.[21] Man, it appeared, was coming unhinged from the rest of nature. Hence the untranslatable perfection of Paul Hazard's title, *La crise de la conscience Européenne (1680–1715)*.[22]

2.2.2 Contemporary Reception of the Mechanical Philosophy

A work that has helped construct the contemporary cultural register on this issue is Carolyn Merchant's *The Death of Nature* (1980). She writes: "The rejection and removal of organic and animistic features and the substitution of mechanically

---

[20] The irreligious implications were drawn out by Diderot in the mid-eighteenth century *Pensées philosophiques* (1998).

[21] "Thus the world of science – the real world – became estranged and utterly divorced from the world of life, which science has been unable to explain – not even to explain away by calling it 'subjective' …" (Koyré, 1965, 24–25).

[22] Translated as *The European Mind: The Critical Years 1680–1715* (1953).

describable components would become the most significant and far-reaching effect of the Scientific Revolution" (Merchant, 1980, 125). Her argument is that there is a "normative import [to] descriptive statements about nature," such that "as the descriptive metaphors and images of nature change, a behavioral constraint can be changed into a sanction" (4). In the sixteenth and seventeen centuries, there clearly emerged a "transition from the organism to the machine as the dominant metaphor" (xxii). And the result was that "two new ideas, those of mechanism and of the domination and mastery of nature became core concepts of the modern world" (2). Mechanism neutralized the inhibitions that organicism had created *vis à vis* exploitation of the natural world: "the passivity of matter, externality of motion, and elimination of the female world soul altered the character of cosmology and its associated normative constraints" (11). Merchant's account minimizes the diversity that persisted into the modern era of science even as it proposes a vision of the pre-modern that proves somewhat over-idealized (Newman, 1998, 216–225). While the balance of our discussion will explore the contesting and complicating features of later thought, it is useful to follow Merchant in her stark characterization of the implications of mechanism, for this tradition did have a profound impact on subsequent Western culture.

It is Merchant's view that "the historical interconnections between women and nature" constitute a decisive vantage for assessing the value-dimension of the formation of the "modern scientific and economic world" (Merchant, 1980, xx–xxi). Indeed, she maintains "in both Western and non-Western cultures, nature was traditionally feminine" (xxiii). What she wishes to expose is "a worldview and a science that, by reconceptualizing reality as a machine rather than a living organism, sanctioned the domination of both nature and women" (xxi). Aggressive human endeavors in the world – both cognitive and technical – took on the tenor of sexual penetration and violation. Merchant invokes an allegory by the medieval author Alain de Lille from 1160 which charged that "in aggressively penetrating the secrets of heaven, [men] tear Natura's undergarments, exposing her to the view of the vulgar" (10). She draws links to wider and more pervasive anxieties and taboos in the epoch: "the sixteenth and seventeenth-century imagination perceived a direct correlation between mining and digging into the nooks and crannies of a woman's body," since it cherished an "image of Mother Earth and her generative role in the production of metals" in the womb of the earth (39, 30).[23]

---

[23] One aspect of Merchant's argument which needs critical reexamination, even on feminist premises, is that the radical innovation that assuredly marked the "mechanization of the world picture" was accompanied by and causally enmeshed with an accentuation of misogyny. The argument here is not against the association of nature with the female, nor against the link with misogyny. It is rather that both of these associations, and their mutual constitution, belong to the long sweep of Western civilization, not uniquely to the modern rupture. Merchant's general claim is that "the image of nature that became important in the early modern period was that of a disorderly and chaotic realm to be subdued and controlled," and this "wild uncontrollable nature was associated with the female" (127). Yet Merchant does not provide nearly sufficient evidence to warrant the claim that this particular view of nature gained in its importance at that point in history. Assuredly it was an image with a much older pedigree. Nor is the association of such wildness with the female new. If Merchant is correct that the domination of nature works in tandem with the domination of women, it remains that the seventeenth century's new determination to dominate nature and its particular method were not caused or sufficiently specified by the determination to dominate women.

For Merchant – and in her train for Keller, Harding, and many other feminist critics of Western science – sexually aggressive metaphors expose the latent value-dimension of ostensibly neutral cognitive actions (Keller, 1985; Harding, 1986). The foundational exponent, in their view, was Francis Bacon. In paradigmatic form, Bacon "sets forth the need for prying into nature's nooks and crannies in searching out her secrets for human improvement" (Merchant, 1980, 33).[24] Merchant accuses Bacon's "vivid metaphor" of creating a "language that legitimates the exploitation and 'rape' of nature for human good" (171). Far from simply enunciating a cognitive procedure (method), "Bacon fashioned a new *ethic* sanctioning the exploitation of nature" (Merchant, 1980, 164, italics added).

This argument about the tenor and import of Bacon's language found earlier expression in William Leiss's 1972 study, *The Domination of Nature*. Leiss saw a different but equally ominous linkage: an "inextricable bond between the domination of nature and the domination of man" (1972, xiv). He stressed class rather than gender as the social concomitant of Bacon's new science. Leiss, too, linked the metaphorical field to that of mining and metallurgy in the medieval tradition. And, in a manner Merchant would echo, Leiss located Bacon's project in the larger question of the refashioning of magic in the Renaissance world. It was in terms of magic that the Renaissance manifested "an intense upsurge of interest in the workings of nature, in its 'secrets' and its 'treasures'" (35). The established Christian tradition had deep misgivings about the demonry in magic, and thus the Renaissance struggled to defend a "natural magic."[25] That is, "Renaissance writers made magic more acceptable by contending that it never violates the principles of nature ... magic is the 'servant of nature,' nature's imitator and assistant" (Leiss, 1972, based on Rossi, 1968). Leiss suggests that this was to a considerable degree disingenuous. William Newman provides massive evidence for this and argues that Merchant's view of the premodern, especially with reference to the alchemical tradition, suffers from nostalgia. Premodern alchemy had many of the very manipulative characteristics that Merchant wants to believe uniquely modern (Newman, 1998, 216–225).

In any event, Bacon undertook revisionism on both scores. He sought to rescue aggressive manipulation of nature from the scruples of tradition and religion by setting

---

[24] Especially in its feminist form, this construal of the centrality and virulence of Bacon's sexually aggressive metaphors has provoked a significant controversy. Alan Soble has presented some compelling criticism, arguing persuasively that in assessing the intellectual genesis of modern science it is methodologically sounder to view Bacon's metaphors "as 'literary embellishments' than as a 'substantive part of science'" (Soble, 1998, 210). While Soble is persuasive that some feminist interpretation has gone too far with Bacon's language, the fact that the same argument was made earlier by William Leiss without any particular gender axe to grind should alert us to the prominence of some of these metaphors. They did have a significant imaginative force in the genesis of modern science, and they did betoken a pervasive concern with domination through inquiry which lies at the methodological heart of the new science of the seventeenth century.

[25] "[T]he notion of natural magic was a fully warranted conception: it was a recognition of the marvellous [sic] character of natural phenomena which could not logically be deduced, combined with the recognition that the investigation of these phenomena is not a sinful attempt to penetrate into realms forbidden to human thinking, but a perfectly legitimate gratification of a natural desire for knowledge" (Dijksterhuis, 1961, 158).

it free from the wrong magical associations while, at the same time, he wished to assert not dependence but dominance. Merchant registers this: Bacon "transformed the magus from nature's servant to its exploiter, and nature from a teacher to a slave" (Merchant, 1980, 169). Leiss, on the other hand, stresses Bacon's affiliation with the Judeo-Christian ideas of divine transcendence and of human participation in this transcendence over against a balance of creation utterly lacking in commensurate dignity.[26] Bacon always viewed dominion over nature through science within "a larger religious scheme" which led him to believe science would be "governed by sound reason and true religion" (Leiss, 1972, 51). He was deliberately and sincerely Christian in his reconstruction: "to master nature through scientific progress did not violate God's plan," as Leiss puts it (49). Bacon consistently argued that the "domination" of nature was to be held within theologically-defined restraints. The new science was to be "restorative" of the knowledge lost at the Fall, but was not to exceed it. To do so was to transcend divine limits. But in the tradition subsequently established in his name, "when the concept of mastery over nature is thoroughly secularized, the ethical limitations implicit in the pact between God and man, whereby the human race was granted partial dominion over the earth, lose their efficacy" (54). In the absence of this external, religious guarantee, "in the influential intellectual tradition originating with Bacon it has been assumed that mastery over nature considered as scientific-technological progress would be 'automatically' transformed into mastery of nature considered as social progress (a reduction in the source of social conflict)" (140). That, Leiss asserts, is blatant ideology because it masks the fact that "mastery of nature has been and remains a social task, not the appurtenance of an abstract scientific methodology ... and the overriding feature of that context is bitter social conflict" (154). "The abstract idea of man (in the phrase 'man's conquest of nature') hides the fact that the actual agents in this process are individuals and societies in violent conflict among themselves" (189). C.S. Lewis put that precisely: "What we call Man's power over Nature turns out to be a power exercised by some men over other men with Nature as its instrument" (Lewis, 1996, 39–40, cited in Leiss, 1972).

2.2.3 Conclusion

MacIntyre, Leiss, and Merchant (and the list could easily be extended) clearly discern an epochal rupture in the establishment of the mechanistic worldview.[27] Method and motive coincided to accentuate an alienation of human will from natural order.

---

[26] "The idea that man stands apart from nature and rightfully exercises a kind of authority over the natural world was thus a prominent feature of the doctrine that has dominated the ethical consciousness of Western civilization" (Leiss, 1972, 32). For a classic environmentalist critique of this Judeo-Christian legacy, see White (1968). For richly detailed accounts of the cultural context, see especially the writings of Thomas (1971, 1996).

[27] Two decisive additions to the list would be Alexandre Koyré and E. A. Burtt. See Koyré (1957) and Burtt (1949).

What both the methodological and the motivational impulses implied was that it was a category error, a "fallacy," to look to nature for normative guidance. Nature was "dead," an object to be manipulated simultaneously for comprehension and control. The values upon which humankind would carry out this endeavor had to be sought (and justified) *elsewhere*. Kant thought this emergence from the "tutelage of nature" the mark of "Enlightenment," the historical advent of human *maturity* in freedom (Kant, 1784, in Reiss, 1970, 54–60). "The proper study of mankind is man," Alexander Pope proclaimed in 1733 (Pope, 1994). David Hume proposed that a "science of man" serve as the essential prolegomenon to all pursuit of knowledge (1739) (Hume, 1888). But man's nature proved no more transparent than the "real" world, and there was the further dilemma, ostensibly pronounced by the very same David Hume, that even if one discerns what "human nature" *is*, that gives no purchase on what it *ought* to be (MacIntyre, 1969). Neither Hume nor his rivals succeeded in becoming "the Newton of the moral sciences."[28] Instead, the very idea that there could be a "moral" *science* came to seem increasingly incoherent. MacIntyre claims that though the philosophers of the Enlightenment essayed preeminently "to find a rational basis for their moral beliefs in a particular understanding of human nature," this was doomed from the start by another of their paramount premises, namely the denial that one could make *any* inference from fact to value. "The meaning assigned to moral and indeed to other key evaluative expressions so changed during the late seventeenth and the eighteenth centuries that what are by then commonly allowed to be factual premises cannot entail what are by then commonly taken to be evaluative or moral conclusions" (MacIntyre, 1984, 77).

The change at the origin of the modern Western scientific worldview was real and epochal, though that does not mean it was instant, total or irreversible. Aristotelianism survived in the non-mathematical sciences for a considerable span (Grant, 1978), and it informed a variety of challenges to the mechanical model over the balance of Western thought. Traditional religious, ethical and political ideas moved in a slower cadence, though they soon felt the shock of the new. Something unquestionably had changed, something so epochal that it came to overwhelm these other currents in the cultural memory and the cultural expectations of the modern West. William Leiss sets it in the relevant frame for our purposes: "Precapitalist societies share almost universally a common feature, namely, a reliance on various 'naturalistic' categories as a basis for social organization" (Leiss, 1972, 180). In the premodern West, such a normative naturalism exercised a palpable restraint upon human conduct, but "the final 'despiritualization' of nature, the steady weakening of the prescriptive force carried by the concept of nature, creates [an] ethical and ideational void" in which "the concept of nature ceases to be a basis for limitations on the scope of human behavior" (Leiss, 1972, 185). In the balance of this chapter we will register the resources that nonetheless persisted or emerged for a normativity in nature for the modern West, and how these tally with some non-Western approaches.

---

[28] Kant thought Rousseau a better candidate than Hume, and one suspects he thought of himself as a better one still. Nobody has won that prize, and now it seems no one even aspires to.

2.3 Nature Reborn

The introduction to this chapter poses the question of how "nature" can be used as a philosophical resource in ways relevant to current issues in the ethics of biotechnology, and situates this problematic in a historical analysis of the changes in the conception of nature that accompanied main aspects of the Scientific Revolution, and particularly the revolution in the physical and mathematical sciences. If nature is only a system of inert matter moving according to fixed mathematical laws, and is, in Descartes's phrase from the posthumous *Le Monde*, "not some Goddess, or some other sort of imaginary power, but ... matter itself," there is little reason to see "nature" as having some ethical significance, or as being worthy of some intrinsic respect.[29] Overemphasis on this mechanistic tradition of the philosophy of nature is, however, to be guarded against, and as this section will develop, there persisted alongside early modern mechanism traditions that were in continuity with Greek and Renaissance philosophies of nature that in the life sciences formed the foundations for a return of a substantive conception of nature in the latter decades of the eighteenth century.

The recognition of the existence of conflicting historical traditions in the philosophy of nature from the eighteenth century is necessary to understand both ethical and religious debates over modern biotechnology. Modern life science has incorporated aspects of both of these historical traditions and illustrates Alasdair MacIntyre's observation that contemporary debates, particularly those involving ethical disputes, often can be analyzed as incoherent discussions drawing upon a fragmentary heritage of forgotten past discussions (MacIntyre, 1984, ch. 1). To see any relation between philosophical reflections on "nature" and contemporary biotechnology requires that some deeper analysis be supplied of these competing traditions of usage that are currently involved in an incommensurable debate.

The analyses of the history of the concept of nature available in the literature illustrate some of the complexity of the problem.[30] The term "nature" has been used with a wide range of meanings in the tradition.[31] Some of these are more relevant than others to current debates over biotechnology, but ingredients of others are also involved.[32] The picture that emerges from a historical review is not that of a simple schema in which a unitary notion has been replaced historically in the early modern period by another settled concept that governs contemporary discourse. Instead we find a plurality of concepts that continue to interact in contemporary discussions. One illustration of this is found in the differing images of the normative value of the "natural" that one can draw from contemporary environmentalism, and those

---

[29] Descartes, *Le Monde* (1630), in Descartes, 1971, 143–144.

[30] See, for example, Ehrard's classic *L'Idée de Nature en France dans la première moitié du xviiie siècle*, new edition (1994).

[31] See the taxonomy in Lovejoy (1927, 444–450).

[32] For a classical reference, Aristotle's definitions at *Metaphysics* V.1015a.13 form a useful starting point. This discussion gives primacy to the concept of an inner principle of a growing thing.

derivable from the discourse current in the physical sciences, biophysics, and much of the philosophy of science derived from these sources.

If it is indeed the case, as argued in the first section of this chapter, that a dominant theme in the modern period was the development of a non-teleological, "mechanistic" conception of nature that rendered it a passive domain upon which human domination could be exerted, it is equally true that this notion has not occupied the conceptual landscape in modern discussions without significant challenge. Nor are these challenges only of recent origin. Rather, these opposing viewpoints have resulted in important traditions of discussion that have profoundly affected important areas of the life and medical sciences over a long expanse of recent history. As a consequence, it is a claim of this section that our contemporary debates over the "natural" involve persistent historical strands of discussion that have substantial historical roots.

To appreciate this, we must step back two centuries to examine briefly the transformations in the concept of nature that occurred in the latter half of the eighteenth and the early decades of the nineteenth century in response to the "mechanization" of nature described in the first section. This reveals stages by which "nature" was reintroduced into scientific and philosophical discussions at this time in a form that has affected at least some traditions of discussion up to the present.

2.3.1 The Rebirth of Nature

The first section of this chapter has developed the framework by which the natural philosophy inherited in the early-modern period from Aristotelianism, Stoicism and Scholasticism was undermined by the new mechanical philosophy. Robert Boyle, for example, in a discussion that proved to be influential down into the 1840s,[33] defined nature as,

> … the aggregate of the bodies that make up the world, framed as it is, considered as a principle by virtue whereof they act and suffer according to the laws of motion prescribed by the author of things. Which description may be thus paraphrased: that nature, in general, is the result of the universal matter or corporeal substance of the universe, considered as it is contrived into the present structure and constitution of the world, whereby all the bodies that compose it are enabled to act upon, and fitted to suffer from, one another, according to the settled laws of motion (Boyle, 1996, 36).

---

[33] Boyle's discussion forms the substance of Ephriam Chamber's discussion of the concept of nature in his *Cyclopaedia: Or an Universal Dictionary of Arts and Sciences* (first edition 1728). Via Chambers, it then was incorporated into the article "Nature," in Diderot's *Encyclopédie ou dictionnaire raisonnée* (1765), vol. 2, 40–41. It is also the basis of the articles in William Nicholson's *The British Encyclopedia or Dictionary of the Arts and Sciences* (1809); Abraham Rees' *Cyclopedia, or Universal Dictionary of Arts, Sciences, and Literature* of 1819; and the entries in the *Encyclopedia Brittannica* from the first edition of 1771 through the fourth edition of 1810, and in altered form it still formed the foundation of the entry in the seventh edition of 1842.

2 Philosophical Approaches to Nature

As natural philosophers, medical theorists, and naturalists interested in the living domain developed their work in the eighteenth century, however, the adequacy of models for the explanation of phenomena in the life sciences was seriously questioned, and finally rejected in the latter decades of the eighteenth century. Accompanying this was a trans-national revival of the conception of nature as an explanatory category, with substantive, dynamic, vital, and even moral properties. In some respects this represented a selective appropriation of classical traditions found in Roman Stoicism, Scholasticism, the tradition of natural law, and the reflections of Renaissance nature philosophers, positions which had been for a time driven into retreat by the rise of mathematical physics and its accompanying mechanical philosophy of nature. Drawing in selective ways upon empirical researches in the life sciences, religious developments (such as German Pietism), literary and artistic movements broadly characterized as "Romanticism," and the systematic philosophical reflections of Spinoza, Leibniz, Christian Wolff, Kant, Herder and Schelling, "nature" reentered philosophical and scientific discussion as a primary category in the discourse of the life sciences around 1800.[34]

This "rebirth" of nature is directly allied to what can be termed the "vitalist" revolution of the latter eighteenth century. By this is meant a broad move within the medical and biological sciences away from mechanical conceptions of life, and the replacement of mechanical models and metaphors by vitalistic alternatives.[35] The causes of this remarkable development are complicated and can only be sketched here. In the biomedical sciences, this transition was in part generated by the persistent failure of mechanistic analysis to explain disease, irritability, animal heat, regeneration, and embryological development as these questions were more deeply explored in the early decades of the eighteenth century (Reill, 2005; Roger, 1997, esp. 335 ff.).

On another level, these transformations were motivated by more general philosophical and theological concerns: the revival of a substantive conception of "nature" was an important ingredient in deistic theologies developed in the eighteenth century. "Nature" could serve as an intermediary creative agency that carried out the divinely imposed laws of the creator without the need for further interventions by the deity. The development of atheistic accounts of the world in the Enlightenment required that an even more active role be given to "nature."

One insight into some of these transformations can be gained from the influential discussions of the concept of nature in the widely read work of the French naturalist and *philosophe*, Georges Louis LeClerc, comte de Buffon (1707–1788). In two lengthy essays prefacing the 14th (1764) and 15th (1765) volumes of his *Histoire naturelle, générale et particulière*, Buffon presented his contemporaries

---

[34] See Reill (2005), Zammito (2002, ch. 8), and Richards (2002). For the post-Boyle developments, see also McGuire (1972, esp. 526–528).

[35] For background see Roger's landmark study, *Les Sciences de la vie dans la pensée française du xviiie siècle*, 3rd edition (1993), partially translated as *The Life Sciences in Eighteenth-Century French Thought* translated by R. Ellrich (1997, 336–353).

with two of the most explicit discussions available in the literature of the late eighteenth century on the concept of nature. Drawing together ingredients from Newton, Leibniz, Aristotle, and the English Deists, Buffon defined nature as a "system of laws established by the Creator." But "nature" for Buffon, was not a thing, nor was it a replacement for God. At the same time, it was conceived as a "living, immense power, which embraces all things, which animates all," and which has its own domain of activity (Buffon, 1954, author's translation).

There is, however, a Janus-faced character to Buffon's characterization of nature in these discussions. In one respect Buffon's conception of nature looks back to Newton and to the Deistic tradition. Nature remains a created system governed by imposed natural laws that act to bring about effects in accord with divinely-established purpose. Nature is also limited in its action by these laws. But facing in another direction, Buffon points toward a vitalization of nature. Nature is a "living" worker. It employs "organic molecules" that are metaphysically distinguished from the "brute matter" that composes inorganic materials. Nature has its own domain of action that can bring about historical change in species (Reill, 2005, ch. 1).

Buffon's grand synthesis of his theory of nature, bringing together cosmology, geology, and biology in his *Epoques de la nature* (1778/9), supplied his contemporaries with a framework against which many could react. In this he outlined a history of nature beginning from the first creation of the planets from matter attracted from the sun by the passing of a comet. From this point the earth and the solar system have arisen as the solar matter cools and consolidates. The creatures of the earth emerge in time as the world cools, formed of matter clumping together by the action of natural Newtonian microforces (Buffon, 1988).

As this schema was laid out in the *Epoques*, Buffon also shared with Newton a "historical pessimism" with regard to the course of the world's history. The world system is cooling down and degenerating. Unlike Newton, however, Buffon did not assume periodic divine interventions to renew the system. In the revolutionary chronology of the world's total history supplied by the *Epoques*, based on Buffon's calculations of the cooling rate of an originally molten earth, 70,000 years have elapsed since the first formation of the world to the appearance of the first human beings approximately 5,000 years ago in north central Europe. The earth's predictable continued cooling meant, however, that at a definable point in the future – 168,000 years from the earth's origin – all life would cease.[36] Buffon's "nature," confined within the limits of the laws of matter with which it works, cannot resist this inevitable death.

As Buffon developed the consequences of his conception of nature, this supplied in his writings a new meaning to the Baconian-Cartesian "domination" thematic discussed in the opening section of this chapter. Within his secularized world history, Buffon's historical pessimism implied a new urgency for human domination over nature. Human civilization is pitted in a struggle against the inevitable decay

---

[36] In the manuscripts, this figure was expanded to 7,000,000 years. See Roger (1997, "Introduction").

of the natural order. Rational science and technology are the means to delay, if not ultimately prevent, this inevitable decline. The natural world, left to itself, inevitably succumbs to the running down of the natural order, resulting in increasing disorder. Even the human aboriginals of Africa and the Americas, more subject to the direct impact of nature, have become brutish and have declined in intelligence. In an almost Faustian paean to human domination, he continues:

> Brute Nature is hideous and dying; it is I, I alone, who can make it agreeable and living. Let us drain these marshes, enliven these dead waters, and in making them flow, form streams and canals. Let us employ the devouring and active element [fire] ...putting to flame the superfluous foliage, and to these ancient forests already half consumed. Let us destroy with iron that which fire could not consume (Buffon, 1954, 34).

The consequence of such a view of nature for the ethics of domination should be evident. Buffon's nature, if not the "machine" of the previous century, is nonetheless an autonomous realm that operates through imposed natural laws that result in a decaying system that at least to the degree possible must be mastered and controlled by human action. The Baconian technological mastery of nature within divinely sanctioned limits has become a technological imperative to dominate nature in a constant struggle for human survival.

Buffon's importance has been highlighted for two reasons. First, as one of the major writers of the late French Enlightenment and from his position as a gatekeeper over the domain of natural historical sciences, Buffon exerted considerable influence over the reflections of his contemporaries, either negatively or positively. Second, he occupied an important transitional point between the tradition of the mechanical philosophy and the new vitalistic views of nature that emerged in his wake. Those who followed him in time – proximately several influential natural historians at the Paris *Muséum national d'histoire naturelle* – and more remotely those who picked up his grand schema and developed similar scenarios, such as Johann Gottfried Herder, generally broke with him by replacing a degenerative account of nature with one that was creative, vital, and even endowed with teleological and moral purpose.

2.3.2 Vitalized Nature

The development of a new conception of nature at the close of the eighteenth century also owed a considerable debt to the philosophies of Kant, Fichte, Goethe and especially Friederich Schelling. These authors were able to bring together transformations in underlying epistemology with the internal developments in the new "vital" life sciences (Reill, 2005; Richards, 2002, esp. ch. 3).[37] Particularly in the Germanies, this combination created a new framework for a "rebirth" of nature. As systematized by Schelling in the program of transcendental *Naturphilosophie*,

---

[37] For brief introductions, see Heidelberger (online) and Jardine (1996).

which he formulated in a series of important statements between 1797–1803, this program deeply affected the inquiries into biology, anatomy, electricity, chemistry, and even psychology for a period of the early nineteenth century, achieving widespread adherence in academic circles in the German university system. Through the association that developed between *Naturphilosophie* and Goethe's brand of Romanticism, it broadly affected the arts and literature in the period.

Schelling's *Naturphilosophie* defies brief summary, but a few select points can be cited of relevance. First, it assumes a dialectical construction of the world from the original positive and negative forces that constitute matter. The construction of nature also involves an intimate interplay between the subjectivity of consciousness and the objectivity of matter. This provides a dynamic historical construction of the world with both a rational and material aspect. In time this gives rise to inorganic matter, the organic world, and finally to consciousness and human freedom (Schelling, 1867).[38] The system is inherently teleological and rational. Schelling claimed by this philosophical program to overcome the mind-matter, subjectivity-objectivity dichotomies. Nature is a dialectically-developing system reflecting a double aspect. Utilizing language first established by Spinoza, but now employed in a dynamic, creative sense, nature can be distinguished both as product (*natura naturata*) and as productive (*natura naturans*). Seen from one side, nature is external. From the other it is subjective:

> In so far as we regard the totality of objects not merely as product, but at the same time necessarily as productive, it rises into *Nature* for us, and this *identity of the product and the productivity*.... Nature as a mere product (*natura naturata*) we call Nature as object (with this alone all empiricism [*Empirie*] deals). Nature as productivity (*natura naturans*) we call Nature as subject (with this alone all theory [*Theorie*] deals) (Schelling, 1867, 199).

Schelling's complex nature philosophy had significant, if still not fully understood, extensions into the English speaking world through Samuel Taylor Coleridge and his disciples (Sloan, 2007). It can also be traced into the writings of American writers who have deeply affected the philosophy of nature embodied in contemporary environmentalism. Ralph Waldo Emerson, for example, develops in his essays on nature a perspective that presents it as immanent, vitalizing, beautiful, and moral. Similar to themes found in Schelling's writings, Emerson also speaks of nature as involving the interplay of an external nature of concrete productions – *natura naturata* – and an inner working causal principle – *natura naturans* – that "publishes itself in creatures, reaching from particles and spiculae through transformation on transformation to the highest symmetries" (Emerson, 1903, 179). Emerson's disciple, Henry David Thoreau, developed similar views in his *Walden* of 1854. There Thoreau spoke of nature as manifest "not [as] a fossil earth, but a living earth; compared to whose great central life all animal and vegetable life is only parasitic" (Thoreau, 1992, 206). As he reflected on this meaning of nature, even its harsher manifestations of predation and prey were to be seen as manifestations of a "universal

---

[38] For further elaborations see the introduction to Schelling 2004 and Richards 2002, chs. 3, 8. The 1867 text still is the only translation of Schelling's separately-issued *Introduction* to his 1799 *First Outline of a System of the Philosophy of Nature* (Schelling 2004).

innocence." Nature, for Emerson and Thoreau, was good, ethical, and alive. Human relationship to nature acquires in Thoreau's writings a mystical, quasi-religious dimension. Violation of the natural order, even fishing and meat-eating, become occasions for guilt (Thoreau, 1992, 212, 143). Thoreau's awareness of the writings of the Oriental tradition, particularly of the *Analects* of Confucius and the *Bhagavad-Gita*, also suggest ties of some of these reflections of Thoreau to the Eastern traditions referred to later in this chapter.

2.3.3 The Ambiguity of Evolutionary Ethics

It is within the framework supplied by the "vital" sciences and these new dynamic conceptions of nature that we see the emergence of the first modern "evolutionary" theories of nature, first articulated in the natural scientific domain by Buffon's onetime understudy, Jean Baptiste Lamarck (*Philosophie zoologique*, 1809). For Lamarck, nature functioned as the creative agency that brought about the transformation of species over time. Nature "is neither a body, nor any being whatever, nor an ensemble of beings, nor a compound of passive objects... [It is] an order of particular things, constituting an ever-active power which is, nevertheless, present in all its actions. It is, effectively, nature which gives existence, not matter..." (Lamarck, 1991, 307–308).

In the British context from which Darwinism emerged, one can detect several different manifestations of the developments in the conception of nature outlined above. Well before the publication of Darwin's *Origin* in 1859, evolutionary developmentalism had dramatically affected general anglophone discussions in the 1840s and 1850s through the enigmatic *Vestiges of the Natural History of Creation*, issued anonymously in 1844 by the Scottish publisher Robert Chambers (Secord, 2000). In this work, Chambers sketched out a historical scenario similar in several respects to that of Buffon's *Epoques*, but with the addition of the assumption of a gradual and complex teleological development of nature over time under a general "law of development."[39] Because this process was taking place under the guidance of Divine Providence, Chambers' "nature" realized an ethical good over the long span of history, if not in each immediate circumstance (Chambers, 1994, esp. 378 ff.). Similar views are to be found in the progressive evolutionism of Herbert Spencer later in the century. In Spencer's formulations, natural laws, if left without interference, would work out to an eventual order of society in which natural benevolence would reign: "Evolution can end only in the establishment of the greatest perfection and the most complete happiness" (Spencer, 1896, 530).

Darwin's omnivorous reading in English, French and German sources during the most creative period of his intellectual formation between 1831 and 1842, when he

---

[39] Chambers did not exactly endorse a genuine transformism of species except in a few rare instances and criticized Lamarck for proposing evolutionism as a general theory of life's origins. See Chambers, 1994, esp. 221–233.

developed his views of the evolution of species by natural selection, brought him in contact with the themes we have developed. We can follow his views on nature through at least three stages. There is the first set of reflections of the *Beagle* years, in which his thought seems to have been most heavily dominated by the romantic conception of nature derived from the writings of Alexander von Humboldt, who in turn had drawn extensively on Goethe, Herder and Schelling. In this period, "nature" for Darwin formed a benevolent, vital, and unifying ground of being. It is within this period that he first formulated his reflections on the transformation of species.[40]

A second phase covers the years from 1837–1859 in which Darwin worked out his theory of natural selection as the means by which species were transformed in time. The writings of this period employ in some form a conception of a wise and providential *demiurge* who selects beings for ends more perfect than those discerned by human art. In these manuscripts, Darwin originally introduced the concept of nature as subordinate to a separate creative demiurge. This notion was succeeded by the replacement of this creative demiurge by "nature" itself, now endowed with the capacity to make wise selections and whose "productions bear the stamp of a far higher perfection than man's product by artificial selection. With nature the most gradual, steady, unerring, deep-sighted selection – perfect adaption [sic] to the conditions of existence…" (Stauffer, 1975, 224). The formulations of this period lead directly into the metaphors of many sections of the first edition of the *Origin of Species*, in which the analogy of art and nature forms the framework for the primary argument.

Of greatest importance for the subsequent interpretations of the relation of evolutionary theory to the concept of nature was the development in Darwin's writings of a third "mechanistic" and anti-teleological phase. In this mature period of his work, nature becomes generally a metaphor for the mechanical action of selective processes that are generally working to no end except that of differential survival.

We can illustrate this development through a revision Darwin inserted in the third edition of the *Origin* in 1861 that was to set the tone for all subsequent editions of the *Origin* and his later writings. Responding to critics, he dropped intentional and teleological metaphors, embracing instead a mechanical conception of nature that endowed it neither with vital properties nor intentional agency: "… it is difficult to avoid personifying the word Nature; but I mean by Nature, only the aggregate action and product of many natural laws, and by laws the sequence of events as ascertained by us. With a little familiarity such superficial objections will be forgotten" (Darwin in Peckham, 2006, 165).[41] The subsequent adoption of Spencer's phrase "survival of the fittest" as a synonym for "natural selection" in the fifth and sixth editions of the *Origin* is but one manifestation of this change.[42]

---

[40] These points are developed in greater depth in Sloan (in Hösle and Illies, 2005, ch.7). See also Richards (2002, esp. ch. 14).

[41] This passage remains unchanged through all subsequent editions. For some of the complexities in Darwin's position on these questions see Sloan (2001).

[42] In Spencer's writings, however, this phrase was employed within the framework of a teleologically-developing system leading to moral perfection.

This latter theme, generally absent from the manuscript history of Darwin's theory, recovered another strand of reflection that accompanied the eighteenth-century "rebirth" of a vitalistic conception of nature we have discussed previously. The re-emergence of a dynamic conception of nature here underpinned the development of an atheistic world view. A historical commencement of this interpretation can conveniently be traced to the later writings of Denis Diderot. Originally an adherent to a deistic conception of nature in some ways similar to that of Buffon, his subsequent contacts with the "medical" vitalism on the rise in the 1760s in France allowed him to develop a very different notion. Most easily accessed through his dialogue *Rêve de D'Alembert*, only published in 1830, but written in the summer of 1769 and circulated in manuscript, Diderot portrays a discussion between himself, Jean D'Alembert, Julie de L'Espinasse, and the medical vitalist physician, Théophile de Bordeu. In the character of Bordeu, Diderot develops a vital materialist view of life that explains all vital activity by fibers endowed with an inherent property of *sensibilité*. The phenomena of vitality, consciousness, generation, and mental function are explained by the action of such sensible fibers in combination. This pan-vitalism is also extended to all of nature, resulting in a universal flux of species. At the same time, and unlike the situation we find in the evolutionary theories of Lamarck, for example, there is no underlying purpose affirmed to govern this process. Nature is indeed dynamic and vital, but it works to no end (Diderot, 1964, esp. 122–129).[43]

Diderot's anti-teleological vitalism can be compared with some similar claims that are to be found in David Hume's late *Dialogues on Natural Religion*, a work that was read by Darwin in the late 1830s.[44] In the interchanges between the interlocutors Philo and Demea, both critics of the providential design arguments of Cleanthes, the natural world does not present the appearance of harmony and benevolence, but rather that of cruelty, disease and death. As Demea writes:

> The whole earth, believe me, Philo, is cursed and polluted. A perpetual war is kindled amongst all living creature. Necessity, hunger, want, stimulate the strong and courageous: Fear, anxiety, terror, agitate the weak and infirm.... Weakness, impotence, distress, attend each stage of that life: And it is at last finished in agony and horror (Hume, 1947, 185).[45]

Philo also responds to Cleanthes by suggesting that the universe is indeed vital and fecund, but without purpose:

> Look round this universe. What an immense profusion of beings, animated and organized, sensible and active!....How hostile and destructive to each other!.... How contemptible or odious to the spectator! The whole presents nothing but the idea of a blind nature, impregnated by a great vivifying principle, and pouring forth from the lap, without discernment or parental care, her maimed and abortive children (Hume, 1947, 211).

---

[43] On Diderot's development of this non-teleological vitalism, see Roger (1997, 614 ff.). This section is missing in the English translation of Roger, *Life Sciences*.

[44] Hume spent from 1760–1762 in Paris and was at this time in contact with Diderot. It is beyond the range of this essay to work out the relations between Diderot and Hume in this period.

[45] Darwin's reading of this is documented in his reading notebooks in Darwin (1988, 444).

This is not a "mechanical" conception of nature, and such a bleak view of the natural world formed no part of traditional mechanical philosophy of the seventeenth century. The "nature" employed here is vital and creative, but creative to no end. Rather than basing moral value on the natural, such a view seems to require a denial of all moral meaning to nature.

Although forming only a minor thematic in Darwin's writings until the 1860s, a similar concept of a dynamic nature, endlessly producing beings without purpose, and displaying profound "natural" evil, became a prominent theme in Darwin's later writings. It has been this latter reading of Darwin that has deeply affected the views of modern neo-Darwinism to the extent it can be said to have developed an explicit philosophy of nature. Nature is rendered a highly ambiguous category within evolutionary biology as a result. The heritage of later Darwinism suggested to many of Darwin's readers that teleological purpose, in the sense assumed by classical writers and the more recent tradition of *Naturphilosophie*, must be denied to nature. Consistent with this conclusion, there is therefore no "moral" to be drawn from nature and natural process. Instead, facts and values must be radically separated.

The complex relation of Darwinism to ethics is a direct result of this ambiguity, although as developed in the final section of this chapter, there have been substantial efforts to reconnect evolution and ethics in recent sociobiology. Critics of "evolutionary" ethics in the late nineteenth century drew out this ambiguity. Thomas Henry Huxley, and on a more technical philosophical level, G. E. Moore, both attacked evolutionary ethics on these grounds. Both explicitly denied that nature, especially Darwinian nature, could be the source of any moral imperatives. For Huxley, "nature" displayed an amoral, and even an immoral, character. As he writes, "Social progress means a checking of the cosmic process at every step and the substitution for it of another, which may be called the ethical process" (Thomas Huxley, quoted in Appelman, 2001, 502).[46] G. E. Moore's attack on evolutionary ethics, particularly in its Spencerian form, but even more telling when directed against Darwinian evolutionism, was the immediate background for his definition of the "naturalistic fallacy – the fallacy which consists in identifying the simple notion which we mean by 'good' with some other notion" (Moore, 1968). As discussed later in this chapter, since Moore, this argument has been directed at any form of ethics that seeks to derive moral norms from descriptive facts, particularly from facts drawn from biology.

2.3.4 Contemporary Biological Reductionism and the Challenge of Bioethics

The foregoing brief overview of some selected themes in the philosophy of nature since the Enlightenment has been intended to illuminate the fragments of discourse

---

[46] On Huxley's 1893 Romanes lecture and its reception, see Farber (1994, ch. 4).

that currently are in play in contemporary discussions of bioethics and biotechnology. From this historical survey we can detect at least three strands. The first is that of a tradition of classical mechanical philosophy of nature that conceptualized matter as inert, governed by natural laws, and "nature" as the system of matter, laws and forces that is a domain open to free human domination without a moral significance attributed to this natural order. The second strand is the dynamic and teleological conception of nature, showing some continuities with Classical and Renaissance traditions, but that rose to new prominence around 1800. This attributed to "nature" both an ethical value and a teleological purposiveness. We can follow this tradition into the roots of environmentalism (Emerson and Thoreau) and beyond. Finally, we have followed a strand that came to prominence with Darwinian evolution. This in its origins was built upon a vital and dynamic conception of nature, but one that denied to it an evident teleological purposiveness. The relation of this tradition to ethics is ambiguous, and it lies at the root of the modern distinction of "is" and "ought."

As we move to more immediate discussions, one of the major developments in the foundations of modern biotechnology has been the turn toward philosophical reductionism and biophysical conceptions of life. First developed in detail by the German biophysical school associated with the names of Hermann von Helmholtz, Emil DuBois-Reymond, and Karl Ludwig who began their pursuit of this program around 1850, this set of explanatory ideals pursued an overtly "mechanistic" conception of organic beings. The metaphors employed were no longer those of a machine, as they had been in the older mechanical philosophy of the seventeenth century, but rather those of a determinate chemical-physical system governed by natural laws that left no room for special forces, principles, or souls in the biological domain. Drawing particularly on the principle of the conservation of force (*Kraft*), this development of extreme analytic and reductive views of life opened the door to technological manipulation of living beings to a degree not seen previously. Biophysical studies on nerves, muscles, senses, and embryological development suggested to many that the conquest of the living state was finally possible. This German biophysical program, expounded in the public sphere by such individuals as emigré American physiologist Jacques Loeb, put forth a research agenda in the life sciences that pursued a complete reductionism of life to its chemical and physical states (Loeb, 1912; on Loeb see Pauly, 1987).

These and similar claims drew a strong response from those who came to be termed "vitalists." The most prominent counterattack was launched by the one-time developmental embryologist, Hans Driesch. He commenced his critique of the mechanistic conception of life in embryological studies of the 1890s in which he argued that the ability of the developing embryo to maintain harmony and even compensate for damage could only be explained by some kind of inherent dynamic force. This claim Driesch developed through many works as he migrated from a career as a zoologist to the holder of a chair in philosophy at the University of Leipzig. For Driesch, the phenomena of development required the assumption of an unknown organizing principle that could not be reduced to physical and chemical causes. Only on the assumption of vital properties could the organic being adequately be explained. Through a massive series of books and articles that extended until his death in 1941, Driesch became the main international spokesperson for "neovitalism."[47]

As will be developed in the next section, the mechanism-vitalism dispute of the early decades of the last century, and the appeals made by Driesch and his followers to non-empirically accessible agencies, formed an easy target for the anti-metaphysical stand of early logical empiricism. If the sharp knife of the Vienna Circle was to cut in both directions on this dispute – both metaphysical materialism and Drieschian vitalism fell into the category of meaningless discourse – the use of the concept of nature as a basis for some kind of ethical stance was presumably also gutted by the same blade.

2.3.5 Concluding Reflections

The function of this section of the chapter has been primarily descriptive and bridging. The combined historical and philosophical analysis supplied here is primarily intended to sort out some of the strands of discourse that are often incoherently blended in contemporary discussions. From this discussion, we have separated out at least four different options.

The first of these, reaching back to Enlightenment discussions, but also behind them to classical sources, accepts nature as substantive, normative, and teleologically directed. The argument presented here is that this conception of nature was not unequivocally overthrown in the seventeenth century, but instead it returned with new vigor in the late eighteenth and early nineteenth centuries.

A second position, related to the first in some formulations, is that of progressive evolutionism or historical developmentalism, illustrated by the views of Lamarck, neo-Lamarckianism, Chambers, Spencer and more recent evolutionary progressivists. By positing a teleological direction to the evolutionary process, it is presumed that some value can be determined from nature, at least in the long run. Those holding to this view need not oppose biotechnological interventions as long as these are in keeping with the progressive course of evolutionary history.

A third view seems most intimately connected to the heritage of Romanticism and *Naturphilosophie*, as this was developed by Goethe, Schelling, and Alexander von Humboldt in the German tradition, and in the writings on nature by Emerson and Thoreau in America. Nature is assumed to be vital, benevolent, normative, and worthy of respect. Human beings must see themselves as an intimate part of this ecosystem, without special rights or privileges. Because the "natural" is also the "good," those holding to strong versions of this position are

---

[47] This was most fully developed in English in his Gifford Lectures, *Science and Philosophy of the Organism* first published in 1908. I have used the second edition of 1929.

typically opposed to biotechnology, genetically-modified organisms, and human germ-line modification. Efforts to link this third position with environmentalism have led to recent revivals of interest in Schelling's philosophy (Schelling 2004, "Introduction").

A fourth position clusters around a denial of the moral normativity of nature. There are many strands that are seen to compose this position – the legacy of enlightenment vitalism, the heritage of the anti-teleology of neo-selectionist evolutionary theory, the reductive conceptions of life of modern biophysics, and the impact of philosophical critiques of metaphysics generally by various forms of empiricism.

It seems safe to argue that the historical remainders of these discussions, often more as incoherent fragments rather than deeply systematic positions, implicitly define the framework of much of the standard discourse of bioethics at present. As a result of these developments, "nature" has essentially dropped from sight in most secular conversations of bioethics. The sharp separation of "fact" and "value" that governs most Anglophone discussions of ethics is a practical consequence of these developments.

2.4 Concepts of Nature in Twentieth-Century Philosophy of Science

Nature as such does not much figure in twentieth (and, so far, twenty-first) century philosophy of science. Nevertheless, many philosophical questions about the nature of nature are addressed under other rubrics, making it possible to identify, indirectly, substantive conceptions of nature operative in or stemming from philosophical thought. In the main, this period has seen a consolidation of the enlarged sphere of the natural and the concomitant elimination of any explanatory role for non- or super-natural elements. At the same time, as the consensual materialist outlook has expanded, new fissures have opened.

Throughout most of the twentieth century, philosophy of science was the site of two major contestations: one between forms of reductionism and forms of antireductionism, and another between, loosely speaking, instrumentalism and realism. Instrumentalism encompasses views that scientific theories should not be taken as literal descriptions of the world but rather as machines for generating predictions of future observable phenomena from descriptions of past observable phenomena. Scientific realism, by contrast, encompasses views holding that scientific theories are to be understood as literal descriptions of the world, including the unobservable structures and processes that underlie what we can observe. Scientific success on the instrumentalist view is empirical adequacy, that is, truth of the observational content of a theory. On the scientific realist view, success is truth of all the content of a theory. To the extent that the nature of nature involves issues beyond the observable, instrumentalists in philosophy of science treat it as beyond the purview of science.[48] The various forms of reductionism and antireductionism bear more directly on the question of the nature of nature. Reductionists want to treat the natural world as fully decomposable into its least parts, whereas anti-reductionists have argued that some part of the natural world resists identification with any set of purported least parts. The form and consequences of this contestation depend upon whether it is framed in instrumentalist or realist terms. Instrumentalists would treat the debates between reductionists and antireductionists as debates about language. Realists, on the other hand, treat it as a debate about, among other things, the nature of nature.

Reflecting the consolidation of a broadly materialist point of view, the presupposition on the part of most philosophers of science is that the nature of nature, whatever it might be, is to be revealed by or inferred from the results of scientific investigation. As seen in earlier sections of this chapter, the success of physical science from Newton on emboldened many thinkers to hope that all natural phenomena could be treated as expressions of the same laws and composed of the same parts as the bodies in motion which constitute the Newtonian ontology. The displacement of Newtonian mechanics by quantum mechanics means that the least parts are differently characterized than Newtonian bodies, but the reductionist thesis simply transfers to whatever the least parts are (waves, particles, wavicles, strings, …). Antireductionists who resisted this way of thought include biologists and philosophers of biology and philosophers concerned with the status of the mental.

2.4.1 The Impact of Logical Positivism

2.4.1.1 The Linguistic Turn

The immense panoply of life forms are distinguished from non-living objects (like rocks, sand, or water) by their capacities of matter absorption and transformation (through ingestion, digestion, respiration, and excretion), by their capacities of patterned growth and reproduction, and by their integration of multiple components (molecules, tissues, organs, and cell types) and different levels of organization of those components into singular wholes. All life forms, from mammals, birds and fishes, to trees and grasses, to bacteria and other single-celled organisms, possess

---

[48] As should be expected, the terms of this debate have changed as it has proceeded. Bas van Fraassen's constructive empiricism understands theoretical statements literally, but proposes that scientific success consists not in the truth of those statements, but only in empirical adequacy, i.e. truth of the observational content of a theory. See van Fraassen (1980). Arthur Fine's natural ontological attitude also understands the theoretical content literally, but eschews any metaphysical interpretation of scientific practice. See Fine (1984). Richard Boyd, on the other hand, advances a robust form of scientific realism employing a form of inference to the best explanation. See Boyd (1984).

these capacities. Advances in microscopy (largely in service to pathology) have, in addition, revealed categories of entity whose status is ambiguous or borderline between life and non-life. Viruses and prions, for example, are borderline in that they possess some but not all of the capacities distinguishing life from non-life.

The nineteenth century debate about the nature of life and life forms continued into the early decades of the twentieth. Vitalism and organicism continued to be advocated as alternatives to the mechanistic conceptions found in the work of Bernard, Helmoltz, and others previously mentioned.[49] These, however, along with other metaphysical debates and doctrines, were swept away by the austerity of logical positivism and logical empiricism. Logical positivism as a school of thought originated in 1920s Vienna, and its intellectual sibling, logical empiricism, in Berlin. The positivists were scientists and philosophers seeking to absorb the lessons of relativity physics. Logical positivism was carried into English-speaking philosophical circles initially by visitors from England and the United States, but cemented there by the escape to England and the US from Nazism by the Austrian and German philosophers. Its hallmark was a principle of (cognitive) meaningfulness. Only those statements that could be directly verified or falsified (later confirmed or disconfirmed) by observation statements were meaningful. Since the claims at issue in the debates about the nature of life failed the tests of observational verifiability or confirmability, they lacked cognitive significance. Now doctrines of reducibility or non-reducibility were to be treated as matters of language – of the derivability of a theory in one domain from a theory in the reducing domain. Explanation was explicated as logical derivability of a description of the phenomenon to be explained from statements of general law and initial conditions, as in the "covering law" model. Theories were understood as deductive systems. The meaning of any non-logical expression occurring in a theory or hypothesis depended on its being translatable, in whole or in part, into an observational expression. Thus, the logical positivist and logical empiricist conception of theories is often described as instrumentalist. While members of the original logical positivist circle articulated different versions of the unity of science doctrine it came to be interpreted as the derivability of all genuine scientific theories from the principles of physics (which themselves were assumed to be paradigms of meaningfulness).[50]

The positivist's criterion of cognitive meaningfulness had consequences for normative as well as metaphysical statements. Statements of value were to be understood as expressions of feeling, not as saying anything substantive about the world.

---

[49] Representative texts include Bergson, *L'Evolution Creatrice* (1907); Driesch, *The Science and Philosophy of the Organism* (1908); Loeb, *The Mechanistic Conception of Life* (1912); and Chardin, *La Phénomène Humaine* (1956).

[50] Some of the key logical empiricist texts include Carnap, *Logische Aufbau der Welt* (Berlin: Weltkreis, 1928) and *Meaning and Necessity* (1947); Frank, *Modern Science and Its Philosophy* (1949); Schlick, "Meaning and Verification" (1936); Reichenbach, *Experience and Prediction* (1938); Hempel, *Aspects of Scientific Explanation* (1965). What we describe here is now associated with the so-called right wing of the Vienna Circle. Scholars have recently been presenting a more nuanced picture of logical positivism and logical empiricism (cf. Friedman, 1999; Giere and Richardson, 1996).

From the positivist point of view, then, there can be no connection between assertions of fact and normative assertions because no logical relations can be drawn between them and *a fortiori*, no implicative relations established between conceptions of nature (to the extent these are cognitively meaningful) and claims about what ought to be done. Several kinds of issue in biology after mid-century produced challenges to this philosophical vision: the widespread use of functional explanation, which was prima facie untranslatable into a deductive argument, as required by the derivability or covering law model, and the apparent explanation of patterns of inheritance by reference to autonomous genetic principles. While from an instrumentalist or linguistic perspective the issues are framed as a matter of the autonomy of the science of biology from physics and physical chemistry, from a realist perspective the issue concerns the relation of life or living organic nature to nonliving, inorganic nature.

2.4.1.2 Functions, Teleology, and Logical Empiricism

Functional explanation typically explains a process or constituent part of an organism by elaborating the contribution that process or part makes to the survival and/or reproduction of the organism. For example, peas have pods because the pods provide nourishment to the pea plant's seeds, which are the germs of the plant's next generation. One problem has to do with the whiff of teleology (and proscribed backwards causation) that attends functional explanation. Another has to do with the association of goal directedness with intention. But even if one abjures thoughts of the future causing the past or present, or the ascription of mentality to plants and organs, such explanation fails the criterion of derivability specified by the covering law model. It does not follow from the contribution of a four-chambered heart to an organism's survival that it does have a four-chambered heart, as the contribution could be made by a functional equivalent of the heart. Philosophers working in the logical empiricist tradition rejected functional explanation as a distinct pattern of explanation. Any talk of functions that was integral to science could and should be translated into causal-mechanical terms or into covering law explanations.[51] Whatever could not be so translated was treated as scientifically illegitimate.

It has often been remarked that philosophy of science from the 1950s through the 1960s and 1970s took physics as the paradigm science.[52] Philosophers attentive to the details of biological investigation were more likely to be persuaded that the translations into explanatory forms modeled on certain kinds of explanation in physics could not be accomplished without remainder, i.e., without loss of crucial explanatory strength. For example, one can know everything about the molecular constitution of a heart, of a portion of the DNA molecule, or of a seed pod without

---

[51] For causal mechanism, see Nagel, E. (1961); for covering law see Hempel (1965).

[52] The "*Ur*text" for this view is Oppenheim and Putnam (1968). But see also Earman and Roberts (1999).

knowing what these contribute to the systems of which they are a part. But a crucial part of the evolutionary account of any of these is the function they play in the system of which they are a part, the contribution they make to the organism's fitness. Thus functions seem both necessary and irreducible. The debate about functional explanation has continued into the present, and is characterized by arguments on the part of philosophers of biology to the effect that function talk in biology is important for biological understanding, but does not carry problematic metaphysical commitments to some alternative form of causal force.[53]

2.4.1.3 Genetic Reductionism

After the discovery (in 1953) by James Watson and Francis Crick of the structure of the DNA molecule, it seemed that a uniform (physical-mechanical) characterization of the natural world (or to stay within the linguistic strictures of Logical Positivism, of the total universe of discourse of the natural sciences) might actually be forthcoming. The double-helical structure of the molecule and the position of constituent nucleotides, together with principles of chemical valence and bonding, were enough to explain some of the deepest mysteries of living systems. As the (in)famous Central Dogma had it: DNA makes RNA, RNA makes protein, protein makes us. Information flows *from* DNA to the rest of the organism, not *to* DNA (Crick, 1970). In philosophy of biology, the question of the reducibility of classical genetics to molecular genetics assumed the same, if not a greater urgency as had debates about the proper understanding of functional explanation.[54] Other areas of biology climbed onto the molecular bandwagon. As physiological psychology, driven by varieties of animal experimentation, morphed into behavioral endocrinology, more and more behavior seemed explicable as the expression of hormonal function, which would in turn be explicable as the expression of genetic information. Advances in neuroscience supported hopes that all mental phenomena could be accounted for in terms of neuronal activity, which again, in line with the reductionist program, would be accounted for in terms of molecular constituents and processes (Churchland, 1986). As long as biological processes could be accounted for in molecular terms, it was assumed that they could ultimately be understood in the language of physics.

2.4.1.4 Modern Mechanism

For the logical positivists, discussion of nature was to be understood as discussion of language (or as meaningless). Nevertheless, their conception of the proper

---

[53] For a sampling of views on functions and functional explanation, see Brandon (1981), Rosenberg (1985), Wright (1976), and Mitchell (1995).
[54] Discussions of reduction in biology include Hull (1972), Schaffner (1967), Darden and Maull (1977), Rosenberg (1985), Kitcher (1984), and Waters (1994).

language for science was based on their understanding of the language of physics (Oppenheim and Putnam, 1968). Despite the banishing of metaphysics, their philosophy contains an implicit commitment to the ontology of physics, and thus to a conception of nature as bodies in motion, as Newtonian (or quantum) mechanism. The spectacular successes of physical science in the first half of the twentieth century contributed to cementing this as the default position in philosophy of science. Philosophers of biology felt they had to defend the scientific status of their domain of interest, which accounts for the concentration of debate on the issues of functional explanation and the reducibility of classical genetics to molecular genetics. As just noted, this defensiveness was only exacerbated by the success of molecular biology in the second half of the century. One consequence of this attitude in philosophy of science is that other philosophers simply assumed the default position as the correct understanding of the natural world. To be part of the natural world, of nature, was to be explicable in (narrowly) mechanical terms.

There are several ways to incorporate this understanding in a wider philosophical approach that seeks to integrate conceptions of nature with conceptions of human life. Physicist Steven Weinberg takes the lesson to be the final disenchantment of nature, forcing upon us a mature acceptance of the universe's supreme indifference to human hopes, fears, and ideals (Weinberg, 2001). At the opposite end, moral philosophers accepting the default characterization of nature have argued that the phenomena of human life that are their concern exceed that narrow frame (Taylor, 1989; Held, 1996). Philosophers of mind and psychology debate whether mental phenomena, intentionality, and action can be incorporated into a causal or causal-mechanical framework. Outside mainstream philosophy of science, however, whether in science itself or in upstart or marginalized movements of thought, challenges to the default position suggest other ways of thinking about nature and by extension of our place in it.

2.4.2 New Directions in Philosophy of Science

2.4.2.1 Kuhn and Post-positivist Philosophy

The most influential book of twentieth century philosophy of science was Thomas Kuhn's *Structure of Scientific Revolutions*, published in 1962 and reissued in a second edition in 1970 (Kuhn, 1970). Logical positivism and logical empiricism were already on the wane, as philosophers found that the formalist reconstructions of concepts such as explanation and theory were unsatisfactory, and for some, beyond remediation. Kuhn's notions of theory-laden meaning and observation and of revolutionary science were embraced by many thinkers who applied them in ways that went well beyond the history of physics that was Kuhn's concern. Both Norwood Russell Hanson and Paul Feyerabend also articulated versions of the theory dependence of meaning and of observation, and, like Kuhn, focused on physics (Hanson, 1958; Feyerabend, 1962). The work of these three provided the other end of a polarity with logical empiricism. It also resonated with the anti-foundationalism of philosophers like Ludwig Wittgenstein and John Austin, although very different from them in the details of their views (Wittgenstein, 1958; Austin, 1961).

The conceptions of theory-dependence, of paradigm and world views, of revolutionary and normal science, were developed by philosophers of science like Larry Laudan (1977), by sociologists of science like David Edge and members of the so-called Strong Programme in Sociology of Science (Barnes and Edge, 1982), by literary scholars, by sociologists, and by feminist and other critics of the sciences.[55] What they saw varied according to their interests, but included freedom from the constraints of a narrow empiricism, from an assumption that current scientific theories were grounded in unquestionable foundations, from the reductionism associated with the positivists' unity of science ideal, and from the notion that the sciences were epistemically independent of their social contexts.

The priority of theory meant that theoretical principles themselves could not be regarded as independently supported by experiential data. Instead data (their identification and description) presuppose theory. This means that no general concept of nature can be treated as independently supported by observation and measurement. Such a concept would form part of what Kuhn memorably called a "world view." How world views, paradigms, theoretical principles are supported, if not by independent observation, is a matter on which those developing Kuhnian and Feyerabendian themes differ. The central point, however, is that some such view or principles are required to get the scientific enterprise going. Even if the sciences are granted authoritative status regarding the knowledge of the nature of nature, however, the message then is that world views are historical and mutable, so that a conception of nature regnant at any given time cannot be regarded as timelessly true. Instead it serves to ground investigation of, interaction with, and intervention in the natural world. Although to say so goes well beyond Kuhn and beyond philosophy of science, as an overarching framework, such a conception would at least place some constraints on what sorts of ethical claims can plausibly be made at any given time.

2.4.2.2 Biology: Complexity and Self-Organization

One consequence of Kuhn's and Feyerabend's work in philosophy of science was a shift from accounts of science as a whole to investigations of the foundations and practices of the individual sciences. In this atmosphere, and in response to the increasing reach of biological investigation, philosophy of biology flourished. In the 1980s and 1990s biological research seemed to be reaching the limits of the reductionist program. Models of complex interaction and self-organization began to replace models that tried to identify a single causal factor responsible for a given

---

[55] Some recent volumes taking stock of Kuhn's legacy include Gutting (1980), Horwich (1993), Favretti et al. (1999), and Nickles (2002).

biological process. These models assumed the same material composition as reductionist models, i.e., no mysterious "vital forces" or "entelechies", but proposed that the organization of those elements in certain patterns resulted in non-reducible, "emergent", properties and processes (Kauffman, 1993). New developments in computer modeling and mathematics made these processes representable and calculable, and, therefore, scientifically respectable.[56] Research in a number of biological arenas supported the turn to complexity. In behavioral endocrinology, researchers found that patterns of hormonal secretion previously thought to be determined genetically were malleable by environmental manipulation (Silver, 1992). It became clear that genes on their own initiated nothing, but required activation by protein complexes.[57] Models of brain and nervous system development incorporated the notion that the neuronal selection setting up functional connectivity in the brain is not genetically controlled, but is responsive to environmental input (Edelman, 1987). In addition, neuronal networks and gene networks exhibit processes of interaction within and among groups that are not traceable to any individual molecule. In ecology, researchers, while varying widely on the details, stress the interdependence of the elements of an ecosystem.

The response in science and philosophy of science to these and other demonstrations of complexity has been bifurcated. In understanding the response of scientists it is important to distinguish between methodological reductionism and metaphysical or ontological reductionism.[58] Methodological reductionism seeks knowledge of a system by observation and manipulation of its least units, e.g., understanding organic processes by investigation of their molecular constituents. Metaphysical/ontological reductionism would claim that all systems can be decomposed without remainder into their least units and that all causally significant processes occur at the level of those least units. Scientists can be methodological reductionists without also adopting a reductionist metaphysics or ontology.[59]

Philosophers, on the other hand, are explicitly concerned with the metaphysical and ontological dimensions of reductionism.[60] As biology outgrew the reductionist

---

[56] For an example, see Segal and Keller (1970).

[57] Some of the research leading to the more complex view of gene activation is summarized in Keller (2000, esp. chs. 1 and 2).

[58] For further discussion, see Longino (1990).

[59] Edelman (1987), for example, is methodologically reductionist in his focus on the role of cellular adhesion molecules in neural system development, but his account of the development of neuronal connectivity as well as his views on the limitations of a neurophysiological account of cognitive process suggest a metaphysical antireductionist attitude. E. O Wilson, on the other hand, investigates systems by observing their behavior at a relatively high level of organization (the ant colony, for example, rather than the molecular composition of individual ants) while seeming to embrace a genetically reductive metaphysics. See Wilson (1980).

[60] While this chapter concentrates on philosophical questions, it is worth noting that many social analysts of the sciences view the contemporary salience of reductionist views in biology as a function of economic interests. Certainly the corporate funding of research on the genetics of disease and of genetic engineering in agriculture is a function of the anticipated economic benefit to industry of that knowledge. See Mirowski and Sent (2003).

hopes that first attended the discovery of the molecular structure of the DNA molecule, philosophers had to revisit the ways in which the problem was configured. Some philosophers (e.g., John Earman) have insisted on a distinction between science directed at discovering the fundamental laws of nature (a science that turns out to be physics) and the so-called special sciences whose range is more limited (from biology to the social sciences) (Earman and Roberts, 1999; see also Earman et al., 2002; Sklar, 2003). Fundamentalists see nature as uniformly characterizable by a small set of fundamental laws, while the laws of the special sciences are short hand for very complicated formulations in the language of the fundamental laws. Anti-fundamentalists comprise several groups. Some philosophers, like Nancy Cartwright, stress the autonomy and nonderivability of mid-level laws in physics from fundamental ones, as well as the highly idealized nature of the fundamental laws (Cartwright, 1983, 1999). Other philosophers, like John Dupré, take an anti-essentialist attitude toward any natural kind talk. Both ordinary language and scientific language reveal multiple non-reducible systems of categorizing. Any kinds appearing in any system of categorizing that serves a purpose are to be treated as being as real and as causally efficacious as those of any other such system (Dupré, 1993). Whatever the degree of pluralism advocated, for pluralists, it is reasonable to accept that there are irreducible laws or principles operating at higher levels of organizations relating complexly organized systems. Pluralists argue that the complexity of some natural processes means that no single investigative strategy will be able to produce a comprehensive account of them (Waters et al., 2006).[61] Instead, multiple accounts drawing on noncongruent models and strategies are required. Other pluralists stress the multiple ways key concepts are and can be defined and used (Ereshevsky, 1998; Teller, 2004). This multiplicity, too, supports a pluralist interpretation of scientific knowledge.

Fundamentalism is a form of reductionism, to the extent that it treats all scientific knowledge as reducible to knowledge of the behavior of the few kinds of entity whose behavior directly exemplifies the fundamental laws. Anti-fundamentalists, however, could resist reductionism, but still hold that for any given level or organization there will be one and only one theory that completely and correctly describes it. Pluralists, on the other hand, hold that there can be different approaches even at the same level of organization, prompted by different cognitive and practical interests, which issue in different and incommensurable or non-congruent models of the processes at work at that level. While the pluralist and the antifundamentalist philosophers might disagree on the scope of disunity, they do agree that no single comprehensive characterization of the natural world is forthcoming from the sciences. At the same time, they would agree with earlier philosophers of science that any meaningful characterization of the nature of the natural world must be grounded in empirical data, and is not to be obtained by *a priori* reflection. Consistent with the pluralist spirit, there may be multiple nature concepts, each consistent with what

---

[61] Other pluralists maintain the requirement for convergence that Kellert et al. reject (cf. Mitchell, 2004).

we know, but each appropriate for different purposes. Given the complexity of the natural world and the multiplicity of possible accounts of aspects of the world, some pluralists would even countenance the reliance on continuity or congruence with high level principles or social values (such as universal individual self-determination, or community well-being) to serve along with empirical adequacy as a constraint on selection of a nature concept.

In spite of the protestations of thinkers like Steven Weinberg, then, contemporary scientific inquiry, especially, but not exclusively in biology, makes available much richer conceptions of the natural world than are available within the framework of a metaphysical reductionism. On such conceptions, the natural world is complex, composed of entities that are both in complex relations with each other and internally complex. New (non-reductionist) ideas about mechanisms have been advanced that are better suited to the new conceptions of nature and natural process (Machamer et al., 2000). The acceptance of such complexity carries with it some loss of certainty. Although the mathematics for representing complex relations has improved, one lesson is that our ability to predict outcomes that are dependent on the intersection of many more factors than those whose status we can identify is seriously circumscribed.

2.4.2.3 Feminist Philosophy of Science

Feminist philosophy of science drew its inspiration and strength from the political movement for women's liberation. The very concept of "nature" is suspect for feminists sensitive to the ways in which concepts of what's natural, such as different roles for women and men, have supported gender oppressive socio-political formations. Feminist scientists and historians and philosophers of science are divided as to whether to treat "nature" as one term of a politically valenced conventional dichotomy or whether to develop alternative conceptions of nature with which to challenge reductionism and androcentric social biology. They are united in challenging the uncritical delegation of the work of characterizing the natural world to the sciences because of the ways in which past science and the nature it purported to reveal contributed to sustaining gender injustice. As a consequence, they have also agreed that scientific research cannot be assumed to be value-neutral. This common point is treated differently in different feminist epistemological approaches.

Among feminists seeking alternative conceptions of particular structures and processes in nature, feminist psychologists, ethologists, and primatologists identified ways in which models of animal behavior wrote stereotypes of culturally and historically specific human gender relations into nature (Hubbard et al., 1979; Fausto-Sterling, 1985; Tavris, 1992). They criticized work focusing on animal analogues of selected human behaviors, e.g., populations characterized by male dominance, and work that saw selected human relations in animal behaviors that could be otherwise described and interpreted. Non-androcentric researchers in ethology, primatology, and psychology produced different data and appealed to different organizing principles than the old androcentric approaches (Zihlman, 1981; see also Haraway, 1978). The new view of animal populations meant that nature could no longer be seen as an ally of male social supremacy.

Feminist thinkers also identified and critiqued typical reductionist explanatory strategies, arguing that attempts to identify a single molecule as the initiator of given biological processes were in many cases supported at least in part by commitment to patriarchal and/or hierarchical social relations (Bleier, 1984; Keller, 1985). These scholars tried to show patriarchal ideology at work in specific cases (Biology and Gender Study Group, 1988). More importantly, they identified alternative explanatory approaches, as in Keller's work on slime mold aggregation and her study of the work of geneticist Barbara McClintock (Keller, 1983a, b). Keller's model of slime mold stressed the role of environment and diffusion gradients in initiating the formation of a fruiting body, in contrast to the dominant model, which took genetic difference among the amoebae to determine the timing and location of this formation. McClintock emphasized the interaction of an organism's genome with various levels of the genome's environment (from intra-cellular to the environment external to the organism). Philosopher of science Nancy Tuana suggested replacing the old nature-nurture dichotomy with an interactive model that stresses the ongoing interaction and mutual modification of structures and processes internal to the organism and external to it (Tuana, 1990). Like advocates of Developmental Systems Theory (Oyama, 2000), Tuana urges that we understand the internal and external as aspects of an integrated system. What these approaches share is an effort to reinvest nature with a certain kind of liveliness, or inherent dynamism, stressing the diversity of natural forms and the capacities of self-organization in contrast to the reductive uniformity and determinism the feminist critics associate with mainstream philosophical approaches to nature.

Feminist thinkers have also raised warnings about the conflations of "normal" and "natural" by advocates of the Human Genome Project (Keller, 1994; Lloyd, 1994). Given the genetic variability among humans, selections must be made in characterizing *the* human genome, even if some variants will be included in the final map. What criteria will govern those selections? And what will be the consequences of selection in and selection out? Both Keller and Lloyd decry the possible consignment of "natural for humans" to "normal (or predominant) in the genome."

Finally, ecofeminists from Carolyn Merchant through Karen Warren emphasize human continuity with the natural world (Merchant, 1980; Warren, 2000). They reject characterizations of nature, such as the mechanistic one, that require a bifurcation between humans (as beings capable of intention) and nature. The world is constituted not just of particles or fields on the one hand and minds on the other, but of organisms in environments. Ecofeminists vary in their treatment of the proper relationship between humans and the rest of nature and in their interpretation of the "feminism" in ecofeminism. Some have argued for a characterization of nature and the earth as female or for a characterization of women as possessing a special relationship to nature (Griffin, 1978). To the extent these treat motherhood as essential or even central to womanhood, these are distinctly minority positions among contemporary feminists. Ecofeminism might most generally be characterized as a rejection of distinctions between human and nonhuman on grounds that it partakes of the same oppressive relations as do traditional conceptions of gender difference, a rejection of reductionist models of natural process in favor of conceptions attributing complexity and integrity to the natural world, and a commitment to extend models of human responsibility and caring to non-human coinhabitants of the same environment that sustains all (Warren, 1997).[62]

Among feminists spurning the concept of nature, perhaps Donna Haraway's work, which features lists of structuring dichotomies that have mutated in response to changes in social, political, military, economic, and technological configurations, is best known (see esp. Haraway, 1991). For these feminists, "nature" and cognates are always part of contrastive dichotomies serving particular rhetorical or ideological purposes. "Natural" contrasted with "unnatural" is often used to reinforce local prescriptions about behavior. (As in the condemnation of homosexuality.) "Nature" can be contrasted with "culture," a dichotomy that can be put to use in defense of wilderness as much as in support of human domination and exploitation of the natural world. It can also support characterizations of gender difference that associate women with nature and men with culture. A parallel contrast between organism and machine serves similar purposes, promoting untenable contrasts between what is natural and unnatural in human life.

Taken together, contemporary feminist thinking about the sciences promotes suspicion toward any use of "nature" or the "natural" to support claims with social or political consequences. Even those who celebrate the identification of women with nature, do so under a radical re-evaluation of the meaning of nature, for they affirm accounts that delegitimate the impulse to domination of either nature or women. For feminists it is crucial to identify the political underpinnings of any appeal to nature, even those perceived to be woman- or feminist-friendly.

2.4.3 Conclusion

Twentieth century philosophy of science, with few exceptions, has not advanced particular conceptions of nature that would guide its thinking about the sciences. It has instead taken any substantive concepts of nature and natural process from the sciences themselves. While concepts are thereby implicitly endorsed, philosophy of science not concerned with the foundations of particular sciences has been primarily concerned with epistemological and methodological issues. As the sciences do not present a single consistent and comprehensive notion of nature, philosophers are divided as to whether any characterization of nature, as a whole contrasted with what can be observed of it, is forthcoming. On the other hand, the sciences themselves are the potential source of conceptions of nature richer than those associated with a narrow Newtonian mechanism. Once the Darwinian revolution placed the

---

[62] For a critical review of much ecofeminist literature, see Agarwal (1998).

living world firmly within the investigative sphere of the empirical sciences, the relation of biology to physics continued to pose analytical challenges. In the biological sciences, the natural world is not initially framed as a collection of intrinsically inert objects set in motion by an external force. While some investigations seemed for a time to support understanding life processes in terms of the action of their least constituents, some of the entities in nature seem better understood as self-organized and possessed of an internal dynamism. That such a dynamism is a function of the complex arrangement of material substances does not gainsay that the temper of these views is not narrowly mechanistic. Thus, while it seemed that a unitary anti-teleological, evolutionary, mechanistic materialism had triumphed over rival conceptions of nature, it contained within itself the seeds of another dialectic. Within the broadly materialist viewpoint, one can discern in the investigation of self-organization echoes of the Enlightenment's non-teleological vitalism alongside the biophysicists' mechanism preferred by those reductionists for whom all explanation eventually ends in fundamental laws of bodies characterized only by position and momentum.

Given the diversity of views, the lesson that can be drawn from contemporary philosophy of science is a cautionary one. There is no universally endorsed conception of nature to be found either in the sciences or in their philosophies. Indeed some conception of nature either explicitly or implicitly underlies the presuppositions that generate and inform the interpretations of empirical research. The lesson of Kuhn is that such conceptions are in an important sense provisional – guiding research and informing its interpretation only as long as they are fruitful and consonant with other deeply held cultural values and assumptions. Thinkers seeking to draw moral or ethical consequences from characterizations of nature or the natural issuing from the sciences would do well to heed the warnings of feminist thinkers. Such characterizations serve different political purposes at different times and ought not be taken at face value. Understandings of nature are bound up with the more particular philosophical, scientific, and social views in which they play a part.

2.5 Modern Moral Philosophies and East Asian Perspectives

The aim of this section is to provide a broad overview of current ethical theories and their engagement with "nature." In an effort to present a more complete, textured, and nuanced account of these modern Western views, where appropriate, we compare and relate them to some early, indigenous Chinese ethical theories and their views on nature. We begin with a short description of five types of ethical theory and the role of nature in each. We then provide more extended descriptions of each theory. In the conclusion, we assess their primary strengths and weaknesses.

As will be clear in the following discussion, contemporary ethical theories offer a range of views on the importance and role that conceptions of nature – both the natural world as a whole and human nature in particular – play. The two most influential theories among Anglo-American philosophers, *utilitarianism* and *deontology*, tend to place the least emphasis on nature. This reflects the "death of nature" or "world we have lost" perspective discussed in earlier sections of this chapter. Under the influence of the scientific revolution and the Cartesian turn in philosophy, ethical theorists have sought to provide an objective account or what Bernard Williams has called "absolute" descriptions of what is valuable and good (Williams, 1985). Normative appeals to nature are regarded by a number of philosophers as a form of error, examples of what G. E. Moore dubbed the "Naturalistic Fallacy" (Moore, 1968). The turn away from nature has been reinforced by several prominent Feminist ethical theorists (Simone, 1953; Jaggar, 1983; Okin, 1989) who have provided devastating criticisms of traditional, gendered conceptions of the natural world and human nature as not only false but profoundly oppressive to women.

Rather than dwell on fundamental principles or theories, a good deal of excellent work in contemporary ethics is done in the area of "applied ethics," focusing on practical problems and the broader issues of policy. Questions about human nature or nature in general rarely play any overt role in such work, and some philosophers have even argued that "applied ethics" offers evidence that there are no unifying principles or theories behind our ethical deliberations. That each of the various domains of applied ethics rests as it were upon its own bottom, however, seems at the least quite controversial. Different realms of applied ethics such as business, law, or medicine may in fact have different understandings of how general ethical principles apply or which should play a leading role in certain types of deliberations, but in either case, general principles and theories play a critical role in deciding ethical issues within such disciplines. And so while theoretical concerns are not in the foreground in the field of practical ethics, they surely lie not very far beneath the surface. The same is true about conceptions of human nature and nature. While a great deal of practical and policy-focused philosophy does not explicitly engage issues concerning human nature and the rest of the natural world, such questions inform many of the assumptions upon which such reflections rest.

While widespread and powerful, the tendency to constrict or eliminate conceptions of nature in the development of ethical theory is not universal. *Virtue ethics* offers a revival or perhaps more accurately a reinterpretation of traditional appeals to human nature as a foundation for claims about what is valuable and good. Gendered versions of such an appeal also can be found among contemporary Feminist Ethicists, for example, Nel Noddings (1984), Mary Daly (1978), and on a certain reading Carol Gilligan (1993). Chinese Confucians offered a wide range of views that are best described as forms of virtue ethics (Lau, 1970, 1979; Knoblock, 1988–1994; Ivanhoe, 1991; Liu and Ivanhoe, 2002; Kline and Ivanhoe, 2000; Tucker and Berthrong, 1998). In addition to this family of theories, a number of contemporary thinkers advocate forms of *intuitionism* that establish a central and important place for nature within ethical theory. Intuitionism can be used to describe a wide range of ethical theories that justify their central claims by appealing to some fundamental, guiding sense or insight. Such views also find strong advocates among Daoist thinkers in early China. Our fifth and final type of ethical theory, what we call *natural science accounts*, insist that their view is strictly scientific and objective. However, in certain prominent and influential cases such theories claim or have been interpreted as claiming a central role for the value of nature.

2.5.1 Five Types of Ethical Theory

Utilitarianism appeals to the widespread belief that ethical decisions should be impartial and for the best, where the latter is defined as the greatest good for the greatest number. In its classical forms, the good was described in terms of feelings of pleasure and pain. And so the right action to perform is that which, from an objective point of view, can reasonably be expected to produce the greatest amount of aggregate pleasure over pain. Some utilitarians insist that we pay attention to general policies rather than individual acts, which distinguishes "rule" from "act" utilitarianism. Others, called preference utilitarians, argue that the good that should be maximized is the satisfaction of individual preferences rather than aggregate pleasure over pain. In any of its forms, utilitarianism denies that the natural world has any ethical value apart from its ability to produce the good whose maximization is being sought. On such a view, the natural world can be viewed as a rich source of good, but its worth will always be understood instrumentally, in terms of the good it can produce for sentient creatures.

Another significant feature of several versions of utilitarianism is their tendency to avoid discriminating between the good of human and non-human animals. For example, one influential utilitarian, Peter Singer (1979), insists that to count the well-being of non-human animals as less important than human beings is to commit the ethical wrong of "speciesism" which is akin to racism and sexism. Singer is very clear that most human beings are capable of much richer forms of both pleasure and pain and that this must be taken into account when calculating what actions one should perform. Nevertheless, on his view, neither the natural world nor human nature occupies a privileged place on the ethical landscape.

Deontology includes a broad range of related ethical theories. Their common feature is that the morally correct thing to do can be codified in certain rules or maxims that by their very nature command our attention, allegiance, and respect. Divine Command Theory is one example of a deontological theory. According to Divine Command Theory, the authority of certain rules or commandments is thought to lie in the fact that God reveals them. Contemporary ethicists who find inspiration in the writings of Immanuel Kant are among the most influential advocates of deontological forms of ethics. Such thinkers believe that certain rational maxims for action are thought by their very nature to command the attention, allegiance, and respect of any rational, free, and autonomous agent.

In the case of Divine Command forms of deontology, the value of the natural world or human nature ultimately is justified in terms of the will of God. So, for example, if one reads the *Book of Genesis* as describing the things of the world as given to human beings by God for their use, edification, and enjoyment, then one finds little warrant for protecting the natural world from human exploitation and manipulation. However, if one focuses on those parts of the text that describe human beings as stewards of God's creation, each part of which He considers "good," then one can derive a strong imperative to protect and nurture the natural realm. The value of human nature is similarly judged by the degree to which it is understood as obeying or defying the will of God.

The deontological theories inspired by Kant appear to offer few resources for seeing ethical value in the natural world itself. Since the source of morality lies in the nature of rational, free, and autonomous agents, the natural world and its non-human inhabitants do not qualify as proper objects of moral concern. Some Kantian-inspired ethicists have tried to enlarge the moral realm to include non-human animals. For example, Tom Regan argues that many non-human animals are "subjects of a life" and purposively pursue their own well-being which makes them "moral patients" if not full "moral agents" (Regan, 1983). Other deontologists, for example, Thomas Hill, have argued that respecting and caring for the natural world and non-human animals are important aspects of "imperfect duties" as defined by Kant (Hill, 1994). Roughly put, the reason we should behave well toward other creatures and even inanimate parts of nature is that it would be an affront to the dignity of our free, autonomous, and rational nature to squander our lives pursuing a cruel, insensitive, and brutish existence.

Kantian-style ethics is well known and rightly admired for the central role it accords the respect of other moral agents. In virtue of their autonomy, freedom, and rationality, moral agents command respect and cannot be treated as mere instruments of individual or collective desire. However, the features that establish the basis for such respect are not in any way distinctively human and so human nature as a whole does not play any central role in founding ethical claims. Any creature that exhibited autonomy, freedom, and rationality would be on an equal moral footing. Angels, highly evolved robots, or intelligent extraterrestrials all would be members of the moral kingdom.

Traditional versions of virtue ethics, for example, those of Aristotle in the West and early Confucians like Kongzi ("Confucius"), Mengzi ("Mencius"), and Xunzi in East Asia, seek to describe the ethically good life in terms of human flourishing (Aristotle, 1985). Such views maintain that human nature has inherent and distinctive capacities which when fully developed lead us toward the best life for creatures like us. Both Aristotle and early Confucians describe paradigmatic individuals who serve as the ideal of and model for human flourishing. Proper action is described largely in terms of what such agents would do, offering what has come to be called the "good person criterion."

According to Aristotle, our most characteristic capacity is rationality and the best kind of life is one in which this capacity plays a central role in all of our activities. The ideal life, as exemplified by the *phronimos* ("man of practical wisdom"), is one in which *practical* reasoning helps us exercise the various excellences or virtues needed to carry out our affairs, and *theoretical* reasoning helps us to understand and appreciate the nature of the world. The ideal life that Aristotle describes is not open to everyone. In fact few have the range of natural abilities and external means needed to realize the best kind of life. In this respect, Aristotle's conception of the virtues shares certain similarities to the ideal described by modern thinkers such as Nietzsche.

Kongzi and his followers are distinguished from Aristotle by their tendency to place much greater emphasis on certain basic human needs and affections, particularly those found within the family. For Confucians, the best kind of life as exemplified by the *junzi* ("gentleman") enables us to develop and enjoy the basic goods of familial well-being and extend these out to society at large. For Confucian thinkers, practical reason plays an important role in helping us to understand, appreciate, and order our lives. At least some Confucians also value an understanding and appreciation of the *dao* ("Way") itself and thereby describe a role for a type of theoretical reason as well. Another important respect in which Confucians differ from classical Aristotelians is their insistence that the virtues are accessible to all human beings and serve as the basis for a shared conception of the ideal life.

Francis Hutcheson, David Hume, and other "sensibility theorists" offer different but equally important versions of virtue ethics.[63] Like traditional virtue ethicists, they see the good life for human beings as closely tied to our particular natures and to the development of certain virtues. Hutcheson argued that moral good lies in the possession of a general benevolent disposition to promote the well-being and happiness of people and other sentient creatures. He thought that such a virtuous state of character will elicit a special sense of moral approbation from any reflective human being who contemplates it. Hume described virtues as traits of character that are "useful or agreeable" to those who possess them or to others. He too believed that such traits have the power to generate a special sense of esteem in reflective individuals. Hutcheson, Hume, and other "sensibility theorists" are more like Confucians than classical Aristotelians in that they believed the virtues they describe are accessible to all and serve as the basis for a shared conception of the ideal life.

Contemporary virtue ethicists such as Alasdair MacIntyre and Martha Nussbaum have attempted, in very different ways, to preserve some of the best aspects of virtue ethics while avoiding some of its clear liabilities (MacIntyre, 1984; Nussbaum, 1993). Specifically, both have sought, in different ways, to sever an account of the virtues from traditional theories about human nature and human flourishing. MacIntyre sees the good life for human beings as the search for the good life guided by a set of virtues that enable people to live meaningful, rich, and effective lives in community with others. Nussbaum seeks to describe and justify the virtues needed for a good life in terms of what she calls "spheres of activity" – for example, dealing with danger or coordinating one's actions with others – that define any recognizably human life. This has led her to focus more on basic issues concerning human well-being and less on specific ideals and virtues.

We shall focus on three versions of intuitionism in which nature serves as a central focus or theme. Our first two theories, the Land Ethic and Deep Ecology, have

---

[63] See the classic collection of texts: Selby-Bigge (1897, reprint, 1964).

been particularly influential among contemporary environmental ethicists in the West. The third, early Chinese Daoism, has attracted the attention of some Western ethicists interested in the ethical status of nature.

The Land Ethic was first described by Aldo Leopold (1970). Leopold's view is complex and not systematically and concisely presented, but the heart of his view is easy to discern. He argues that humans must abandon the long-held view that land is a commodity and come to see themselves not as conquerors of nature but citizens within it. We must come to regard the land, by which he means, "soils, waters, plants, and animals," as part of a larger community. The non-human members of this community have a "right to continued existence, and at least in spots, their continued existence in a natural state." Leopold does not disavow the human use and management of nature – indeed these are central parts of his view as well as his life-long practice – however, he insists on something like the inherent value of nature, both its individual parts and as a larger system. He offers sustained and eloquent testimony for this basic intuition throughout his writings, and argues convincingly that such an appreciation of nature can only be acquired through informed and reflective acquaintance. One could argue that Leopold's view is best understood as a form of virtue ethics. On such a reading, his guiding sense of the value of nature is at least partially a reflectively acquired and rationally endorsed taste, albeit one that is firmly grounded in certain aspects of human nature.

The Deep Ecology view was established by Arne Naess (1973, 1989). Its most significant and distinctive claim is that each organism and ecosystem is of *equal* value and hence every one should be protected and preserved as it is. This fundamental intuition distinguishes Deep Ecology from the Land Ethic described above. Since other organisms and ecosystems are actively being harmed and in many cases irreparably damaged by current human behaviors, and their habitats are continually being invaded by excessive human populations, deep ecologists insist that we must dramatically alter our behavior and reduce our population. For the deep ecologist, human beings have reached the point of being a strongly destabilizing infestation on the planet and are regularly and systematically violating and harming the non-human parts of nature.

Early Chinese Daoists, such as Laozi and Zhuangzi, express the profound appreciation for nature common to both the Land Ethic and Deep Ecology (Ivanhoe, 2003; Graham, 2001; Cskiszentmihalyi and Ivanhoe, 1998; Kjellberg and Ivanhoe, 1996). They also believe that most human beings have a grossly inflated sense of self-importance that distorts their view of and action within the world in ways that tend to undermine natural harmony. Early Daoists further argue that from the perspective of the *dao* all things are equal in value and have a place within the great "Way." In these respects they are much like deep ecologists. Early Daoism differs from either of the Western theories discussed above in its explicit faith in the benign character of both human nature and the natural world. Daoists describe the origin of human problems as a kind of fall from an original, uncomplicated, and harmonious form of life. The cause of this decline in natural well being is an excess of human cleverness, which gives rise to deceit, hypocrisy, unbridled desire and greed, as well as a host of other human failings.

Early Daoism, like Leopold's Land Ethic, also can be understood as a kind of virtue ethics. In certain respects, the case for doing so is even stronger. For example, early Daoist writings describe a type of self-cultivation and insist that most people must engage in a kind of spiritual therapy in order to understand the Daoist view. One can appreciate and endorse many aspects of such an interpretation of early Daoism from a modern critical perspective. For example, one might believe that human nature strongly inclines us to adopt certain innate intuitions and tendencies in regard to nature that are important constituents of a good human life. One might further believe that life in modern industrialized society often effaces or distorts these parts of our nature and thus tends to leave us alienated from important sources of satisfaction. People in such a benighted state might then be led to inflict additional harm upon themselves, others, and the rest of the natural world. At least parts of such a reading of early Daoism finds support in the natural science accounts of thinkers like E. O. Wilson, described below. However, early Daoists tend to give much greater weight to the satisfaction and joy one feels when one realizes a harmonious relationship with the rest of the Natural world.

By natural science accounts we mean views that claim to draw ethical prescriptions from scientific descriptions of human beings and the world in which they live. We will look at two representative examples in the writings of E. O. Wilson and James E. Lovelock. In the course of discussing Wilson's views, we will also offer some comments on the nature and value of evolutionary ethics in general.

E. O. Wilson has produced a wide range of writings that contribute either explicitly or indirectly to ethics. In *Sociobiology: The New Synthesis* and *On Human Nature*, Wilson argues that human behavior is primarily the result of evolution and that there is no purpose to human life beyond the "imperatives" our genetic history provides (Wilson, 1980). Given these "facts," ethics should be handed over to biologists who will endeavor to provide an account of human social behavior in terms of neural activity. This more "fundamental" explanation will then serve as the basis for decisions about the future of the human species, now that we have the capacity to alter our nature.

Wilson also has written a fascinating and provocative book called *Biophilia* (1984; Kellert and Wilson, 1993). While not expressly concerned with ethical philosophy, this work has tremendous potential to contribute to both ethics and aesthetics. Mustering his characteristic appeals to the role that evolution has played in the development of human nature, Wilson argues that human beings have a deep, complex, and persistent relationship with and need for living things – what he calls *biophilia*. This need gets expressed in a variety of ways – being channeled and shaped by culture – but it remains and is discernable beneath the play of a wide range of human activities. It is so strong and intricately interwoven into our nature that it constitutes an ineradicable feature of humanity and one that we cannot live without. If Wilson is right, an understanding of *biophilia* would offer important insights into a broad spectrum of human behavior and value. It would also raise questions about those, including Wilson himself, who advocate the plausibility and desirability of altering human nature in fundamental ways.

The British atmospheric chemist James E. Lovelock and his Gaia Hypothesis offer a second, influential natural science account (Lovelock, 1979, 1995).

Lovelock's original view, presented in the late 1960s, arose out of his observation that phenomena like mean global temperature and the salinity or alkalinity of the oceans are not fixed but rather move around a roving set point. That is, these phenomena are *regulated* over certain ranges by the combined interactions of the earth's air, water, surface soil, and living things (the "biota"). This resembles the kind of *self*-regulation we see in other organisms and that many regard as a central feature of any living organism. Lovelock took this to imply that all of the various constituents of the biota are parts of a comprehensive system that operates more like a living organism than a mechanical system.

At times, Lovelock alternates between the more modest claim that the Gaia hypothesis is a useful model for thinking about global systems and the stronger claim that the earth itself *is* a single living organism. We understand the latter as his primary position. Because he accepts the fact that life on earth is continually evolving, this single organism, Gaia, keeps changing. No species, no particular part of Gaia is, in principle, indispensable to its "health" by which Lovelock means its viability as a self-regulating system. In its ongoing effort to maintain the ideal conditions for life, Gaia occasionally destroys and sloughs off parts of itself. Human beings may in fact be on the way out, purged by Gaia in order to keep the conditions for life from being degraded to a dangerous point. A wide range of later writers, inspired by Lovelock, embrace and develop stronger and in some cases overtly religious versions of the Gaia hypothesis. However, we will not explore these here.

2.5.2 Critical Commentary

Utilitarianism has been criticized for providing too narrow a conception of human value. Simply put, while pleasure and pain are important aspects of many good human lives, they do not begin to exhaust the range of things that we value. They appear more as markers or at best constituents of what is good rather than representing the good itself. Contemporary defenders of utilitarian-type theories have attempted to expand the scope or "basket" of goods used to calculate "utility." While such "consequentialist" theories help to make utilitarianism more plausible, they also tend to make it more complicated and less distinctive. For example, the more diverse forms of good one includes, the more difficult the calculation of overall utility becomes. Including psychological goods like a sense of shame or honor tends to blur the distinction between utilitarianism and virtue ethics.

As noted earlier, utilitarianism sees no inherent moral value in nature in general or human nature in particular. Any good that may be found in such phenomena reflects their ability to produce certain valuable states in sentient creatures. According to such a view, there is nothing inherently wrong with altering nature. Indeed there is clear warrant for any alteration that can reasonably be expected to increase the aggregate amount of pleasure over pain. It might make good sense to appeal to such a view in order to justify and plan how to ameliorate human damage to the non-human world or to manage parts of it in ways that facilitate their flourishing.

Utilitarianism also is helpful as a justification and guide for eliminating certain painful defects or enabling the enjoyment of certain basic pleasures in human beings. However, it is much less clear that such a theory can provide a warrant for more wide-ranging manipulation of nature. As noted above, human beings value many things that may well be put at risk through the widespread alteration of human nature.

Except for certain versions of divine command theory, deontological theories don't have much to say directly about the value of the natural world. According to Kantian style forms of deontology, any moral good that may be found in nature is a reflection of the inherent value of certain kinds of agents. There are plausible Kantian arguments, in terms of "imperfect duties," against the wanton destruction of nature and even against a failure to appreciate nature. However, these don't offer any reason not to alter nature with an eye toward improving it for human use and appreciation.

Deontological theories do offer strong constraints against any attempt to alter human nature in ways that undermine human rationality, freedom, or autonomy. The wide range of prohibitions restricting experimentation on human subjects expresses the sense that such work often constitutes an affront against a person's fundamental dignity. Regardless of the good consequences that may result, even for the person involved, human beings must never be treated merely as means. Kantian style theories can mount compelling arguments against genetic manipulation when the aim is to produce something like a "better" soldier, for the person being modified is being treated solely as a means. However, such arguments do not seem to offer similar constraints in cases where a person freely chooses to undergo a procedure designed to increase her basic capacities for rationality, freedom, or autonomy.

Regardless of its particular form, virtue ethics appeals to some kind of theory about human nature in order to describe and justify its conception of the good life. At least implicit in such theories are views about how human beings relate to the greater natural realm as well. For example, in early Confucianism, there are a range of views describing how the *dao* ("Way") establishes a proper order both among human beings and between human beings and nature.

The central role that theories of human nature and nature play in virtue ethics places certain constraints on their views of the good life. These also seem to limit the range of legitimate ways in which one may alter either human nature or the natural realm. This does not mean that the ideal human life does not require us to develop or reform our natures, but it does appear to rule out altering the basic content of human nature or nature. Fundamentally changing human nature would preclude realizing the *human* good that such theories describe, while dramatically altering nature would deny the possibility of locating good human lives harmoniously within the larger natural order. If there is any substantial content to human nature as a result of its evolutionary past, the idea that certain kinds of lives in certain kinds of environments will prove decisively more satisfying to most human beings seems quite plausible. Such a view provides support for the virtue ethics approach.

All three forms of intuitionism discussed earlier offer firm guidance regarding how to decide certain issues concerning both human nature and the natural world.

In one way or another, they all argue that human beings have mistakenly arrogated to themselves a position of preeminence and dominance within the Natural order. They warn that this unwarranted claim of superiority injures many other creatures and organisms, threatens the well being of nature as a system, and harms and threatens human beings in a variety of ways as well. One might agree with many of the claims these theories make but seek to justify them with reasons that go beyond appeals to intuitions.

Some of the strongest claims made by such theories stand or fall on whether one shares or can and would choose to cultivate their founding intuitions. There is a spectrum of views within this category and considerable latitude in how one interprets different expressions of what we have called intuitionism. The "intuitions" that define such ethical perspectives can play greater or lesser roles in defining what is good for creatures like us. They also can be more or less "raw" in the sense of being simple and direct. Earlier, we pointed out how both the Land Ethic and early Daoism can be understood as forms of virtue ethics and how both can find support in E. O. Wilson's theory of *biophilia*. The guiding sensibilities of these two ethical perspectives could also play important roles in forms of utilitarianism.

Natural science accounts offer a number of insights but also leave us with several important unanswered questions. While E. O. Wilson's account of sociobiology is a brilliant achievement in many respects, it is also plagued by a number of difficulties. Consider this representative statement of his position:

> Human beings inherit a propensity to acquire behavior and social structures, a propensity that is shared by enough people to be called human nature. The defining traits include division of labor between the sexes, bonding between parents and children, heightened altruism toward closest kin, incest avoidance, other forms of ethical behavior, suspicion of strangers, tribalism, dominance orders within groups, male dominance overall, and territorial aggression over limiting resources (Wilson, 1994, 332, quoted in Arnhart, 1998, 5).

Wilson's approach provides at best a description of – not a prescription for – human behavior. While no one should doubt that evolution has brought us to where we are, Wilson provides no argument for his claim that biology sets our ends or purposes. We may judge that evolution has in fact brought us to an ethically better state. However, that would be a contingent fact about our particular history and such a judgment can only rest on a standard independent of an appeal to the course of evolution. Reaching back to the views of Darwin and Spencer, and emphasizing the evolutionary importance of the emergence of altruism, ethical "oughts" are presumed to be derivable from "nature." The difficulty is that the "nature" invoked by Darwinian evolution and its neo-selectionist heirs has ethically ambiguous consequences. On the one hand, evolutionary biology reports how life requires predators devouring prey in order to survive. When we base a social ethic on this aspect of nature, we end up with Herbert Spencer's survival-of-the-fittest as our moral guide (Spencer, 1896). On the other hand, today's sociobiologists and evolutionary psychologists look for examples in animal communities of reciprocal altruism; and, finding such examples, they argue that our social ethic should emphasize altruism and cooperation (Wilson, 1998, 171; Barkow et al., 1992, 167; Wright, 1994, 189–209). Both precedents can be found in nature. Which should become our moral guide:

survival-of-the-fittest or reciprocal altruism?[64] "The choice between these attitudes is a moral choice," writes Mary Midgley, "so it cannot be determined by science" (Midgley, 2003, 24). At Wilson's ideal level of description, it is difficult to see where one could find the resources to engage in critical evaluation. As important as evolution is for understanding ourselves, it offers no clear guidance on what we might become. Evolution is not necessarily going anywhere "higher" or "better." "Humankind's progress has nothing to do with unfettered evolution, which is always unpredictable and not necessarily upward bound," writes bioethicist Lee Silver (1997, 122). There is nothing inherent in evolutionary theory that ensures we *will* in fact evolve beyond what we presently are. Wilson's own discussion of our ability to choose how – or whether – to alter our nature presumes an ability to imagine and pursue ends other than those set by biology.

Wilson's suggestion that an account of human behavior on the level of cells is more "fundamental" and better seems misguided as well. First, in this case, reduction seems to eliminate much of the power of the resulting analysis. If one were to provide an account of aesthetic appreciation in terms of neural activity, one would know something, but it is not clear that one would have an explanation of aesthetics, much less a better one than what one can provide on the level of human experience. As Mary Midgley has argued, the same holds true for mathematical knowledge as well (Midgley, 1995).

There are several questions one might raise about Lovelock's Gaia Hypothesis. One of the first concerns his claim that "Gaia" is an organism and not just *like* one in certain ways. If Gaia is *not* an organism, it is more difficult to regard it as a proper object of ethical concern. If we grant that Gaia *is* an organism, then what might this tell us about how we ought to live? Minimally, it would provide us with a counsel of prudence: we must not threaten "Gaia" in ways that will cause it to slough *us* off. Beyond this, if Gaia can be shown to be an agent worthy of moral consideration, this might be used to argue that we must not harm or interfere with Gaia's on-going process of self-regulation. However, this would require much more than what Lovelock has established. Gaia would need to be sentient, capable at least of feeling pain if not disapproval, in order to qualify as a proper object of moral concern. Much like Wilson's sociobiology, Lovelock's theory is fascinating as a description of important features of our world but has difficulty providing prescriptions to guide us.

2.5.3   *Conclusion*

Utilitarianism and deontology remain the dominant ethical theories among Anglo-American philosophers. Unfortunately, as noted in our discussion, they seem to provide little normative guidance concerning the deepest ethical challenges arising

---

[64] Actually, this is a false dichotomy. Within circles of evolutionary conversation, these two complement one another. Reciprocal altruism becomes the means for establishing reproductive fitness. The alternatives appear contradictory only at the level of ethics.

from our ability to fundamentally alter nature. Contemporary or future philosophers may find ways to draw upon these theories in new and insightful ways, however the very nature of these views seems to limit their application to this particular realm of ethics. Utilitarianism grounds value in certain psychological states of sentient creatures, while deontology defends a particular conception of rational agency. Since neither sees ethical value in nature or human nature *per se*, both appear amenable to altering nature in quite radical ways as long as such changes result in greater amounts or finer qualities of their respective fundamental goods. There is a widespread sense, though, that such unbridled manipulation of our selves and the rest of the natural world is not a proper guide for how to alter nature but rather an expression of what is ethically wrong with doing so.

Virtue ethics is experiencing a significant revival and directly links its conception of value with various appeals to human nature. Some contemporary virtue ethicists, such as Rosalind Hursthouse, also are beginning to work out views about the role that nature plays in good human lives (Hursthouse, 1999). As noted in our earlier discussion, these aspects of virtue ethics suggest that such theories might provide important resources for addressing the challenges of altering nature. However, virtue ethicists are still coming to grips with their theory's historical association with discredited conceptions of biology, anthropology, and psychology. The future importance of virtue ethics largely hangs on the success of such efforts.

"Intuitionism" and "natural science accounts," include a range of theories that argue for a direct link between conceptions of human nature or nature and ethical value. Intuitionist theories have attracted significant attention. However, such views face difficulties persuading the unconverted. It is not evident that the purported "intuitions" used to justify such views are shared by every reasonable and caring person. This leads to the suspicion that they are more acquired tastes than innate dispositions. If acquired, then such "intuitions" quickly come to resemble the *cultivated* tastes of virtue ethics. In any case, if intuitions are acquired then they are in need of some kind of additional justification explaining *why* one might choose to cultivate and follow them. Natural science views attempt to link descriptions of nature and human nature with prescriptions for human behavior. However, as noted in our earlier discussion, while their efforts often succeed in revealing important aspects of both nature and human nature, their attempts to derive prescriptive principles or norms from such descriptions are not equally successful.

These last two types of ethical theory at best remain on the fringes of orthodox Anglo-American philosophy. Nevertheless, most attempts to discuss the kinds of ethical challenges that we face in light of our ability to alter nature fundamentally are found among proponents of these types of theories. Perhaps this points to what is missing in mainstream ethics. Intuitionism grants human nature an important place in ethics by appealing directly to human sensibilities while natural science accounts seek to explain human nature and its sensibilities in terms of more comprehensive theories about the structure and processes of nature. Part of the appeal of such theories may lie in their attempts to ground ethical theory in conceptions of human nature and views about how such nature fits into a larger natural order. In other words, such views focus attention upon what many feel is missing in utilitarianism and deontology. From a different angle, these same concerns can be understood as attempts to provide a more solid and reliable foundation for some of the more appealing but less well-supported aspects of traditional virtue ethics.

Part of the remarkable success of utilitarianism and deontology resulted from their commitment to provide more "scientific" or objective accounts of ethics. These theories offer an "absolute" view of morality, one not tied specifically to human nature or any larger conception of the natural world. However, this very strength seems to leave these theories with little to say about the central problems arising from our ability to alter nature fundamentally. Moreover, the natural sciences, while aiming for objectivity, are inextricably tied to the actual world in which we live and many philosophers have begun to hold philosophical theory to a similar standard. In an effort to avoid dubious metaphysical commitments, contemporary philosophers have sought to "naturalize" philosophy and this movement has begun to influence ethical theory. Recent work such as Elliott Sober and David Sloan Wilson's *Unto Others: The Evolution and Psychology of Unselfish Behavior*, which brings ethical theory together with contemporary work in psychology and evolutionary ethics, offers a powerful and promising paradigm for such efforts (Sober and Sloan Wilson, 1998). In order adequately to address the challenges that arise from our ability to alter nature, we will need work of this kind that also employs and incorporates a broad approach to value theory – one that draws upon modern insights as well as traditional western and non-western sources.

If ethics involves the search for principles and policies that can guide human beings to the most edifying and satisfying kinds of lives, then it is difficult to see how one could altogether ignore questions of human nature and the value of the natural world in the course of ethical reflection. Eliot Sober and David Sloan Wilson argue that altruistic desires constitute an important component of human nature; this means that most human beings have deep and abiding interests in the interests of others. While we cannot directly read off of this fact some raw natural value or virtue, the existence of such distinctive desires must inform our views about what kinds of lives we should seek for ourselves and others. For example, if we should find that our moral or political theories and practices work against the exercise and fulfillment of such widely shared and deeply held desires, this constitutes a good *prima facie* reason for questioning such theories and practices.

E. O. Wilson offers a revolutionary and radical set of claims about the human need for nature that, like the work of Sober and Wilson, is grounded in evolutionary biology. He argues not only that humans tend to have deep and powerful preferences and aversions to certain natural organisms, objects, and scenes but also that our fascination, attentiveness, curiosity, wonder, and desire to understand the world arose out of and are sustained by our interactions with nature. Even if one questions some or all of the particular claims that Wilson advances, his approach is based upon a powerful idea that would be foolish to deny. Natural selection has operated on what are now human beings for vast stretches of time and for all but a thin slice of this history understanding and responding to nature in certain ways were critical for our survival. It is implausible that such experience has not fashioned and bequeathed to us significant and characteristic dispositions to understand and respond to the natural world. While we certainly do not want to be bound by our natural tendencies, it seems equally clear that what we value and enjoy is inextricably tied up with such inclinations. If our ethical reflection is in any way aimed at leading us to edifying and satisfying lives, if part of such lives includes an understanding of our selves and the world around us, then it cannot hope to succeed without a broad, intricate, and critically informed understanding of human nature and the natural world.

2.6 General Conclusion

For much of the Western scientific and philosophical world by the end of the nineteenth century, the notion of nature as normative had simply dropped from serious discussion.[65] The rise of logical empiricism as the dominant philosophy of science of the twentieth century served to seal this tomb. For the architects of the Vienna Circle, late nineteenth-century Neo-vitalism and *Lebensphilosophie*, the remnants of Schelling's *Naturphilosophie*, became primary examples of the metaphysical errors into which misleading language could lead.[66] Either the "mechanism-vitalism" dispute of the early decades of the nineteenth century seemed resolved in favor of the claims of mechanism (meaning a chemical-physical determinism of organic actions), rendering such traditional concepts as the soul problematic at best, or the issues generally were consigned to the level of meaningless metaphysical discourse. Not only metaphysics but also ethics lost logical solvency: both were reduced to *mere* assertions, non-rational and arbitrary. Simultaneously, the ambition for complete domination of nature, set out at the opening of this chapter as one of the ideals of at least the Cartesian variety of the mechanical philosophy, has become virtually hegemonic, and the "right" of human domination over the natural ironically presents itself as a virtually "moral" imperative. Concomitantly, ethical discussion today holds human moral norms autonomous. Within this isolated ("autonomous") ethics, the standard debate has been between those advocating maximization of social good according to some version of utilitarianism, and those holding a deontological position, primarily some form of Kantianism, based on the moral dictates of autonomous reason. Thus the dominant ethical approaches in modern Western philosophy, utilitarianism and deontology, like the dominant philosophy of science, positivism, prove relatively inhospitable to any normative role for nature.

Yet the persistence of variants of virtue ethics, whether traditionally Aristotelian or grounded in the "moral sense" tradition, and the emergence of environmentalist

---

[65] Symptomatically, the entry "Nature" disappeared from the famous ninth edition of the *Encyclopedia Brittanica* of 1875–1889. See also note 33.

[66] An example is Wittgenstein's comment in the *Tractatus* that Darwin's theory has no relevance for philosophy. *Tractatus logico-philosophicus*, 4.112.

ethics (the "Land ethic," "deep ecology," the "Gaia hypothesis") and of evolutionary ethics (in either its sociobiological or its naturalist formulations) offer considerable scope for a normative role for nature. These endeavors appear to flout one of the most entrenched of philosophical pieties in the modern era, the distinction of facts from values, of *is* from *ought*, the prohibition of the "naturalistic fallacy" (Woolcock, 1999; Meyer, 2001; for the counterargument see Callicott, 1982; Warren, 1985).

2.6.1 The "Fact/Value Dichotomy"

Many scientists and philosophers consider this prohibition of the "naturalistic fallacy" decisive, and they uphold a sharp separation between "facts" and "values." Enforcers of the piety acknowledge that there may be *some* constitutive relation between human experience and the natural world that envelops it, but they deny that anything *prescriptive* can be established from nature to bear upon human conduct. Whether the intervening variable be construed as freedom or culture or politics, it renders any direct normativity of nature problematic. Nature can teach us "facts" about the world around us, to be sure. Yet on this understanding, nature cannot provide ethical guidance. What "is" is not necessarily what "ought" to be. The attempt to move from nature to ethics stands accused of perpetrating the "naturalistic fallacy."

It is critical to both this chapter and the larger project of which it is a part to insist that there is nothing *philosophically conclusive* about the "fact-value dichotomy," such that the realm of values and ethics *must* appear cognitively arbitrary and the natural order *cannot* affect the ethical order. The "fact-value dichotomy" has come under considerable criticism in recent thought. In articulating these criticisms, we distinguish the broader question of the "fact-value dichotomy" from the commonly invoked notion of the "naturalistic fallacy." First, let us clarify this latter category, then turn to the more fundamental issue of the "fact-value dichotomy."

The "naturalistic fallacy" is an ambiguous and controversial notion. It is usually associated with G. E. Moore, though he was not the first to raise the issue or use the term (Moore, 1968). Moore in fact raised *two* distinct issues. One is whether ethical terms such as "good" or "right" can be *equated* with or *reduced* to natural terms, such as "conducive to growth" or "admired by everyone." The other issue is whether judgments about what is "good" or "right" can be *derived* from "facts" – i.e., whether they can be inferred using inductive or deductive methods (Hancock, 1974, 14). Moore's concern was primarily with the *first* of these two issues: the question of definition of terms.[67] It arose in a quite specific context and against a quite specific target. Nineteenth-century Naturalists (e.g., Herbert Spencer) maintained

---

[67] As G. J. Warnock notes, "although the naturalistic fallacy has played an important and colorful part in more recent writings, in Moore's book it amounts only to the bald assertion that the quality of goodness is *not* as other qualities are – that, in particular, its presence is not to be detected by any ordinary species of observation, experience, or investigation" (Warnock, 1967, 7).

"good" could be defined in natural terms – "conducive to life of self or others," for example. Moore replied that "good" cannot be defined in other terms and therefore is not reducible to some natural phenomenon (Frankena, 1939). In what is termed the "open question" argument, Moore invoked the simple fact that we can say "X is conducive to life" and still ask "is X good?" Hence "good" cannot simply mean "conducive to life." The question "is X good?" and hence the meaning of "good" remains open (Hancock, 1974, 29–31). To be sure, Moore was reduced to entirely *stipulative* assertions of what "good" meant: "intuitions" as ungrounded as those we have considered in other approaches. In practice, his ethic fell back to utilitarian inferences. He held that the "right" thing to do was to maximize "good." Moore found himself involved in judgments and predictions from factual statements about what does the most "good."

The fundamental issue in the "is/ought" debate is the question of whether one can *derive* ethical judgments from factual judgments. The objection is that ethics does not depend *logically* on facts about human nature and the world. William Frankena argues ethical judgment comes about because of an additional, non-factual premise. There is always a *tacit* value postulation (the definition of "good"). Thus, from the mere *fact* that X injures someone, we do not get an ethical judgment. If we say "X is wrong" because X injures someone, there is a tacit premise or "bridge principle" in the logic – that injuring someone is wrong. In this sense, Frankena concludes, "those who insist that we cannot go from Is to Ought or from Fact to Value are correct" (Frankena, 1973, 97). The conclusion is, then, ethical norms and values cannot be justified by grounding them in "the nature of things."[68]

In the words of the logician Gerhard Schurz: "the situation in ethics is *basically different* from the situation in science. *Nature* cannot tell us which ethical concept is the right one: *we* have to decide. Intersubjective agreement in ethics … is basically the result of a *common culture* which enables mutual understanding and a common life practice" (Schurz, 1997, 285). If, however, "the conflicting parties differ in their conceptions of 'good,' their conflict can never be solved," because "the possibility of an intersubjective justification in ethics is (quite seriously) limited" (Schurz, 1997, 236). Given these "serious restrictions on objectivity in ethics," in which "there exist a plurality of competing ethical concepts of what is good and what ought to be done, among which it is not possible to decide in the way empirical science decides between competing theories," Schurz opines, the fact/value dichotomy "advances tolerance and respect in ethical discourse" (Schurz, 1997, 31). This is, however, too optimistic a judgment. Hilary Putnam has a far better handle on the implications of this dichotomy: "the worst thing about the fact/value dichotomy is that in practice it functions as a discussion-stopper, and not just a

---

[68] Frankena suggests that there may be another sense – a non-logical sense – in which "our basic norms and value judgments can be justified by appeal to the nature of things" (Frankena, 1973, 101). For example, if human nature is so constituted that each of us "by nature" pursues our own good, then what we "ought" to do could indeed be derived by an examination of human nature. This has resonance with the Roman Catholic understanding of natural law, although Frankena does not extend his argument into that arena.

discussion-stopper, but a thought-stopper" (Putnam, 2002, 44). The force of this dichotomy has been to marginalize ethical discourse from the sphere of "knowledge" and to reinforce unbridgeable boundaries between entrenched value communities (by invocation of "incommensurability" or relativism). This has been one of the crucial features of a modern Western culture dominated by scientism, provoking the disquiet of such thinkers as MacIntyre noted in our introduction.

Putnam is apt here as well: it was the logical positivists who "were the most influential marketers of the fact/value dichotomy" (Putnam, 2002, 23). As he elaborates, "their real target is the supposed objectivity or rationality of ethics," and "Carnap's purpose was to *expel* ethics from the domain of knowledge, not to *reconstruct* it" (Putnam, 2002, 19–20). This sets the question of the fact/value dichotomy in a different light. In suggesting that the dichotomy ought not stand as an indefeasible barrier, we propose to ask again, in the words of Marinus Doeser, "Is it possible to render intelligible – at least somewhat intelligible – how we choose our values in relation to the factual conditions attending the choices?" (Doeser, 1986, 16). In Putnam's terms, the question is "whether there is a notion of rationality applicable to normative questions" (Putnam, 2002, 2).

In the twentieth century, two extreme positions proved deeply invested in the fact/value dichotomy: those committed to the "autonomy of ethics" and those committed to the "purity" of scientific knowledge. It is striking that neither of these commitments seems sensibly ascribable to the historical David Hume, from whom the "law" severing "is" from "ought" is conventionally derived. The famous passage in Hume's *Treatise of Human Nature* notes the unconsidered transition from observations about the nature of things to notions of right and then goes on to suggest that "is" judgments are "entirely different" from "ought" judgments (Hume, 1888, 3.1.2). How to understand his remark, however, as Schurz observes, "has been the subject of continuing philosophical debate" (Schurz, 1997, 4). In a crucial intervention in 1959, Alasdair MacIntyre noted that "for the tradition which upholds the autonomy of ethics from Kant to Moore to Hare, moral principles are somehow self-explicable; they are logically independent of any assertions about human nature" (MacIntyre, 1969, 50). MacIntyre advanced a strong argument that this could not be reconciled with the historical Hume's cultural context, theory of knowledge or moral philosophy. He urged that interpreters could not take Hume to be using "deduction" in the same sense we use it today, and that Hume did not intend to deny the inference from "is" to "ought" but to problematize it and to suggest more cogent approaches to that problem, i.e., "to preserve morality as something psychologically intelligible" (MacIntyre, 1969, 50). Situating Hume's statement within Hume's overall philosophy of knowledge and of morals, MacIntyre insisted that the modern reception of "Hume's Law" simply did not fit the historical Hume. A lively debate ensued. In an excellent review of the debate, R. E. Creel considers the core of the issue to be the claim that "Hume has forever destroyed the presumption that what we ought to do can be determined by an examination of what is. And since there is no bridge between 'is' and 'ought,' morality must be grounded on extra-empirical foundations" (Creel, 1995, 517). Considering the ebb and flow of decades of debate and based on his own close reading of Hume's text, Creel concludes that Hume "cannot be appropriated with justification by rationalistic intuitionists" (Creel, 1995, 517; see also Millgram, 1995, 1997).

If these *twentieth-century* preoccupations have a key progenitor, it is not Hume but Immanuel Kant. Kant's concerns for the autonomy of ethics and for the purity of theoretical reason are overweening both for his system and for his legacy.[69] The logical positivists – both in their philosophy of science and in their views on ethics – represent a moment in the history of philosophy in which Kant's *a priori* warrants for ethics had suffered a substantial loss of credibility just as his *a priori* structures of theoretical knowledge shattered under the impact of new mathematical and scientific developments. In the triage operation that the logical positivists undertook, it was the integrity of scientific knowledge they chose to save. And this bore directly upon the manner in which they took up the fact/value dichotomy. They cared nothing for a discriminating inquiry into value. Their concern was entirely with *fact*. But if we ask, with Putnam, "what *exactly* did the logical positivists ... understand by *fact*," we come to the answer that "the logical positivist fact/value dichotomy was defended on the basis of a narrowly scientistic picture of what a 'fact' might be" (Putnam, 2002, 23, 26). It hinged, to be concrete, on Carnap's project to hold "observation terms" distinct from "theoretical terms," and more generally, to uphold the analytic-synthetic distinction (Carnap, 1956). But these were precisely what Willard Quine dubbed and drubbed as "Two Dogmas of Empiricism" (Quine, 1980).

The logical positivist idea of "fact" is irredeemable.[70] "But if the whole idea that there is a *clear* notion of fact collapsed with the hopelessly restrictive empiricist picture that gave rise to it, *what happens to the fact/value dichotomy?*" (Putnam, 2002, 30). Putnam argues very persuasively that the fact/value dichotomy cannot really survive the demolition of the other "dogmas of empiricism," because *fact* must now be recognized to be thoroughly impregnated with theory and with values.[71] Science cannot be "value-free" since "science itself presupposes values [which] are in the same boat as ethical values with respect to objectivity" (Putnam, 2002, 4). That means that the idea of *objectivity* needs to be reinterpreted. "It is time we stopped equating *objectivity* with *description*" in a manner which ascribes to the description/prescription distinction an ontological grounding in separate "natural kinds" (Putnam, 2002, 33). We are compelled, Putnam argues, to follow pragmatism in "developing a less scientific account of rationality [by reflecting on] how normative judgments are presupposed in all reasoning" (viii). Putnam holds that the pragmatists, exemplified by John Dewey, emphasized that "value and normativity permeate *all* of experience" (30).

---

[69] The authority and autonomy of "pure reason" is the core of the "critical philosophy." See Kant, *Critique of Pure Reason*, *Critique of Practical Reason*. For a prominent instance of the Kantian legacy, see Korsgaard (1996).

[70] For a penetrating and path-setting account of the rise of "post-positivism" in philosophy of science, see Hacking (1983). For the views of authors of this chapter, see Longino (1990, 2002) and Zammito (2004).

[71] Thus he favors "rethinking the whole dogma (the last dogma of empiricism?)" and urges that coming to "think without these dogmas is to enter upon a *genuine* 'post-modernism'..." (2002, 145, 9).

If the "purism" of scientific knowledge has collapsed, what of the commitment to the "autonomy of ethics"? (see Prior, 1960; Jackson, 1974; Pidgen, 1989). The embrace of "Hume's Law," as it has been called, had to do for its advocates not simply with the prestige of scientism, though this is not to be underestimated. It had to do as well with the conviction that ethical concern must not be *reduced* to determination by the physical order. The alternative, here, as Kant expressed it, was "heteronomy" – the compromise and, at the extreme, the foreclosure of human moral freedom (Kant, 1964). For such thinkers, rationalists and intuitionists alike, it seemed imperative to bar any determination by physical cause in order to create a space for ethical freedom. Logical rigor, in that light, had a special place: if one could establish that no ethical judgment could follow from *merely* descriptive judgments, no "ought" from any "is," then this made space, somewhere in the chain of logic, for a *primordially ethical posit*. But what was that supposed to be? Putnam is not alone in expressing skepticism about "any special 'Platonic' realm of 'ethical properties'" (Putnam, 2002, 3). This is the other front facing the proponents of the "autonomy of ethics" – not only must they defend the ethical domain from physical determinism, but they must also somehow explain the primordial ethical posit.

For non-cognitivists, it is simply an arbitrary preference. This means, of course, that rational examination of it is foreclosed. But what is the alternative? It would appear, as Putnam puts it, that we cannot come up with "a *metaphysical explanation of the possibility of ethical knowledge*" (2002, 44). For Putnam this is an instance of a far more general realization: the failure of "foundationalism" across the board, and he maintains, further, "nothing much follows from the failure of philosophy to come up with an explanation of *anything* in 'absolute terms'" (2002, 45). The "entanglement of facts and values" simply is the condition in which humans find themselves, and it is palpable in ordinary language, which throngs with "*thick* ethical concepts" (Putnam, 2002, 29).[72] The recourse of advocates of ethical autonomy faced with this unquestionable feature of ordinary language is to suggest that ordinary language can always be "rationally reconstructed." That is, as the ethicist R. M. Hare puts it, it always *in principle* possible to disaggregate any "thick" ethical concept, any "mixed" (descriptive/prescriptive) statement into the strictly descriptive and the strictly prescriptive (Hare, 1952). But a whole series of critics have questioned the plausibility and the good sense of this move (i.e., Murdoch, Anscombe and Foot). As Putnam writes: "the attempt of noncognitivists to split thick ethical concepts into a 'descriptive meaning component' and a 'prescriptive meaning component' founders on the impossibility of saying what the 'descriptive meaning' of, say, 'cruel' is without using the word 'cruel' or a synonym" (2002, 38). As Doeser puts it, "Is it really true that our ordinary use of language requires correction on this point? If fact and subjective response or evaluation can always be distinguished completely, then why do we make use of statements in which reference is made to both in mutual coherence?" (1986). As Alasdsair MacIntyre put it in

---

[72] The notion of "thick ethical concepts" is developed by Iris Murdoch, G. E. Anscombe, and Philippa Foot. See the discussion in Hudson (1969).

a classic essay, "this belief arises out of accepting formal calculi as models for argument and then looking for entailment relations in nonformal discourses" (MacIntyre, 1969, 45).

The logician Schurz observes that "the primitive symbols of natural language may have both descriptive and normative meanings," and therefore arguments "in natural language [are] not really conclusive in the strict logical sense" (Schurz, 1997, 9, 4). Schurz chooses to believe that Hare is right that one can always "at least in principle" distinguish between descriptive and prescriptive (elements in) statements, and hence the matter can be brought under the domain of rigorous ("scientific") logical analysis, via "alethic-deontic modal logic." While Schurz makes clear that hitherto "the is-ought problem has to be regarded as basically open and unsolved" (which is all that we believe needs to be established), he undertakes to remedy this irresolution by a rigorous logical analysis (Schurz, 1997, 4 and passim). But this is to suppose that the issue really has always been "entailment" in some strict sense of logical necessity. As we have noted, it goes vastly beyond this to epistemological and ontological questions which are at the very least as primordial as the force of the logical symbols that Schurz employs. The scope of philosophical necessity – of logic – has proven remarkably restricted in accounting for human knowledge – in science or in ethics (Kitcher, 1992). There is no "value-free" (foundational) perspective anywhere, no "neutral" rationality (Nagel, 1986). As Doeser puts it, "the choice for a specific form of rationality always takes place on the basis of what people consider relevant and important" (Doeser, 1986, 8).

This bears upon Max Weber's famous effort to discriminate *Zweckrationalität* – means/end or "instrumental" rationality – from *Wertrationalität*, the rational examination of ultimate values (Weber, 1946.)[73] It does not preempt rational examination to renounce absolute grounding, for the presumption that without absolute grounding there is no platform for investigation is simply unfounded (Doeser, 1986, 10; Putnam, 2002, 63). That is the point of a pragmatic holism. In the words of Jack Nelson, "To ask what there is is to ask what posits we must make for our theories to come out true," and these posits include recognizing our own presence as interested inquirers, for whom, moreover, knowledge is always only what we come to *share* (Nelson, 1996, 65). In that sense, "values and value theory are important parts of the holistic web" (Nelson, 1996, 66). The point of breaking the fact/value dichotomy is not, Nelson argues, to collapse into relativism, but to recognize and pursue adjudication.

How the world is, and what humans are, can, in this limited sense, directly impact ethical judgment. If we are always situated in a cultural/historical community, this does not cage us into "incommensurable" systems of Rortian relativism in which a decisionist "solidarity" totally displaces "objectivity" (Rorty, 1991). But we must reconstitute what objectivity means. It makes no sense to uphold a notion of objectivity as "neutral description," or as the attempt to describe "the world as it

---

[73] On the enormous impact of Weber's concern with "value-fee" social science, see Stedman-Jones (1998).

is in itself, independent of all observers" (Putnam, 2002, 40). Instead, we must begin with a notion of experience rich enough to take account of the fact, as John Dewey argued, that it is always already value-infused (Dewey, 1925). Thus rationality is a developmental practice, not a timeless norm, and "objectivity" is its contingent and fallible outcome, wherein not only are the facts "theory-laden," and the theory "value-laden," but facts and theory can occasion moments when the values themselves need to be revised.[74] Obviously not all values can be reevaluated simultaneously, any more than all theories can. But that suggests we invoke Quine's favorite metaphor not only for our theory of knowledge but for our understanding of the relation of facts to values, of description to prescription: we are like sailors on Neurath's boat: always making our necessary repairs while at sea.

From a post-positivist vantage the fact-value dichotomy seems little more than the last dogma of positivism, as dubious as any of the other dogmas, e.g., the analytic-synthetic distinction, the distinction of observation from theory, or of a context of discovery from a context of justification (Nelson, 1996). The core of this criticism bears on the notion of "fact." All "facts" are constructed in and through a theoretical language, and all theoretical languages are embedded in values which organize the question, the procedures of investigation, and the signification of warrant. This cuts the argument away at the ground. Once it is seen that the domain of "fact" is pervaded by value, the issue comes squarely to bear upon the putative authority of *logic* on its own: the grand, austere normative necessity of the *a priori*. But there are even stronger reasons to believe that we have come to a point in the history of philosophical thought where we can grant the *a priori* precious little of the necessary authority by which it has reigned so long in the tradition. In a sense, this rounds us back to Hume, and to his insistent naturalism. Hume *qua* naturalist supplemented Hume *qua* skeptic decisively and consistently across his whole opus. If his insight as a skeptic was profoundly deflationary for the claims of transcendental necessity, of the *a priori*, his insight as a naturalist was to relocate the sought-after order in custom, culture and history. He saw well that these fall short of the Platonic ambition of timeless truth, but he insisted that we need to take courage and come to terms with the limits of logic, settling for what we can learn about ourselves and our world in the inevitably blurred genres of value-laden experience (Hume, 1977). This does not disempower rational discrimination in science or in ethics; it humbles and pluralizes it. Contingency and fallibility are as central – if not more central – to ethical discourse as they have proven to be to scientific discourse. And all of this suggests that there is no categorical divide between these two discourses after all.

---

[74] "For [Dewey] 'inquiry' in the widest sense, that is human dealings with problematical situations, involves incessant reconsideration of both means *and* ends... Any inquiry has both 'factual' presuppositions ... and 'value' presuppositions, and ... changing one's values is not only a legitimate way of solving a problem, but frequently the only way of solving a problem" (Putnam, 2002, 97–98). "If Dewey does not believe that inquiry requires 'criteria,' in the sense of algorithms or decision procedures, either in the sciences or in daily life, he does believe that there are some things that we have *learned* about inquiry in general *from* the conduct of inquiry" (104). "Objective value arises, not from a special 'sense organ,' but from the *criticism of our valuations*" (103).

While for some this threatens the "autonomy" of ethics, for others it opens the prospect of a larger integration of experience and inquiry. As Baird Callicott has urged, "ethics are embedded in larger conceptual complexes – comprehensive worldviews – that more largely limit and inspire human behavior." Thus, "an ethic – despite the professionally decreed divorce between fact and value, *is* and *ought* – does not exist in a cognitive vacuum… [C]hanging worldviews open up and shut down possibilities, either implicitly or explicitly" (Callicott, 1994, 4–5). It thus becomes important to see how different traditions understand and use the term nature and how they relate it to their ethical judgments and warrants for evaluating biotechnological developments.

2.6.2 *Approaches That Bring Nature Back In*

Against the isolation of facts from values, science from ethics, we have noted at least three positions involving some renewed appeal to the concept of nature. One is the revival of *evolutionary ethics*, typically allied with some forms of sociobiology. This position seems to carry strong appeal for some empirical scientists. Within the larger non-teleological interpretation of the evolutionary process endorsed by mainstream evolutionary theory, the theoretical struggle over evolutionary ethics centers on the degree to which an ethical adaptationism, displaying relative degrees of fitness to conditions of life, sufficiently supplies the foundation for the robust ethics demanded by moral philosophers.[75] With respect to issues in biotechnology, even if there is some way of connecting biology and ethics together through a concept of the "natural," Darwinian evolution itself suggests little reason to prohibit technological intervention in normal biological process as long as these further human evolutionary fitness.

Another stance in these discussions, modern *environmentalism*, can be traced to the heritage of Romanticism and *Naturphilosophie*. This nineteenth-century heritage, reworked through the writings of Thoreau, Leopold, and many others, re-values the "natural" as normative, demanding moral respect. Some have even characterized this movement as a new religious faith (see, for example, Hitt, 2003). Unquestionably, it has become broad and deep enough to indicate that "nature" is still a viable category on which to build ethical norms, at least in some domains. The revival of the "natural" as moral in modern environmentalism opens up several opportunities for a reconnection of religious thought with a concept of nature, although how this plays out on the terrain of practical ethics remains to be seen. From vegetarianism to opposition to genetically modified foods to the more extreme forms of "deep" ecology and the *Gaia* hypothesis, it represents a powerful set of attitudes that at present play a major role, at times even a destructive one, in the debates over biotechnology.

---

[75] See, for example, Rosenberg (2000, ch. 6). Several perspectives on the broad debate over sociobiology are summarized in Maienschein and Ruse (1999).

A third stance might best be characterized as neo-Aristotelian. Advocates of this position do indeed allow values to be derived from "nature." Furthermore, they embrace a universalist conception of human nature, and from this derive precepts that can legitimately restrain technological interventions. Within this tradition would also stand the "moral sense" ethics of the early Hume, Hutcheson, and Adam Smith, and the more recent "moral sense" theories found in the writings of Leon Kass and others who appeal to our instinctive distaste for biotechnological change – what Mary Midgley has called the "yuk" factor – as worthy of moral consideration (Midgley, 2000). The natural law tradition – important for, but not confined to, Catholic ethical teaching – also seems best located in this category. If the "right" thing to do is to seek our proper *telos* or end, and if this end is already "built into" our nature – which is what natural law theories generally hold – then we should be able to discern what is right by examining human nature. Natural law theory claims to be rationally warranted, i.e., its premises are presumably open to discovery by human reason, even if this requires educated reason and reflection. Natural law theories are not reducible to inadmissible religious premises.[76]

Relating the various interpretations of the concept of nature developed in this chapter to religious traditions requires a fine-grained analysis of the differential importance of the concept of nature, or its equivalent, in the tradition-by-tradition examination carried out in other portions of this book. From the analyses offered in this chapter, one might be able to draw at least a few summary conclusions on the relations of philosophical views on nature to religious traditions. One issue concerns the relation of theological views to the development of the mechanical philosophy. Within specific historical periods examined in this chapter, operative conceptions of nature – mechanism, organicism, vitalism – have had connections to underlying metaphysical views and, at least in some traditions, to underlying theologies.[77] There is also considerable linkage in the early modern period between conceptions of the passivity of matter, voluntarist theology, and mechanical philosophy (see Deason, 1986; McGuire, 1972, esp. 526–528; Roger, 1997, esp. ch. 2). Although this might be seen as leading away from religious conceptions, and indeed even to the replacement of God by nature-as-machine, this development of "theistic" mechanism historically proved congenial to the assumptions of mechanical philosophy. It also served as a restraint upon the will to domination that otherwise might be seen as a consequence of mechanistic views of nature. The subsequent development of vitalistic conceptions of nature in the late Enlightenment occasioned several extensions that could either lead to a self-sufficient, purposeless world order, as seen in Diderot's vital materialism, or support forms of religious interpretation. One such

---

[76] For discussion of the historical affiliations between these two positions, see Sloan (1999).

[77] For example, in the early developments of mechanistic dualism, as found in Descartes, the traditional theological soul was distinct in essence from the body, working by mechanical action, and had no role in vitalization of organisms. The equation of a vital principle and the theological concept of a soul is not a necessary one. Some efforts to revive Cartesian dualism in relation to modern mechanistic accounts of the body can be found in the writings of neurophysiologist John Eccles. See, for example, Eccles (1970).

extension was the dynamic pantheism in the *Naturphilosophie* of Goethe and the Romantics that revived in part Spinoza's identification of God and nature.

There are at least two ways in which religious perspectives offer not only helpful but also critical resources for thinking through the "normativity of nature." The first concerns what might be called metaphysical comfort. The idea here is that human beings have a need to explain to themselves and others where they fit into the greater scheme of nature. This can be understood as parallel to the need to explain oneself and one's actions to other human beings. This perspective is missing in analytic ethics and is front and center in most religions. The problem with the latter is that the answers they provide often seem philosophically implausible. The perspective, though, is not in any way implausible. In fact, what is implausible is *not* taking this perspective seriously. Hence focusing attention on this perspective is the first contribution of religious traditions. Religious traditions are rich resources for such inspirational conceptions (world-views, metaphysical visions) of what *we* might be vis-à-vis the world. One can provide secular accounts from this perspective that are not implausible. E. O. Wilson's book *Biophilia* does this to some extent and Sober and Wilson's *Unto Others* can also be understood in this sense.

The second contribution concerns how one responds to this challenge. Here, one can see religious traditions as offering *ways of life* (and the rules, or "casuistries," informing them) that one could endorse even without committing oneself to the metaphysical assumptions that underwrite these views. For example, Christian environmentalists espouse the view that we are caretakers or stewards of nature while Daoists argue for the underlying connection between all parts of the *dao*. One could adopt a version of either of these views as an expression of the kind of human life one deems most satisfying and appealing. It is not necessary to believe in God or a particular conception of the *dao* to want a world in which human beings work to cultivate certain attitudes, practices, and policies toward the natural world. Similarly, it may not be necessary to believe in the intuitions guiding more recent ethical naturalisms in order to subscribe to their proposals for a more measured human relation to the natural order.

Nor is the most recent trend in science and its philosophy adverse to these tendencies. While the austerities of logical positivism purged some quasi-mystical conceptions of nature from science and from scientific philosophy, those austerities themselves later gave way in response to developments in the sciences (biological complexity, ecological complexity) and in philosophy (anti-foundationalism, anti-fundamentalism, anti-reductionism). Contemporary science and philosophy of science seem considerably more hospitable to multiple conceptions of nature's complexity. Nature may be so complex that any conception can only be partial (the point of pluralism). What the sciences can show is what conceptions are *not* viable, or beyond empirical investigation. Thus, even among advocates of emergence, the best mode of access to the nature of nature is through the sciences.

Sciences can also show the limits of acting on our knowledge or understanding. Because consequences extend beyond the reasonably foreseeable, we may draw a pragmatic maxim to give up the ideal of control of nature, and instead work with (what we know of) nature. That some singular conception of nature does not emerge from the sciences does not leave the field open to religion alone. Philosophical approaches that stress the interdependence of human activity and nature (e.g., Dewey's pragmatism) appear to hold some promise. Moreover, critical analysis (e.g., feminism) shows the interrelations of conceptions of nature and political conceptions. Concepts of nature develop in the context of a whole culture/society and serve multiple ends. Intellectuals (and religious thinkers) can advocate particular conceptions, but none has an autonomously *a priori* status. Those that resonate with other cultural needs and ideals are more likely to be taken up.

2.6.3 Summing Up

This brief survey of some important contemporary resources for thinking about the ethical status of nature can be criticized as failing to provide a single definitive theory or standard and as offering only a glimpse of the philosophical resources available in non-Western traditions. However, we regard these failures as pointing the way toward a more adequate account. The degree to which nature, including human nature, can be rightfully manipulated and dominated for human ends is a question that cannot be resolved from any already definitive set of moral, metaphysical, or theological assumptions. Instead several discernible traditions have come into opposition in the present discussions. In view of the contrasting positions, current debate shows little possibility of immediate closure. If historical-philosophical analysis cannot resolve their differences, it can at least help us gain a clearer perspective on the shape of the questions to be resolved. As we reflect about what we've done, it seems that our work points out some dead ends (e.g., our conclusion on the limits of current utilitarianism and deontology) but, on the positive side, we have sought to indicate what resources do exist in the philosophical tradition for thinking constructively about the human relationship with the rest of the natural world, including one another. We do not try to offer a single best direction in which to move. We can see some viable procedures and approaches coming out of the material we've surveyed, possibly needing modification, but at least not stuck in the starting blocks.

Indeed, one might conclude from the diversity of ethical theories and their varying degrees of plausibility regarding different ethical problems that it is simply misguided to insist that a single theory can solve every problem. A pluralistic approach to ethical theory need not result in chaos. The natural sciences show no strong theoretical unity and yet provide us with the most compelling and reliable account of the world around us. The diversity of ethical theory and especially the contributions that other traditions can make to contemporary philosophy should also be understood as pointing to a significant difference between the natural sciences and the humanities. While the former do not offer us a theory-neutral account, they are designed and focused on discovering the structure and function of a world that exists independently of human minds. The humanities and in particular ethical theory are, or at least should be, constrained by facts about the world; however, their central projects involve making as much as finding, creating as much as discovering. Ethics involves a complex type of interpretation that can color the world in a variety of equally satisfying and mutually irreducible ways.

The entire project arose in response to the gathering conflict between modern science and technology and traditional views about human nature and the natural world. This confrontation has generated diverse and often opposing views. Some insist that modern science and technology have discredited and disenchanted our views of human nature and Nature, that this is good, and that the costs of alienation and uncertainty are worth the benefits of overthrowing oppressive conceptions of what we are and where we belong in the great scheme of things. On the other side are those who insist that traditional views about human nature and the natural world provide us with the only reliable standards for understanding ourselves, grounding certain fundamental moral principles, and guiding us in our increasingly complex interactions with the natural world. Our view is that neither of these extreme positions is plausible. We applaud the role that science and technology have played in overturning a range of racist, sexist, and generally oppressive theories and attitudes and defend the need to continue such efforts. Such a critical attitude in the search for firm foundations is an important constitutive good for any human life. No religious or philosophical tradition and no contemporary ideology or movement is exempt from challenge. On the other hand, none must be excluded from the conversation concerning what human life is or what it can or should be.

For reasons that we have explored in earlier sections of this chapter, we believe that a serious conversation between science and a broad range of religious and philosophical traditions as well as contemporary ideologies and movements is the most sensible and productive starting point for genuine understanding. Some traditional views or at least versions of them may prove to offer reliable and important insights about the role that human nature and the natural world can play in ethical deliberation. There is no *a priori* reason to doubt this or to rule out the possibility of building new compelling and significant theories about human nature and the natural world that can fulfill such a role. Contemporary science offers very strong evidence for rejecting the view that human beings come into the world as *tabula rasa* or that they are infinitely plastic and able to be shaped and directed toward any conceivable end. As we noted earlier, evolutionary biology offers strong evidence that substantial features of human nature are important for ethical deliberation and that at least some of these involve our ethical relationship with the greater natural world. Of course, given the view that we have sketched above, the effort to define human nature will remain ongoing and open-ended. As we understand ourselves and the world better, our theories, attitudes, and values will alter and afford us a moving benchmark for ethical deliberation. But a moving benchmark is still a guide and standard.

And so, we conclude, first, that the best account of the ethical role and status of nature will draw upon a variety of ethical theories and that, like the natural sciences, our search for answers should remain open to new as well as familiar theories. Second, any ethical theory that aspires to philosophical plausibility must be consistent with well-established scientific views. Third, we must recognize and come to appreciate that our sense of the value of nature is inextricably intertwined with larger, meaningful interpretations of the world, which help to define different traditions and ways of life. It is at least plausible that several of these will satisfy our first two criteria and yet present distinctive and mutually irreducible views about the value of nature. This does not represent the threat of chaos but rather the chance to expand and enrich our understanding of ourselves, our world, and what is worth valuing in each.

Bibliography

Agarwal, Bina (1998). "Environmental Management, Equity, and Ecofeminism: Debating India's Experience," *Journal of Peasant Studies* 25(4), 55–95.
Appleman, Philip, (ed.) (2001). *Darwin: A Norton Critical Edition*, 3rd ed. New York: W. W. Norton.
Aristotle (1941). "Metaphysics," in Richard McKeon (ed.), *The Basic Works of Aristotle*. New York: Random House, 689–934.
Aristotle (1985). *Nicomachean Ethics*, Terence Irwin (trans.). Indianapolis, IN: Hackett.
Arnhart, Larry (1998). *Darwinian Natural Right: The Biological Ethics of Human Nature*. Albany, NY: SUNY.
Austin, John (1961). *Philosophical Papers*. Oxford: Oxford University Press.
Bacon, Francis (2000). *New Organon*. Cambridge: Cambridge University Press.
Barkow, Jerome H., Leda Cosmides, and John Tooby (1992). *The Adapted Mind*. New York/Oxford: Oxford University Press.
Barnes, Barry, and David Edge (eds.) (1982). *Science in Context*. Cambridge, MA: MIT Press.
Beauvoir, Simone de (1953). *The Second Sex*, H. M. Parshley (ed. and trans.) New York: Vintage.
Bergson, Henri (1907). *L'Evolution Creatrice*. Geneva, Switzerland: Skira.
Biology and Gender Study Group (1988). "The Importance of Feminist Critique for Contemporary Cell Biology," *Hypatia* 3(1), 61–76.
Bleier, Ruth (1984). *Science and Gender*. Elmsford, NY: Pergamon.
Boyd, Richard (1984). "The Current Status of Scientific Realism," in Jarrettt Leplin (ed.), *Scientific Realism*. Berkeley, CA: University of California Press, pp. 41–82.
Boyle, Robert (1664). *Some Considerations Touching the Usefulness of Experimental Natural Philosophy*. Oxford: H. Hall.
Boyle, Robert (1996). *A Free Inquiry into the Vulgarly Received Notion of Nature* (orig. 1686). Cambridge: Cambridge University Press.
Brandon, Robert (1981). "Biological Teleology: Questions and Explanations," *Studies in History and Philosophy of Science* 12, 91–105.
Bray, Michel (1988). *Reasoning with the Infinite: From the Closed World to the Mathematical Universe*. Chicago, IL: University of Chicago Press.
Brown, Theodore (1941). *The Mechanical Philosophy and the 'Animal Oeconomy.'* New York: Arno.
Buchanan, Allen, Dan Brock, Norman Daniels, and Daniel Wikler (2000). *From Chance to Choice: Genetics and Justice*. Cambridge: Cambridge University Press.
Buffon (1954). *Œuvres philosophiques*. Paris: Presses Universitaires de France.
Buffon (1988). *Des Epoques de la nature*, in Jacques Roger (ed.), reprint. Paris: Museum d'histoire naturelle Press.
Burtt, Edwin A. (1949). *Metaphysical Foundations of Modern Physical Science*. London: Routledge & Keegan Paul.
Callicott, J. Baird (1982). "Hume's *Is/Ought* Dichotomy and the Relation of Ecology to Leopold's Land Ethic," *Environmental Ethics* 4, 163–174.

Callicott, J. Baird (1994). *Earth's Insights: A Multicultural Survey of Ecological Ethics from the Mediterranean Basin to the Australian Outback.* Berkeley, CA: University of California Press.
Callicott, J. Baird, and Roger T. Ames (eds.) (1989). *Nature in Asian Traditions of Thought: Essays in Environmental Philosophy.* Albany, NY: SUNY.
Carnap, Rudolf (1928). *Die Logische Aufbau der Welt.* Berlin: Weltkreis.
Carnap, Rudolf (1956). *Meaning and Necessity.* Chicago, IL: University of Chicago Press.
Cartwright, Nancy (1983). *How the Laws of Physics Lie.* New York: Oxford University Press.
Cartwright, Nancy (1999). *The Dappled World.* Cambridge: Cambridge University Press.
Chamber, Ephraim (1728). *Cyclopaedia: Or an Universal Dictionary of Arts and Sciences* (first edition). London J. and J. Knapton.
Chambers, Robert (1994). *Vestiges of the Natural History of Creation and Other Evolutionary Writings,* facsimile of the first (1844) edition, in James A. Secord (ed.). Chicago, IL: University Chicago Press.
Chardin, Pierre Teilhard de (1956). *La Phénomène Humaine.* Completed 1926. Paris: Editions du Seuil.
Churchland, Patricia (1986). *Neurophilosophy.* Cambridge: MIT Press.
Cohen, H. Floris (1994). *The Scientific Revolution: A Historiographical Inquiry.* Chicago, IL: University of Chicago Press.
Comte, Auguste (1998). *Cours de philosophie positive,* new ed. Paris: Hermann.
Cranor, Carl, (ed.) (1994). *Are Genes Us?* New Brunswick, NJ: Rutgers University Press.
Creel, Richard E. (1995). "The 'Is-Ought' Controversy," in Stanley Tweyman (ed.), *David Hume: Critical Assessments, Vol. IV: Ethics, Passions, Sympathy, "Is" and "Ought".* London/New York: Routledge, pp. 517–529.
Crick, Francis (1970). "The Central Dogma of Molecular Biology," *Nature* 227 (8 August), 461–563.
Cskiszentmihalyi, Mark, and Philip J. Ivanhoe (eds.) (1998). *Religious and Philosophical Aspects of the Laozi.* New York: SUNY.
Cunningham, Andrew, and Perry Williams (1993). "De-centring the 'Big Picture': *The Origins of Modern Science* and the Modern Origins of Science," *British Journal of the History of Science* 26, 407–432.
Daly, Mary (1978). *Gyn/Ecology: The Metaethics of Radical Feminism.* Boston, MA: Beacon.
Darden, Lindley, and Nancy Maull (1977). "Interfield Theories," *Philosophy of Science* 41, 43–64.
Darwin, Charles (1988). *The Correspondence of Charles Darwin,* Vol. 4, Frederick Burkhardt et al. (eds.). Cambridge: Cambridge University Press.
Daston, Lorraine (1988). *Classical Probability in the Enlightenment.* Princeton, NJ: Princeton University Press.
Dawkins, Richard (1989). *The Selfish Gene,* new ed. New York: Oxford University Press.
Dear, Peter (ed.) (1995). *The Scientific Enterprise in Early Modern Europe.* Chicago, IL: University of Chicago Press.
Deason, Gary (1986). "Reformation Theology and the Mechanistic Conception of Nature," in David C. Lindberg and Ronald L. Numbers (eds.), *God and Nature: Historical Essays on the Encounter Between Christianity and Science.* Berkeley, CA: University of California Press, pp. 167–191.
Descartes, René (1971). *Descartes Dictionary,* John Morris (trans.). New York: Philosophical Library.
Descartes, René (1985). *Philosophical Writings,* 3 Vols. Cambridge: Cambridge University Press.
Dewey, John (1925). *Experience and Nature.* Chicago, IL: University of Chicago Press.
Diderot, Denis (1964). "D'Alembert's Dream," in Jacques Barzun and Ralph H. Bowen (trans.), *Rameau's Nephew and Other Works.* Indianapolis, IN: Bobbs-Merrill.
Diderot, Denis (1998). "Pensées philosophiques", in P. Vernière (ed.), *Oeuvres philosophiques,* Paris: Garnier.
Dijksterhuis, E. J. (1961). *The Mechanization of the World Picture.* Oxford: Clarendon.
Doeser, Marinus (1986). "Can the Dichotomy of Fact and Value Be Maintained?," in Doeser and John Kraay (eds.), *Facts and Values: Philosophical Reflections from Western and Non-Western Perspectives.* Dordrecht, The Netherlands: Kluwer, pp. 1–20.
Driesch, Hans (1908). *The Science and Philosophy of the Organism.* London: Black.

Dubos, René (1961). *The Dreams of Reason: Science and Metaphysics.* New York: Columbia University Press.
Dupré, John (1993). *The Disorder of Things.* Cambridge, MA: Harvard University Press.
Earman, John, and John Roberts (1999). "Ceteris Paribus. There Are No Provisos," *Synthese* 118, 439–478.
Earman, John, John Roberts, and Sheldon Smith (2002). "*Ceteris Paribus* Lost," *Erkenntnis* 57, 281–301.
Eccles, John (1970). *Facing Reality: Philosophical Adventures by a Brain Scientist.* New York: Springer.
Edelman, Gerald (1987). *Neural Darwinism.* New York: Basic Books.
Ehrard, Jean (1994). *L'Idée de Nature en France dans la première moitié du xviiie siècle*, new ed. Paris: Michel.
Emerson, Ralph W. (1903). "Nature," (orig. 1841), in *The Complete Works of Ralph Waldo Emerson*, Vol. III. Boston, MA: Houghton Mifflin, pp. 169–96.
Ereshevsky, Marc (1998). "Species Pluralism and Anti-Realism," *Philosophy of Science* 65, 103–120.
Farber, Paul L. (1994). *The Temptation of Evolutionary Ethics.* Berkeley, CA: University of California Press.
Fausto-Sterling, Anne (1985). *Myths of Gender.* New York: Basic Books.
Favretti, Rema Rossini, Giorgio Sandri, and Roberto Scazzieri., (eds.) (1999). *Incommensurability and Translation: Kuhnian Perspectives on Scientific Communication and Theory Change.* Northampton, MA: Edward Elgar.
Feyerabend, Paul (1962). "Explanation, Reductionism and Empiricism," in Herbert Feigl and Grover Maxwell (eds.), *Scientific Explanation, Space and Time. Minnesota Studies in Philosophy of Science*, Vol. III. Minneapolis, MN: University of Minnesota Press.
Fine, Arthur (1984). "The Natural Ontological Attitude," in Jarrett Leplin (ed.), *Scientific Realism.* Berkeley, CA: University of California Press, pp. 83–107.
Foucault, Michel (1966). *Les mots et les choses.* Paris: Gallimard.
Fraassen, Bas van (1980). *The Scientific Image.* Oxford: Oxford University Press.
Frank, Philip (1949). *Modern Science and Its Philosophy.* Cambridge, MA: Harvard University Press.
Frankena, William (1939). "The Naturalistic Fallacy," *Mind* (N.S.) 48, 464–477.
Frankena, William (1973). *Ethics*, 2nd ed. Englewood Cliffs, NJ: Prentice-Hall.
Friedman, Michael (1999). *Reconsidering Logical Positivism.* Cambridge: Cambridge University Press.
Giere, Ronald, and Alan Richardson, (eds) (1996). *Origins of Logical Empiricism, Minnesota Studies in the Philosophy of Science*, Vol. XVI. Minneapolis, MN: University of Minnesota Press.
Gilligan, Carol (1993). *In a Different Voice: Psychological Theory and Woman's Development*, reprint. Cambridge, MA: Harvard University Press.
Gillispie, Charles (1960). *The Edge of Objectivity: An Essay in the History of Scientific Ideas.* Princeton, NJ: Princeton University Press, 1960.
Goldmann, Lucien (1964). *The Hidden God.* London: Routledge & Kegan Paul.
Graham, A. C. (trans.) (2001). *Chuang-tzu: The Inner Chapters.* Indianapolis, IN: Hackett. ("Chuang-tzu" is another romanization of "Zhuangzi.")
Grant, Edward (1978). "Aristotelianism and the Longevity of the Medieval World View," *History of Science* 16, 93–106.
Griffin, Susan (1978). *Women and Nature.* San Francisco, CA: Harper & Row.
Gruen, Lori, and Dale Jamieson, (eds). (1994). *Reflecting on Nature: Readings in Environmental Philosophy.* New York: Oxford University Press.
Gutting, Gary, (ed.) (1980). *Paradigms and Revolutions: Appraisals and Applications of Thomas Kuhn's Philosophy of Science.* South Bend, IN: University of Notre Dame Press.
Hacking, Ian (1975). *The Emergence of Probability.* London/New York: Cambridge University Press.
Hacking, Ian (1983). *Representing and Intervening: Introductory Topics in the Philosophy of Natural Science.* Cambridge: Cambridge University Press.
Hall, Mary Boas (1941). *The Mechanical Philosophy.* New York: Arno.

Hancock, Roger N. (1974). *Twentieth Century Ethics*. New York: Columbia University Press.
Hanson, Norwood Russell (1958). *Patterns of Discovery*. Cambridge: Cambridge University Press.
Haraway, Donna (1978). "Animal Sociology and the Body Politic," *Signs: Journal of Women in Culture and Society* 4, 21–60.
Haraway, Donna (1991). "A Manifesto for Cyborgs: Science, Technology, and Socialist Feminist for the 1980s (orig. 1985), *Socialist Review* 8 (1985), 65–108; in Haraway (ed.), *Simians, Cyborgs and Women*, reprint. New York: Routledge, 149–182.
Harding, Sandra (1986). *The Science Question in Feminism*. Ithaca, NY/London: Cornell University Press.
Hare, Richard M. (1952). *The Language of Morals*. Oxford: Oxford University Press.
Hazard, Paul (1953). *The European Mind: The Critical Years 1680–1715*. New Haven, CT: Yale University Press.
Heidelberger, Michael (online). "Naturphilosophie," in von Edward Craig (ed.), *Routledge Encyclopedia of Philosophy*, available at http://www.rep.routledge.com.
Held, Virginia (1996). "Whose Agenda? Ethics Versus Cognitive Science," in Larry May, Marilyn Friedman, and Andy Clark (eds.), *Mind and Morals*. Cambridge, MA: MIT Press, pp. 69–86.
Helden, Albert van (1997). "The Telescope in the Seventeenth Century," in Peter Dear (ed.), *The Scientific Enterprise in Early Modern Europe*. Chicago, IL: University of Chicago Press, pp. 133–153.
Hempel, Carl G. (1965). *Aspects of Scientific Explanation*. New York: Free Press.
Henry, John (1986). "Occult Qualities and the Experimental Philosophy: Active Principles in Pre-Newtonian Matter Theory," *History of Science* 24, 335–381.
Hill, Thomas E., Jr. (1994). "Ideals of Human Excellence and Preserving the Natural Environment," in Lori Gruen and Dale Jamieson (eds.), *Reflecting on Nature: Readings in Environmental Philosophy*. New York: Oxford University Press, pp. 98–110.
Hitt, Jack (2003). "A Gospel According to the Earth," *Harpers* 307 (July), 41–55.
Horwich, Paul (ed.) (1993). *World Changes: Thomas Kuhn and the Nature of Science*. Cambridge, MA: MIT Press.
Hösle, Vittorio, and Christian Illies (eds.) (2005). *Darwinism and Philosophy*. Notre Dame, IN: University of Notre Dame Press.
Hubbard, Ruth, Mary Sue Henifin, and Marian Lowe (eds.) (1979). *Women Look at Biology Looking at Women*. Cambridge, MA: Schenkman.
Hudson, W. D. (ed.) (1969). *The Is/Ought Question*. London: Macmillan.
Hull, David (1972). "Reduction in Genetics – Biology or Philosophy?," *Philosophy of Science* 39, 491–499.
Hume, David (1888). *Treatise of Human Nature*. Oxford: Clarendon.
Hume, David (1947). *Dialogues Concerning Natural Religion*. Indianapolis, IN: Bobbs-Merrill.
Hume, David (1977). *Enquiry Concerning Human Understanding*. Indianapolis, IN: Hackett.
Hursthouse, Rosalind (1999). *On Virtue Ethics*. Oxford: Oxford University Press.
Hutchison, Keith (1983). "Supernaturalism and the Mechanical Philosophy," *History of Science* 21, 297–333.
Hutchison, Keith (1997). "What Happened to Occult Qualities in the Scientific Revolution?," in Peter Dear (ed.), *The Scientific Enterprise in Early Modern Europe*. Chicago, IL/London: University of Chicago Press, 86–106.
Huxley, Thomas (1894). *Evolution and Ethics*. London: Macmillan.
Ivanhoe, Philip J. (1991). "A Happy Symmetry: Xunzi's Ethical Thought," *Journal of the American Academy of Religion* 59(2) (Summer), 309–322.
Ivanhoe, Philip J. (trans.) (2003). *The Laozi or Daodejing*. Indianapolis, IN: Hackett.
Jackson, Frank (1974). "Defining the Autonomy of Ethics," *Philosophical Review* 83, 88–96.
Jaggar, Alison M. (1983). *Feminist Politics and Human Nature*. Totowa, NJ: Rowman & Allanheld.

Jardine, Nicholas (1996). "*Naturphilosophie* and the Kingdoms of Nature," in Nicholas Jardine, James A. Secord, and Emma Spary (eds.), *Cultures of Natural History*. Cambridge: Cambridge University Press, pp. 230–245.

Jardine, Nicholas (1999). "Inner History; or, How to End Enlightenment," in William Clark, Jan Golinski, and Simon Schaffer (eds.), *The Sciences in Enlightened Europe*. Chicago, IL: University of Chicago Press, pp. 477–494.

Kant, Immanuel (1784). "An Answer to the Question, What is Enlightenment?" (1784), in Hans Reiss (ed.), *Kant: Political Writings*. Cambridge; Cambridge University Press, pp. 54–60.

Kant, Immanuel (1786). "Metaphysical Foundations of Natural Science", in Immanuel Kant, *Philosophy of Material Nature*, James Ellington (trans.). Indianapolis, IN: Hackett, Part 2.

Kant, Immanuel (1956). *Critique of Practical Reason*, Lewis W. Beck (trans.). Indianapolis, IN: Bobbs-Merrill.

Kant, Immanuel (1964). *Groundwork of the Metaphysics of Morals*, H. J. Paton (trans.). New York: Harper & Row.

Kant, Immanuel (1965). *Critique of Pure Reason*, Norman Kemp Smith (trans.). New York: St. Martin's.

Kauffman, Stuart (1993). *The Origins of Order*. Oxford: Oxford University Press.

Keller, Evelyn F. (1983a). "The Force of the Pacemaker Concept in Theories of Cellular Slime Mold Aggregation," *Perspectives in Biology and Medicine* 26, 515–521.

Keller, Evelyn F. (1983b). *A Feeling for the Organism*. New York: Freeman.

Keller, Evelyn F. (1985). *Reflections on Gender and Science*. New Haven, CT: Yale University Press.

Keller, Evelyn F. (1994). "Master Molecules," in Carl Cranor (ed.), *Are Genes Us?* New Brunswick, NJ: Rutgers University Press, pp. 89–98.

Keller, Evelyn F. (2000). *The Century of the Gene*. Cambridge, MA: Harvard University Press.

Kellert, Stephen R., and Edward O. Wilson (eds.) (1993). *The Biophilia Hypothesis*. Washington, DC: Island Press.

Kitcher, Philip (1984)."1953 and All That: A Tale of Two Sciences," *Philosophical Review* 93, 335–73.

Kitcher, Philip (1992). "The Naturalists Return," *Philosophical Review* 101, 53–114.

Kjellberg, Paul, and Philip J. Ivanhoe (1996). *Essays on Skepticism, Relativism and Ethics in the Zhuangzi*. Albany, NY: SUNY.

Kline III, T. C., and Philip J. Ivanhoe (eds.) (2000). *Virtue, Nature, and Moral Agency in the Xunzi*. Indianapolis, IN: Hackett.

Knoblock, John (trans.) (1988–1994). *Xunzi: A Translation and Study of the Complete Works*, Vols. 1–3. Stanford, CA: Stanford University Press.

Koertge, Noreta (ed.) (1998). *A House Built on Sand*. New York/Oxford: Oxford University Press.

Korsgaard, Christine (1996). *The Sources of Normativity*. Cambridge/New York: Cambridge University Press.

Koselleck, Reinhard (1985). *Futures Past*. Cambridge, MA: MIT Press.

Koyre, Alexandre (1957). *From the Closed World to the Infinite Universe*. Baltimore, MD: Johns Hopkins University Press.

Koyré, Alexandre (1965). *Newtonian Studies*. Cambridge MA: Harvard University Press.

Koyré, Alexandre (1978). *Galileo Studies*. Atlantic Highlands, NJ: Humanities.

Kuhn, Thomas (1970). *Structure of Scientific Revolutions*, 2nd ed. Chicago, IL: University of Chicago Press.

Lamarck, Jean B. (1809). *Philosophie zoologique*. Paris: Chez Dantu et l'Auteur.

Lamarck, Jean B. (1991). "Nature," in Julien-Joseph Virey (ed.), *Nouveau dictionnaire d'histoire naturelle*, 2nd ed. (1816–1819), in Jacques Roger and Goulven Laurent (eds), *Lamarck: Articles d'histoire naturelle*, reprint. Paris: Belin, pp. 307–308.

Laslett, Peter (1971). *The World We Have Lost*, 2nd ed. London: Methuen.

Lau, D. C. (trans.) (1970). *Mencius*. Harmondsworth, London: Penguin ("Mencius" is another romanization of "Mengzi.").

Lau, D. C. (trans.) (1979). *The Analects*. New York: Dorset (The most reliable source of the teachings of Kongzi or "Confucius").
Laudan, Larry (1966). "The Clock Metaphor and Probabilism: The Impact of Descartes on English Methodological Thought, 1650–65," *Annals of Science* 22, 73–105.
Laudan, Larry (1977). *Progress and Its Problems*. Berkeley, CA: University of California Press.
Leibniz, Gottfried W., and Samuel Clarke (2000). *Correspondence*. Indianapolis, IN: Hackett.
Leiss, William (1972). *The Domination of Nature*. New York: Braziller.
Leopold, Aldo (1970). *A Sand County Almanac*, reprint. New York: Ballantine Books.
Leplin, Jarrett (ed.) (1984). *Scientific Realism*. Berkeley, CA: University of California Press.
Lewis, C. S. (1996). *The Abolition of Man*. New York: Simon & Schuster.
Lindberg, David C., and Robert S. Westman (eds.) (1990). *Reappraisals of the Scientific Revolution*. Cambridge: Cambridge University Press.
Liu, Xiusheng, and Philip J. Ivanhoe (eds.) (2002). *Essays on Mengzi's Moral Philosophy*. Indianapolis, IN: Hackett.
Lloyd, Elisabeth (1994). "Normality and Variation," in Cranor (ed.), *Are Genes Us?*, pp. 99–112.
Locke, John (1975). *Essay Concerning Human Understanding*. Oxford: Clarendon.
Lodge, David, and Christopher Hamlin (eds.) (2006). *Religion and the New Ecology: Environmental Responsibility in a World of Flux*. Notre Dame, IN: University of Notre Dame Press.
Loeb, Jacques (1912). *The Mechanistic Conception of Life*. Chicago, IL: University of Chicago Press.
Longino, Helen E. (1990). *Science as Social Knowledge*. Princeton, NJ: Princeton University Press.
Longino, Helen E. (2002). *The Fate of Knowledge*. Princeton, NJ: Princeton University Press.
Lovejoy, Arthur (1927). "'Nature' as Aesthetic Norm," *Modern Language Notes*, pp. 444–450.
Lovejoy, Arthur (1936). *The Great Chain of Being*. Cambridge, MA: Harvard University Press.
Lovelock, James E. (1979). *Gaia: A New Look at Life on Earth*. Oxford: Oxford University Press.
Lovelock, James E. (1995). *The Ages of Gaia: A Biography of Our Living Earth*, reprint. New York: W. W. Norton.
Machamer, Peter, Lindley Darden, and Carl Craver (2000). "Thinking about Mechanisms," *Philosophy of Science* 67, 1–25.
MacIntyre, Alasdair (1969). "Hume on 'Is' and 'Ought,'" in W. D. Hudson (ed.), *The Is/Ought Question*. London: Macmillan, pp. 35–50.
MacIntyre, Alasdair (1984). *After Virtue*. Notre Dame, IN: University of Notre Dame Press.
MacIntyre, Alasdair (1990). *Three Rival Versions of Moral Inquiry*. Notre Dame, IN: University of Notre Dame Press.
Maienschein, Jane, and Michael Ruse (eds.) (1999). *Biology and the Foundation of Ethics*. Cambridge: Cambridge University Press.
Mancosu, Paolo (1996). *Philosophy of Mathematics and Mathematical Practice in the Seventeenth Century*. New York: Oxford University Press.
Mason, Richard (1997). *The God of Spinoza*. Cambridge: Cambridge University Press.
McGuire, J. E. (1972). "Boyle's Conception of Nature," *Journal of the History of Ideas* 33, 523–542.
Menand, Louis (ed.) (1997). *Pragmatism: A Reader*. New York: Random House.
Merchant, Carolyn (1980). *The Death of Nature*. San Francisco, CA: Harper & Row.
Meyer, James (2001). *Political Nature: Environmentalism and the Interpretation of Western Thought*. Cambridge: MIT Press.
Midgley, Mary (1995). *Beast and Man: The Roots of Human Nature*, rev. ed. London: Routledge.
Midgley, Mary (2000). "Biotechnology and Monstrosity: Why We Should Pay Attention to the 'Yuk' Factor," *Hastings Center Report* 30, 7–15.
Midgley, Mary (2003). "Criticizing the Cosmos," in Willem B. Drees (ed.), *Is Nature Ever Evil?* London/New York: Routledge.
Millgram, Elijah (1995). "Was Hume a Humean?," *Hume Studies* 21, 75–93.

Millgram, Elijah (1997). "Hume on Practical Reasoning," *Iyyun: The Jerusalem Philosophical Quarterly* 46, 235–265.
Mirowski, Philip, and Esther-Mirjam Sent (eds.) (2003). *Science Bought and Sold*. Chicago, IL: University of Chicago Press.
Mitchell, Sandra (1995). "Function, Fitness, and Disposition," *Biology and Philosophy* 10, 39–54.
Mitchell, Sandra (2004). *Biological Complexity and Integrative Pluralism*. Cambridge: Cambridge University Press.
Moore, George E. (1968). *Principia Ethica*, reprint. Cambridge: Cambridge University Press.
Naess, Arne (1973). "The Shallow and the Deep, Long-Range Ecology Movement. A Summary," *Inquiry* 16, 95–100.
Naess, Arne (1989). *Ecology, Community and Lifestyles: Outlines of an Ecosophy*, David Rothenberg (trans. and ed.). New York: Cambridge University Press.
Nagel, Ernest (1961). *The Structure of Science*. New York: Harcourt Brace & World.
Nagel, Thomas (1986). *The View from Nowhere*. New York: Oxford University Press.
Nelson, Jack (1996). "The Last Dogma of Empiricism?," in Lynn Hankinson Nelson and Jack Nelson (eds.), *Feminism, Science, and the Philosophy of Science*. Dordrecht, The Netherlands: Kluwer, pp. 59–78.
Newman, William (1998). "Alchemy, Domination, and Gender," in Noreta Koertge (ed.), *A House Built on Sand*. New York/Oxford: Oxford University Press, pp. 216–225.
Newton, Isaac (1952). *Opiticks*. New York: Dover.
Newton, Isaac (1953). *Newton's Philosophy of Nature: Selections from his Writings*, H. Standish Thayer (ed.). New York: Hafner.
Nickles, Thomas (ed.) (2002). *Thomas Kuhn*. Cambridge: Cambridge University Press.
Nobis, Heribert (1967). "Frühneuzeitliche Verständnisweisen der Natur und ihr Wandel bis zum 18. Jahrhundert," *Archiv für Begriffsgeschichte* 11, 37–58.
Noddings, Nel (1984). *Caring: A Feminist Approach to Ethics and Moral Education*. Berkeley, VA: University of California Press.
Nussbaum, Martha C. (1993). "Non-Relative Virtues: An Aristotelian Approach," in Amartya Sen and Martha C. Nussbaum (eds.), *The Quality of Life*. New York: Oxford University Press, pp. 242–269.
Okin, Susan Moller (1989). *Justice, Gender, and the Family*. New York: Basic Books.
Oppenheim, Paul, and Hilary Putnam (1968). "Unity of Science as a Working Hypothesis," in Herbert Feigl, Michael Scriven, and Grover Maxwell (eds.), *Concepts, Theories and the Mind-Body Problem. Minnesota Studies in the Philosophy of Science*, Vol. II. Minneapolis, MN: University of Minnesota Press, pp. 3–36.
Osler, Margaret (1994). *Divine Will and the Mechanical Philosophy: Gassendi and Descartes on Contingency and Necessity in the Created World*. Cambridge/New York: Cambridge University Press.
Oyama, Susan (2000). *The Ontogeny of Information*, 2nd ed. Durham, NC: Duke University Press.
Pascal, Blaise (1941). *Pensées and Provincial Letters*. New York: Random House.
Pauly, Philip (1987). *Controlling Life: Jacques Loeb and the Engineering Ideal in Biology*. New York: Oxford University Press.
Peckham, Morse (2006). *The Origin of Species: A Variorum Text*. Philadelphia, PA: University of Pennsylvania.
Pidgen, Charles (1989). "Logic and the Autonomy of Ethics," *Australasian Journal of Philosophy* 67, 127–151.
Pope, Alexander (1994). *Essay on Man*. New York: Dover.
Porter, Roy (1986). "The Scientific Revolution: A Spoke in the Wheel?," in Roy Porter and Mikulas Teich (eds.), *Revolution in History*. Cambridge: Cambridge University Press, pp. 290–316.
Prior, Arthur N. (1960). "The Autonomy of Ethics," *Australasian Journal of Philosophy* 38, 199–206.

Putnam, Hilary (2002). *The Collapse of the Fact/Value Dichotomy and Other Essay.* Cambridge: Harvard University Press.
Quine, Willard van Orman (1980). "Two Dogmas of Empiricism" (orig. 1950), in Quine, *From a Logical Point of View*, 2nd, rev. ed., reprint. Cambridge: Harvard University Press, pp. 20–46.
Rees, Abraham (1819). *Cyclopedia, or Universal Dictionary of Arts, Sciences, and Literature.* Philadelphia, PA: Bradford.
Reichenbach, Hans (1938). *Experience and Prediction.* Chicago, IL: University of Chicago Press.
Reill, Peter Hanns (2005). *Vitalizing Nature in the Enlightenment.* Berkeley, CA: University of California Press.
Reiss, Hans (ed.) (1970). *Kant: Political Writings.* Cambridge: Cambridge University Press.
Regan, Thomas (1983). *The Case for Animal Rights.* Berkeley, CA: University of California Press.
Restivo, Sal (1978). "Parallels and Paradoxes in Modern Physics and Eastern Mysticism: I – A Critical Reconnaissance," *Social Studies of Science* 8, 143–181.
Restivo, Sal (1982). "Parallels and Paradoxes in Modern Physics and Eastern Mysticism: II – A Sociological Perspective on Parallelism," *Social Studies of Science* 12, 37–71.
Richards, Robert (2002). *The Romantic Conception of Life.* Chicago, IL: University of Chicago Press.
Roger, Jacques (1997). *Les Sciences de la vie dans la pensée française du xviiie siècle*, 3rd ed. (Paris: Michel, 1997), partially translated as *The Life Sciences in Eighteenth-Century French Thought*, R. Ellrich (trans.) Stanford: Stanford University Press.
Rorty, Richard (1991). "Solidarity or Objectivity?," in Richard Rorty (ed.), *Objectivity, Relativism, and Truth: Philosphical Papers*, Vol. 1. Cambridge: Cambridge University Press, pp. 21–34.
Rosenberg, Alexander (1985). *The Structure of Biological Science.* Cambridge: Cambridge University Press.
Rosenberg, Alexander (2000). *Darwinism in Philosophy, Social Science, and Policy.* Cambridge: Cambridge University Press.
Rossi, Paolo (1968). *Francis Bacon: From Magic to Science.* London: Routledge & Kegan Paul.
Schaffner, Kenneth (1967). "Approaches to Reduction," *Philosophy of Science* 34, 137–147.
Schelling, Friedrich Wilhelm J. (1867). "Introduction to the Outlines of a System of Natural Philosophy," T. Davidson (trans.), *Journal of Speculative Philosophy* 1, 194–220.
Schelling, Friedrich Wilhelm J. (2004). *First Outline to a System of the Philosophy of Nature*, K. R. Peterson (trans.). Albany, NY: State University of New York.
Schlick, Moritz (1936). "Meaning and Verification," *Philosophical Review* 45, 339–369.
Schurz, Gerhard (1997). *The Is-Ought Problem.* Dordrecht, The Netherlands: Kluwer.
Secord, James A. (2000). *Victorian Sensation: The Extraordinary Publication, Reception, and Secret Authorship of Vestiges of the Natural History of Creation.* Chicago, IL: University of Chicago Press.
Segal, Lee, and Evelyn F. Keller (1970). "Slime Mold Aggregation Viewed as an Instability," *Journal of Theoretical Biology* 26, 399–415.
Selby-Bigge, Lewis A. (ed.) (1897). *British Moralists.* Indianapolis, IN: Bobbs-Merrill.
Shapin, Steven, and Simon Schaffer (1985). *Leviathan and the Air Pump.* Princeton, NJ: Princeton University Press.
Silver, Lee M. (1997). *Remaking Eden.* New York: Avon.
Silver, Rae (1992). "Environmental Factors Influencing Hormone Secretion," in Jill Becker S. Marc Breedlove, and David Crews (eds.), *Behavioral Endocrinology.* Cambridge, MA: MIT Press, pp. 401–422.
Singer, Peter (1979). *Practical Ethics.* New York: Cambridge University Press.
Sklar, Lawrence (2003). "Dappled Theories in a Uniform World," *Philosophy of Science* 70, 424–441.
Sloan, Phillip (1999). "From Natural Law to Evolutionary Ethics in Enlightenment French Natural History," in Maienschein, and Ruse (eds.), *Biology and the Foundations of Ethics*, pp. 52–83.
Sloan, Phillip (2001). "'The Sense of Sublimity': Darwin on Nature and Divinity," *Osiris* 16, 251–69.

Sloan, Phillip (2005). "'It Might Be Called Reverence,'" in V. Hösle and C. Illies (eds.), *Darwinism and Philosophy*. Notre Dame, IN: University of Notre Dame Press, pp. 143–165.
Sloan, Phillip (2007). "Kant and British Bioscience," in P. Hunemmann (ed.), *Understanding Purpose: Kant and the Philosophy of Biology*. Rochester, NY: University of Rochester Press, pp. 149–171.
Sober, Elliott, and David Sloan Wilson (1998). *Unto Others: The Evolution and Psychology of Unselfish Behavior*. Cambridge: Cambridge University Press.
Soble, Alan (1998). "In Defense of Bacon," in Noreta Koertge (ed.), *A House Built on Sand*, New York/Oxford: Oxford University Press, pp. 195–215.
Spaemann, Robert (1967). "Genetisches zum Naturbegriff des 18. Jahrhunderts," *Archiv für Begriffsgeschichte* 11, 59–74.
Spencer, Herbert (1896). *First Principles*, 4th ed. New York: Appleton.
Stauffer, Robert (ed.) (1975). *Charles Darwin's Natural Selection*. Cambridge: Cambridge University Press.
Stedman-Jones, Sue (1998). "Fact/Value," in Chris Jenks (ed.), *Core Sociological Dichotomies*. London/Thousand Oaks, CA/New Delhi: Sage, pp. 49–62.
Tavris, Carol (1992). *The Mismeasure of Women*. New York: Simon & Schuster.
Taylor, Charles (1989). *Sources of the Self*. Cambridge, MA: Harvard University Press.
Teller, Paul (2004). "How We Dapple the World," *Philosophy of Science* 71, 730–741.
Thomas, Keith (1971). *Religion and the Decline of Magic*. New York: Scribner.
Thomas, Keith (1996). *Man and the Natural World*. New York: Oxford University Press.
Thoreau, Henry David (1992). *Walden* and *Resistance to Civil Government*, William Rossi (ed.), 2nd ed. New York: W. W. Norton.
Tuana, Nancy (1990). "Re-fusing Nature-Nurture," in Azizah Y al- Hibri and Margaret Simons, (eds.), *Hypatia Reborn*. Bloomington, IN: Indiana University Press, pp. 70–89.
Tucker, Mary Evelyn, and John Berthrong (eds.) (1998). *Confucianism and Ecology: The Interrelation of Heaven, Earth, and Humans*. Cambridge, MA: Harvard University Press.
Warnock, G. J. (1967). *Contemporary Moral Philosophy*. London: Macmillan.
Warren, Anthony (1985). "Beyond Intrinsic Value: Pragmatism in Environmental Ethics," *Environmental Ethics* 7, 321–339.
Warren, Karen (1990). "The Power and Promise of Ecological Feminism," *Environmental Ethics* 12, 125–146.
Warren, Karen (ed.) (1997). *Ecofeminism: Women, Culture, Nature*. Bloomington, IN: Indiana University Press.
Warren, Karen (2000). *Ecofeminist Philosophy: A Western Perspective on What It Is and Why It Matters*. Lanham, MD: Rowman & Littlefield.
Waters, C. Kenneth (1994). "Genes made Molecular," *Philosophy of Science* 61, 163–185.
Waters, C. Kenneth, Helen Longino, and Steven Kellert (2006). "The Pluralist Stance," in Kenneth C. Waters, Helen Longino, and Steven Kellert (eds.), *Scientific Pluralism. Minnesota Studies in Philosophy of Science*, Vol. XIX. Minneapolis, MN: University of Minnesota Press.
Weber, Max (1946). "Politics as a Vocation," and "Science as a Vocation," in Kurt Wolff (ed.), *From Max Weber*. New York: Oxford University Press, 77–156.
Weinberg, Steven (2001). *Facing Up*. Cambridge, MA: Harvard University Press.
Westfall, Richard (1971). *The Construction of Modern Science*. New York: Wiley.
White, Jr., Lynn (1968). "The Historical Roots of Our Ecological Crisis," in *Machina ex Deo: Essays in the Dynamism of Western Culture*. Cambridge, MA: MIT Press.
Williams, Bernard A. O. (1978). *Descartes: The Project of Pure Enquiry*. Sussex, England: Harvester.
Williams, Bernard A. O. (1985). *Ethics and the Limits of Philosophy*. London: Fontana.
Wilson, Catherine (1995). *The Invisible World: Early Modern Philosophy and the Invention of the Microscope*. Princeton, NJ: Princeton University Press.
Wilson, Edward O. (1978). *On Human Nature*. Cambridge, MA: Harvard University Press.
Wilson, Edward O. (1980). *Sociobiology: The New Synthesis*. Cambridge, MA: Harvard University Press.

Wilson, Edward O. (1984). *Biophilia*. Cambridge: Harvard University Press.
Wilson, Edward O. (1994). *Naturalist*. Washington, DC: Island Press.
Wilson, Edward O. (1998). *Consilience*. New York: Knopf.
Wittgenstein, Ludwig (1958). *Philosophical Investigations*, G. E. M. Anscombe (trans.), 3rd ed. New York: Macmillan.
Wittgenstein, Ludwig (1999). *Tractatus logico-philosophicus*. Atlantic Highlands, NJ: Humanities, 1974.
Woolcock, Peter (1999). "The Case against Evolutionary Ethics Today," in Maienschein and Ruse (eds.), *Biology and the Foundation of Ethics*. Cambridge: Cambridge University Press, pp. 276–306.
Wright, Larry (1976). *Teleological Explanations*. Los Angeles, CA: University of California Press.
Wright, Robert (1994). *The Moral Animal*. New York: Pantheon.
Zammito, John (2002). *Kant, Herder, and the Birth of Anthropology*. Chicago, IL/London: University of Chicago Press.
Zammito, John (2004). *A Nice Derangement of Epistemes: Post-Positivism in the Study of Science from Quine to Latour*. Chicago, IL/London: University of Chicago Press.
Zihlman, Adrienne (1981). "Women as Shapers of the Human Adaptation," in Frances Dahlberg (ed.), *Woman the Gatherer*. New Haven, CT: Yale University Press.

Chapter 3
Scientific and Medical Concepts of Nature in the Modern Period in Europe and North America

Laurence B. McCullough, John Caskey, Thomas R. Cole, and Andrew Wear

3.1 Introduction: Appeals to Nature in Science and Medicine in the Modern Period

This chapter addresses scientific and medical concepts of nature in the modern period in Europe and North America, from the seventeenth century to the present. During this period, under the influence of Francis Bacon (1561–1626) and his followers, biology and medicine self-consciously became increasingly scientific, resulting in the biotechnologies that are the focus of this volume. The reach of contemporary biotechnology is global, but its origins are to be found in the science and medicine of early modern Europe and North America, the main focus of this chapter.

Appeals to nature, the natural, and the unnatural in contemporary assessments of biotechnology and policy proposals for its regulation owe an unacknowledged debt to historical concepts of nature. Our choice of the plural, *concepts*, is deliberate. The central theme of this chapter is that since the seventeenth century, there has been no single, canonical concept of nature. Instead, there has been a competition among *concepts* of nature which continues to this day.

Our goal in this chapter is to map three major conceptions of nature in the context of three case studies: food, animal, and plants as pure and impure; medical interventions to correct the deficiencies of nature; and the dynamic and ever-changing conceptions of aging.[1] We therefore do not attempt a comprehensive historical explanation of the social, cultural, and other forces that shaped concepts of nature during these centuries. We do attempt to understand how concepts of and appeals to nature were conceptualized in each of the three case studies, allowing the distinctive voices of the past to be heard, as best we can, on their own terms. On this basis we identify lessons learned from these three case studies for appeals to nature as normative. These lessons emphasize the complexities and challenges of appeals to nature as normative in assessing scientific and technological advances. Responsible management of these complexities can be accomplished by adhering to requirements,

---

[1] Andrew Wear is the author of the first case study, Laurence McCullough of the second, and John Caskey and Thomas Cole of the third.

drawn from the three case studies, for constructing historically informed appeals to nature as normative in the assessment of biotechnologies.

We want, at the outset, to be clear about what we mean by the term, 'normative'. 'Normative' can be used descriptively, to describe what is most commonly the case or most commonly observed. 'Normative' can also be used prescriptively, i.e., to set standards of judgment, decision making, and behavior. For example, descriptively speaking, heterosexuality in the human species appears to be normative: most, but not all, human beings experience a sexual orientation of attractiveness to and for human beings of the "opposite sex." Prescriptively speaking, the normativity of heterosexuality concerns whether we can reliably form the judgment – and then make decisions and implement them – that heterosexual sexual orientation is the only acceptable sexual orientation.

'Nature' is used in both ways throughout the history of various discourses about nature and its normativity. In this chapter we are mainly concerned with the prescriptive sense of 'nature': Does nature itself provide human beings with standards or benchmarks or points of reference by which authoritative judgments about the worth of things or events can be made and implemented? When nature does so, nature becomes normative. Normative judgments based on what is "natural" or "unnatural" concern what should or should not be the case or allowed to occur, what we should or should not do, and what sort of people we should want to become or not become. Normative judgments that invoke nature as the source of standards are thus moral judgments. Such judgments can be both positive and negative. When nature is positively valorized, its observed structures and functions set standards to which human judgment, decision making, and actions should conform. When nature is negatively valorized, then standards from sources other than nature are invoked to support judgments that the given order of nature should be changed.

When nature is valorized, positively or negatively, the concept of nature has become value-laden, not value-neutral. One can therefore reliably reason from nature, the natural, or the unnatural to what ought or ought not to be the case. When one invokes a value-laden concept of nature in this way, there is no "naturalistic fallacy," i.e., the supposed fallacy of deriving morally normative claims from descriptive claims of fact. The naturalistic fallacy applies only to value-neutral conceptions of nature and the natural. As we shall see in the three case studies to which we now turn, 'nature' refers to a value-laden conception. In the first case study, the origins of nature's values in God's act of creation are plain; in the second those origins have become obscured; and in the third they have become contested.

3.2 Appeals to Nature and the Natural as Normative in Contemporary Bioethics Policy Literature

Bioethics developed, in considerable part, in the context of health policy. Since the 1970s there have been a series of national commissions, usually authorized by the U.S. Congress and appointed by the President. These public bioethics bodies have

3 Scientific and Medical Concepts of Nature in the Modern Period in Europe   139

taken up a wide variety of ethical issues and issued reports and policy proposals. Some of these, especially concerning end-of-life decision making and the definition and determination of death have been very influential.

Developments in biotechnology have been a persistent focus of these public bodies. The President's Commission for the Study of Ethical Problems in Medicine and Biomedical and Behavioral Research, authorized during the Carter Administration (1977–1981) and continued into the Reagan Administration (1981–1989), issued a report on genetic engineering in 1982. Almost two decades later, the National Bioethics Advisory Commission (NBAC), authorized during the Clinton Administration (1993–2001), issued reports on human stem cell research and on cloning.

In its report on genetic engineering, *Splicing Life*, the President's Commission identified as one major area of concern what it labeled "deeper anxieties," framed in terms of the "Frankenstein factor." The Commission explained this in terms of "the notion that gene splicing might change the nature of human beings" (President's Commission, 1982, 14). The Commission went on to consider whether there were ethical objections, including moral theological objections, to "interfering with nature" in the sense of violating God-given natural law. The Commission also noted the "moral revulsion" of some toward crossing species lines, especially "the mixing of human and nonhuman genes" (President's Commission, 1982, 57). While the Commission took up serious consideration of the issues, especially in terms of potential beneficial and harmful consequences and the prudential management of scientific investigation, it did not provide a sustained ethical analysis of appeals to nature as normative.

NBAC explored the successor biotechnologies to recombinant DNA biotechnologies, namely human stem cell research (1999) and human cloning (1997). NBAC undertook its ethical analysis in terms of potential physical harms, the potential impact of reproductive cloning on individuality and self-identity, the potential impact on family structure and function of cloning, potential harm to important social values, treating people as objects, eugenic concerns, personal autonomy, freedom of inquiry, and freedom of reproductive choice. Like the President's Commission before it, NBAC provides no sustained analysis of appeals to nature as normative.

The President's Council on Bioethics, authorized during the George W. Bush Administration (2001–2009), should be credited with being the first public bioethics commission in the United States to take up the task of providing such a sustained ethical analysis. In its 2004 report, *Reproduction and Responsibility*, the Council articulates a framework for its ethical analysis and policy recommendations regarding assisted reproduction and assisted reproduction technologies (ART). Artificial initiation of fertilization poses three "potential hazards," the first of which is characterized as follows:

> First, ART raises novel possibilities for altering the biological relationships that are central to normal sexual reproduction, and thus for confounding the human relationships that follow from it (President's Council on Bioethics, 2004, 44).

The claim here is more than factual or descriptive, as becomes clear in a later passage:

These procedures, and others like them, raise the possibility that children conceived through ART might be connected to their biological parents in fundamentally different ways than children conceived and born without artificial intervention. In some cases, children conceived with these technologies might be denied the biparental origins that human beings have always taken for granted and that have been the foundation of family relations and generational connections. ART techniques do not have to disrupt such relations, but they might be used in ways that confound parentage, involve more or fewer than two biological parents, or otherwise depart from the biologically grounded parent-child relationship (President's Council on Bioethics, 2004, 45)

The appeal appears at first to be to prudence, that one should not press against biological limits without very great care, if at all (a line of reasoning anticipated by John Gregory (1724–1773) in the eighteenth century; see second case study, below). However, the implicit appeal is to the givenness of biological relationships as providing not just practical but also moral limits on the development of new biotechnologies. It is not just that transgressing these limits would be potentially harmful; the very meaning of parental and transgenerational relationships could be disrupted, relationships that take their foundation and therefore their intellectual and moral authority for judgment from human biology, i.e., from the givenness of human nature.

This implicit appeal to the givenness of nature and its limits become explicit in the Council's 2003, report, *Beyond Therapy: Biotechnology and the Pursuit of Happiness*. One of the central themes of this report is that biotechnology used to go beyond therapy in the attempt to improve or enhance the range of human capacities and therefore achievements involves stepping beyond "natural limits" (President's Council on Bioethics, 2003, 17), a concept and discourse that does not appear in the work of previous U.S. public commissions. Such technology is of moral concern because it reflects and encourages dreams of "overcoming limits of body and soul" which the Council characterizes as "[d]reams of perfection" (President's Council on Bioethics, 2003, 18). The worry arises when such perfection is pursued "at all costs" (President's Council on Bioethics, 2003, 18), i.e., without recognition of and consequent willingness to respect nature's limits on human ambition. The Council suggests that such dreams have deep historic roots, back at last to Bacon's injunction that we should master nature.

The Council frames its ethical concerns about enhancement technologies in terms of a "disquiet" about attempts to improve upon human nature (President's Council on Bioethics, 2003, 286) This language echoes that used by the Council's Chairman, Leon Kass, in an influential essay in *The New Republic* in 1997, "The Wisdom of Repugnance" (Kass, 1997). In this essay Kass appeals to (but nowhere argues for) "the inherent procreative teleology of sexuality itself." That is, human organ systems have not just functions, as modern biology teaches, but also natural ends or purposes, which modern biology rejects. We experience repugnance at some scientific advances or possibilities, Kass goes on to claim, when "we intuit and feel, immediately and without argument, the violation of things we rightly hold dear." Human biological things have their purposes that exist independently of human ambitions and set limits on those ambitions that are intellectually and morally authoritative. There is a "natural way of all mammalian reproduction" and such "biological truths about our origins foretell deep truths about our identity and about our human condition altogether." Among these are "naturally rooted social practices," such as the creation of families. "[F]undamental changes" in human nature also affect "basic human relationships, and what it means to be a human being." This is a full-blown naturalism of moral reasoning, in which one can authoritatively reason from facts to moral judgments on the basis of the built-in, discoverable, and value-laden teleologies of human nature, indeed of biological entities generally.

Along these very same lines *Beyond Therapy* claims that moral judgment and public policy regarding enhancement technologies should be based on "[r]espect for the [g]iven" (President's Council on Bioethics, 2003, 287). This givenness of human nature is built right into it, including its built-in purposes that justify limits on human technological ambition. As the Council puts it, the "naturalness" of the means of human accomplishment "matters" (President's Council on Bioethics, 2003, 292). The Council's moral concern is that using drugs and devices to alter human capacities and therefore accomplishments involves "the danger of violating the nature of human agency and the dignity of the naturally human way of activity" (President's Council on Bioethics, 2003, 292). This danger exists for medicine as well. When focused on therapy, medicine aims at "making whole that which is broken or disabled" (President's Council on Bioethics, 2003, 306). We can reach judgments about disease or disability on the basis of observations of given structure and function. Once medicine starts down the path of enhancement, however, this clear direction will become lost, unhinging medicine from natural limits and making it potentially dangerous rather than a source of healing.

The work of the President's Council, more than public bioethics commissions that preceded it and more than much of the bioethics literature, is important because it takes seriously nature as normative, i.e., as setting intellectually and morally authoritative limits on what human beings should expect and therefore desire or allow. If our ambitions, our desires, become undisciplined, they become disconnected from the givenness of nature. The givenness of nature is the course of its normative, corrective force in human judgment and behavior. Once that givenness is abandoned, there are no remaining constraints on human ambition.

The President's Council's thinking is strongly influenced by the prior work of its Director, Leon Kass. In his book *Toward a More Natural Science*, Kass' move from nature to ethics is not offered as a philosophical argument or a doctrine. In proposing a more "natural" biology, Kass defines "natural" as "'true (or truer) to life as found and lived" (Kass, 1985, xiii). He explicitly states that his view is not rooted in a romantic view of nature or a preconceived doctrine, such as Roman Catholic moral law theory. In suggesting that we look for "a more congenial encounter" between nature and ethics, Kass calls for thinking about how a "more natural science might be useful for ethics" (Kass, 1985, 346). It must be said, however, that Kass's definition of "natural" ("true to life as found and lived") incorporates human history and culture as well as biology and therefore allows for a plurality of interpretations.

Kass's views in turn are strongly influenced by the work of Hans Jonas (1903–1993), in particular *The Phenomenon of Life: Toward a Philosophical Biology*, a neo-Aristotelian work by one of the founders of the field of bioethics that proposed that ethics ought to flow from a philosophy of nature. Jonas called for an ethics "which is grounded in the breadth of being ... an ethics no longer founded on a principle of divine authority must be founded on a principle discoverable in the nature of things, lest it fall victim to subjectivism or other forms of relativity" (Jonas, 2001, 284).

This effort to discover ethics "in the nature of things" immediately runs into the problem that "nature" and "the nature of things" in contemporary life and philosophy are construed in so many contradictory ways that no universally accepted, singular conception of nature is possible. This situation reflects the complexities and challenges of appeals to nature as normative. Hence, deriving moral guidance from "nature" must inevitably include philosophical and public dialogue and debate whose outcomes will be uncertain, tentative, and subject to renegotiation as new discoveries and events emerge. Such dialogue and debate are to be welcomed in the ongoing search for ethical norms adequate to our scientific knowledge and technological capabilities. The following three case studies illustrate the complexities and challenges of appeals to "nature" and the "nature of things" that have been made in the modern history of science and medicine, with important implications for how appeals to nature and normative should be responsibly made.

3.3 Three Case Studies of Appeals to Nature in Science and Medicine

As we pointed out above, nature has been valued in the modern histories of science and medicine along a continuum. At one end of the continuum, nature has been valued positively, as setting standards for human judgment and behavior. On such a view, public policy protects us by encouraging and supporting biotechnology within natural limits. Public policy is cautiously permissive and also restrictive. At the other end of the continuum, nature has been valued negatively, as somehow incomplete or deficient and thus in need of correction. On this view, human judgment appeals to sources other than nature as the justification for such correction. Public policy, on such a view, should support and, indeed, encourage, biotechnological advances that protect us from nature's deficiencies and pathologies.

We have chosen our three case studies because they illustrate both the ends of this continuum and, at times, its sometimes muddled middle. At one end of the continuum, we will see caution about the limits of nature and attempts to alter nature. At the other end of the continuum we will see enthusiasm for altering nature, with sometimes little or no sense that there are ethically justified limits on such altering. We will also see that appeals to nature, the natural, and the unnatural have created discourses about fundamental human concerns, such as food, health, and old age that sustain and enrich such appeals.

One of the distinguishing features of these discourses is their secular nature. In modern Europe and North America, scientists and scientific physicians took the view that they could and should do their work without making any appeals to transcendent reality, sacred texts, or revelation. At the same time, many scientists and physicians shaped a discourse that strove not to be hostile to faith communities and traditions. They mainly did this by transforming "why" questions into "how" questions, with the focus on the quest for causal explanations of observed phenomena and interventions into natural processes based on hypotheses designed to test such explanations. As a consequence, our case examples do not have an explicit religious component. Indeed, there was a general emptying out of a role for religious discourses in the discourses of science and medicine concerning nature. Our case studies illustrate the emergence of the distinctively secular methods, results, and discourses of modern science and medicine. It is precisely their secular character that has allowed the methods, results, and discourses of biotechnology to become global.

Our three case studies also illustrate both change and continuity in how we think about nature and its alteration during the several centuries of the modern period. In our treatment of the three case studies, we will point out both, especially as they bear on contemporary moral assessment of biotechnology that appeal to nature, the natural, and the unnatural.

3.3.1 Food, Animals and Plants as Pure and Impure, and the Environment as Natural and Improved in the Early Modern Period 1500–1750

3.3.1.1 Introduction

The materials for the discussion of these topics are drawn from European and especially English history from the sixteenth and seventeenth centuries. We will have one eye on current debates on genetically modified (GM) crops, etc., in making our argument, but hope to avoid too much "presentist" history. Indeed, one of our aims is to show differences as well as similarities between the past and the present. However, we have made some comments about GM crops in order to show why we have chosen particular historical material and make clear its significance to the GM food debate. In the later part of this volume others may interpret the material in different ways.

The early modern world was a macrocosmic world. True, there was the newly invented microscope which gave promise that it would discover an invisible world. But until the late nineteenth century for lay people and for most writers on medicine, on regimen (rules of health) and on the environment, the world that they sensed constituted reality.[2]

---

[2] This was philosophically expressed by Aristotle, who argued that the qualities perceived by the senses were the building blocks of the world and that hot, cold, dry and wet were the primary qualities. Aristotle, *On Generation and Corruption* 329^{b}7–330^{a}29.

Food was judged to be pure or impure by a combination of preconceived ideas and the senses. Religion could certainly be brought in to define impurity or to declassify animals from the category of impure. In Protestant England the dietary laws of Leviticus were ignored, for 'by the coming of Christ, to the pure all things were pure'; only a few, especially Scots, refrained from eating pork (Thomas, 1984, 53–54). Clearly, pure and impure foods could be culturally defined or constructed, in this case by religion, although, of course, the prohibitions in Leviticus may have been drawn up from pragmatic experience. Cultural construction and the practicalities of food safety are two poles of the GM crop argument today which others can ponder.

We would like to develop another approach centered around received medical ideas. In early modern Europe diet was an integral part of preventive medicine. The Hippocratic writers and Galen had emphasized the importance of lifestyle in the struggle to retain health. Food and drink together with air and climate, exercise and rest, the evacuations, sleep and waking, and the emotions all played a part in preventing or bringing on illness. One's relationship to the environment in the form of what one ate or the air one breathed was especially important.

The teachings of classical medicine were popularized in the vernacular books of regimen that were printed in large numbers in early modern Europe. A pivotal belief in many discussions about diet and food was that there was a unity between the environment and organisms like plants, animals and humans and that one expression of this was the food cycle. Animals were what they ate, and humans were what they ate of animals. As Thomas Cogan (1612) put it: 'Such as the food is, such is the blood: and such as the blood, such is the flesh'. The medical-philosophical underpinning for this view lay in the Galenic system of digestion whereby food was altered into the nature of the different parts of the human body (Galen, 1916, 241–263, 1968, 222–223, 233–236), and more generally in the Aristotelian-Galenic view that human beings shared with plants, animals and, indeed, with the rest of the organic and inorganic world the same constitutions or mixture of qualities. The four primary mixtures of qualities were, declared William Bullein, "Not only in man, but in beastes, fish, foule, serpents, trees, herbes, metals, and everie thing sensible and insensible" (Bullein, 1558, fol. 12 v, in Wear, 2000, 172). This view, given appropriate changes in theories of physics and biochemistry, is still, we think, a powerful underlying factor in how the public or some commentators come to judge genetically produced food. People still believe that what they eat shapes their natures and, more particularly, can help them either to stay healthy or can make them ill. The mountainous mass of medical dietary trials and of advice on food all help to confirm people in this belief.

In the early modern period the purity of food had both absolute and relative meanings. Corruption was a potent marker of pathological states. Many illnesses were explained as being caused by internal putrefaction or corruption, while putrefying wounds and sores in the living and in the changing state of dead bodies attested to the destructive power of corruption. The medical procedures of evacuation such as bleeding, purging, sweating, blistering, were designed to expel corrupt matter from the body and so cure it. Putrefaction also played a role in the most destructive disease of the age, plague. It was believed to originate from God's secondary causes: from poisonous putrefying air rising from swamps, dunghills, cesspits, butchers' waste, and sewers. Corrupt putrefying food was, therefore, not surprisingly, abhorred. It spoke to peoples' deepest pathological fears, as well as offending the senses and causing obvious digestive problems.

It is worth asking if the language by which genetically modified crops are explained also touches upon present day views of pathology. For instance, cancers are explained as being caused by abnormal cell division, so eating something which has been engineered at the microscopic level to be out of the 'normal' may be seen as creating abnormality within the body, given the belief that we are what we eat. Thinking about it, the comparison between putrefaction and an abnormal or engineered cell may sound strained. But, as we pointed out earlier, in the sixteenth century the world of the senses was coterminous with reality; today the micro-world below the unaided human senses is part of our reality. The dimensions of the normal and the pathological were different in the sixteenth century from what they are today as are the theories that explain the make up of the world and of living organisms. Such differences, however, do not invalidate analogies that we are suggesting. Eating something which, because it is putrefying, is pathological, or eating what looks as if it shares pathologically inducing abnormalities may appear dangerous to the public in the sixteenth century and today. The difference is that the public could in the sixteenth century judge the purity and quality of food with their senses. We can still do this today, though "sell by" dates and public health regulation take some of the responsibility away from us. However, the public cannot with its senses judge what we eat at the micro level, and this applies especially to GM foods.

We want to turn now to a set of very long-lasting beliefs centered around the question of what were the best conditions for animals and fish to live in if they were to be food for human beings. Writers on regimen believed that animals that were kept in the open air in the wild provided better, healthier food than those kept in close confinement. Thomas Mouffet (c.1552–1604) wrote that the pig provided good food "if he feed abroad upon sweet grass, good mast and roots; for that which is penn'd up and fed at home with tap droppings, kitchen offal, soure grains and all manner of drosse cannot be wholsom" (Mouffet, 1655, 68). Thomas Tryon (1634–1703) similarly wrote: "That Bacon and Pork, which is fed with Corn and Acorns have their liberty to run, is much sweeter and wholsomer, easier of digestion and breeds better blood that that which is shut up in the Hogg-Sties, such Bacon for want of motion becomes more of a gross phlegmatic Nature" (Tryon, 1683, 67).

Such views were commonplace. They emphasized motion, cleanliness, fresh air and water and condemned stagnation, closed-in spaces, darkness, dirt and pollution. For instance, eating fish could result in "much grosse slimie superfluous flegme" and so produce ill health. One had therefore to select fish that "swimmeth in a pure sea, and is tossed and hoysed with winds and surges: for by reason of continuall agitation it becometh of a purer and less slimie substance, and consequently of easier concotion [digestion], and that of a purer iuyce" (Venner, 1628, 69). Purity and freedom from corruption were linked to cleanliness and motion. In medical writings the visible environment shaped the nature of fish as it did that of humans.

Fresh-water fish was best that "is bred in the deepe waters, running swiftly toward the North, stony in the bottome, cleane from weedes, whereunto runneth no filth nor ordure comming from townes or cities, For, that which is taken in muddie waters, in standing pooles, in fennes, motes and ditches, maketh much fleame and ordure" (Cogan, 1612, 140). Fish were not only shaped by the environment, but also by their responses to it. Andrew Boorde made this explicit: "The fysshe the whiche is in ryvers and brokes be more holsomer that they the which be in pooles, pondes or mootes, or any other standynge waters; for they doth laboure and doth skower [clean] themselves" (Boorde, 1870, 268–269).

What can we take from this small case example? Again, we see an expression of the sense of the unity of the organic and inorganic world. At this level the unity is expressed by environmental determinism. Pigs, humans, and fish are alike shaped by their environment. This is a long-lasting view, still present today and first articulated in the western medical tradition by the Hippocratic writer of *Airs, Waters, Places*.[3] Does this sense of unity apply to GM crops and/or animals? In a very general sense it could be argued that the unity between the organic world at least is preserved. What is happening to plants in terms of genetic manipulation is also happening to animals and is around the corner for humans. This unity is expressed at the levels of knowledge and practice, and is emerging as part of the socio-economic structure of developed countries. However there are differences between the past and present. (Here we would stress, again, that GM crops, etc., are not our area of expertise and those who are experts may well see things differently.) The early modern link between plants, animals and the environment is unrecognizable in the GM world. To early modern eyes the natural environment acts upon the nature of animals and makes them more or less pure, or fit for us to eat. With genetic manipulation human action upon genetic material, rather than nature's action upon an already living organism, is what many see as significant. True, the genes that are being manipulated have survived natural selection, which involved their relationship to the environment. But, that was in the past. Now genetic manipulation produces human-selected and created organisms that may be made "fit" for humans to eat and to make money.

It is worth looking at the last two sentences more closely. In the sixteenth and seventeenth centuries evolution by natural selection was unknown. The nineteenth century introduced a Darwinian discourse/theoretical framework that lies between the early modern period and our own and which very much influences the way we think. In the sixteenth and seventeenth centuries monstrosities that were "beyond nature" were acknowledged, as was the power of the imagination to affect the result of conception.[4] But, by and large, in the writings on preventive medicine the animals that were made "pure" by nature to eat were made so only for their lifetimes. They did not pass on the pure or good characteristics that they had acquired on to their offspring. Caveats, however, need to be added. The development of national

---

[3] This seminal text described how, for instance, people living in the North where cold and heat alternated were hardy and fierce, while those living in Asia where the climate was equable and agriculture easy were tamer in character.

[4] See Daston and Park, 2001.

traits through the action of the environment implies some inheritance of acquired characteristics. Moreover, human selection was involved in the age-old practice of selective breeding. However, the medical discourses on healthy eating were largely devoid of references to changes in heredity either through nature's interaction with an organism or through human selection in breeding.

We have sometimes used 'nature' to refer to an undifferentiated whole. Yet, our examples show that 'nature' was also used to refer to something that was considered to be highly differentiated. This was explained in Christian terms by the Fall of Adam and Eve, when perfect paradisical nature was closed to them, the Garden of Eden being guarded by "Cherubims and a flaming sword which turned every way, to keep the way of the tree of life." Moreover, the "ground" that Adam and Eve would live upon was "cursed" it would bring "Thorns also thistles." Labor and death were now their lot: "In the sweat of thy face shalt thou eat bread, till thou return upon the ground" (Genesis, ch. 3, v. 24, 17–19). Not all parts of nature after the Fall appeared equally cursed and barren. Faint echoes of paradise could be discerned by Europeans in the beauties of the countryside, in gardens and in the paradisical-seeming New World. Medical writers believed that those parts of nature were healthiest which were aesthetically most pleasing, perhaps closest to paradise, having qualities such as untouched open countryside, crystal- clear running water, and healthy sweet smelling breezes. There were also parts of nature which were unhealthy, such as low lying land, marshes, pent-up situations unventilated by breezes, stagnant water, that had their counterparts in the unhealthy crowded dark and stinking city areas.

We want now to turn to less general issues. The ability to select food lay at the heart of the medical advice on what was healthy food. Selection involved knowing the different environments from which the animal or fish had come. That might be difficult in a city market in the sixteenth century, though in the face-to-face society of the time the customer had a better chance of finding this out than today. Many types of food, especially in the countryside, would be reared at home or in the neighborhood, so selection was more feasible. Towards the end of the seventeenth century and in the eighteenth century, the "commercial revolution of the eighteenth century" replaced much of the household economy. Many items including animal foods, breads, and also medicines came to be produced and sold commercially. At first, this was on a local basis and then through national wholesale networks and the process accelerated in the nineteenth and twentieth centuries. Knowledge of the environmental origins of food became more difficult to acquire but not impossible, as the modern debate in Europe about battery hens or intensively reared pigs demonstrates (Hansard, 21 October 1999, columns 558–561; *The Times of London*, 19 March 2007, 9). However, knowledge about environmental conditions is now less personal; instead it is often mediated by governments and interest groups.

What has this to do with the environment, purity of animals as food and reproductive technologies? Well, put simply, we suggest that people still conceptualize what should count as healthy food using the language of "we are what we eat," and of the environment having good, healthy qualities. People do this despite factory-reared animals, etc. In the case of genetically manipulated food, the traditional language of selection that we might want to use appears even less appropriate than for factory products. Nature in all its forms seems singularly absent in shaping GM products. In reality nature may well act upon GM food-animals once they are alive, but – and here, again, we are aware that we write as non-experts – the public mainly perceives that they are laboratory-based products. Both opponents and proponents emphasize the uniformity of the products and their ability to resist the shaping power of nature, the opponents also adding that such products can deform nature.

The emphasis has changed, but the old language of selecting products shaped by a diverse natural (and human) environment is still being employed. In the case of factory farming many have used the values contained in the traditional discourse to argue for change, for instance in the move to organic farming, or in lobbying for legislation for animals to have cleaner conditions and more space. Whether such values are in play in relation to genetically manipulated foods we leave others to argue. Perhaps, though, the inability to apply such values to GM foods creates helplessness, anger, or indeed the sense that such foods are unnatural, or outside of nature, precisely because they are not subject to nature and the values that have been imputed to the different parts of nature.

We also leave the question of whether the values relating nature to animals as healthy food and, indeed, to human health are transcendent universal values that apply across time. We suspect an anthropologist might argue in this way. A hardnosed economist or economic historian might argue that the old language of purity, freedom to move, etc., has had its day, that it reflects conditions that no longer apply and that a new language and values have to take its place. Yet, what would such a language of selection consist of and who would speak it – scientists or the public? Moreover, the old discourse linked personal selection to the conditions in the macro-world, to visible nature. Factory food conditions still relate to the visible environment; genetic manipulation does not, though its products, once created, usually do live in such an environment. The techniques of genetic manipulation require esoteric knowledge and specialized equipment. It may be that only experts and corporations will articulate and put into practice choices and values relating to the food we eat.

3.3.1.2 The Environment as Natural and/or Improved

There is a strand of modern thinking which argues that "the natural is best" that clearly invokes nature as normative. At its most extreme, this judgment is expressed in the belief that a natural environment is one which is untouched and unaltered. Such a view is perhaps a component in some critiques of GM crops. More usually, the criticism begins from the unnaturalness of genetic manipulation and then goes on to condemn its possible effects in the field, in nature. Nature in this sense is often taken as both pristine nature and nature as it has been shaped up to now by humankind. Another way of expressing this is that GM crops are viewed by some as threatening the ecological order of nature. But, again, what is meant by nature is ambiguous. It could be argued that ecologists look both to untouched nature and nature as it has been made by humans. And there is little dispute that nature in the period before the introduction of genetic manipulation had been profoundly altered by human interventions such as forest clearing, farming, building, etc.

The ideology or justification for altering nature has over the centuries been that of utility, improvement, and progress. This, of course, is the justification often made for GM crops. We therefore have the situation that the same ideology is being used for both traditional non-GM farming and for GM farming. Parallel to the continuity in ideology is the claim made by the proponents of GM farming that their activity is a continuation and expansion of previous farming practices; the example from selective breeding is one that is often used.

Historically, the belief that the environment was there to be utilized by human beings recurs in western thinking and in a variety of contexts – religious, national, colonial, and commercial. The reaction to such a view is present from the seventeenth century and gathers strength from the nineteenth century onwards.[5] The view that nature was there for mankind's benefit is strongly expressed early on in the Judaeo-Christian tradition, though, as the philosophy chapter (Chapter two) in this volume points out, it only became influential in Europe in from the sixteenth century on. In the Bible, God famously gave Adam and Eve "dominion over the fish of the sea, and over the fowl of the air and over every herb and over every living thing that moveth upon the earth. And God said, Behold I have given you every herb bearing seed, which is upon the face of all the earth, and every tree in the which is the fruit of a tree yielding seed: to you it shall be for meat [food]" (Genesis 1, v. 28–29). To this sense that everything on earth was for mankind's use there were two additional biblical messages. As well as having dominion, Adam and Eve and their offspring were to "replenish the earth and subdue it" (Genesis 1, v. 28).[6] The notion of controlling natural resources constantly recurs across time to the present. The other biblical theme came, as we saw in the previous section, with the Fall of Adam and Eve. Nature or the environment becomes degraded with the Fall; from then on food is produced from the earth only with a great deal of sweat and toil. Man's mastery and use of the earth is now linked to perpetual hard labor.

In the sixteenth century this link continued to be emphasized. For instance, in Conrad Heresbachius' (1496–1576) popular book on agriculture, which was devoted to the best techniques for growing plants and rearing animals, Genesis 3.19 features prominently on the title page:

> In the sweate of thy face shalt thou eate thy bread, tyll thou be turned agayne into the ground, for out of it wast thou taken: yea, dust thou art, and to dust shalt thou returne (Heresbachius, 1577, title page).

3.3.1.3 Sixteenth-Century Farming: Social Aspects

The status, role, and cultural perception of the producers of food relates to the public's sense of what they are eating. Advertising illustrates this at a true if trite level. Packets of butter may feature idyllic pastoral farmhouses with contented cows grazing

---

[5] See, for instance, the classic study by Glacken, 1967.
[6] The early modern meaning would be "fill."

in the open air, or rosy-cheeked farm maids, brimming full of health carrying milk pails. A romantic view of farming rooted in the past informs many such images. In food production trust is important for the consumer, and there is a need to believe that the producers as well as their techniques are trustworthy. In the past, positive images of farmers and farming were constructed, though, of course, the rapacious farmer was also a familiar figure. Today with GM products the public is faced with a schizoid situation: crops made by scientists in the laboratory and planted and grown by farmers. The public's image of the scientist in some countries such as the United States and Britain, is often a negative one, while farming has positive associations for many, if not all. The upholders of the thesis of continuity between non-GM and GM farming could counter that this dualism is not new, as farming had already come to rely on science. In the nineteenth and especially in the twentieth century chemical fertilizers and pesticides were being increasingly used on farms, and scientific research institutes were providing farmers with new, improved crop and fruit varieties which were the outcome of programs of selective breeding. And, on the converse side of the continuity thesis, some of the opponents of GM crops could argue that GM technology is but the latest step in the scientization of farming. However, in the sixteenth century there was no doubt that farmers or people acting as farmers were the sole producers of animal and plant food.

The worker in agriculture was an essential part of sixteenth- and seventeenth-century European society. Unlike the present-day situation, the land employed a very large proportion of Europe's population. Agriculture did not merely sustain Europe's population, but in the early Middle Ages had allowed it to increase. Although city states like Genoa, Florence, and Venice created more wealth from commerce than larger countries did from agriculture, for countries like England and France agricultural production lay at the heart of their wealth. The numbers in agriculture and the economic significance of what they produced ensured that they had a high cultural and social profile. Farmers or husbandmen, who might often own, lease, or rent land and employ some workers were an occupational group that was rising in status in the sixteenth century, increasingly literate and confident of their position in society. Traditional cultural products such as folk and morality tales reinforced their image as the true successors of Adam and Eve. They represented the Biblical archetypes of human beings as laborers on the land: hardworking and respectable. Sir Anthony Fitzherbert (1470–1530) posed the paradoxical question:

> This is the question, whereunto is everye manne ordeyned. And as Job saythe.... a man is ordeyned and borne to do labour, as a bird is ordeyned to flye. And the Apostle sayethe ... he that laboureth not, shulde not eate, and he ought to labour and doo goddess warke, that wyll eate of his goodes or gyftes. The whiche is an harde texte after the lyterall sense. For by the letter, the kynge, the quene, nor all other lordes spirituall and temporal shuld not eate, without they shuld labour, the whiche were uncomely, and not convenyente for such estates to labour (Fitzherbert, 1534, Prologue 1).

Fitzherbert answered his question by writing that labor does not need to involve manual labor and can be taken to be equivalent to an occupation. Yet, he wrote, though agricultural workers were similar to chess pawns being of low rank, nevertheless, they were important "in so moche the yomen [pawns] in the sayde moralytyes and game of the chesse be set to labour, defende and maynteyne all the other hyer [their] estates ..." (Fitzherbert, 1534, Prologue, pp. 1–2 citing Caxton's *Book of the Chess*).

The value placed on farmers made it easier to trust the goodness of their products. This was reinforced by the high degree of manual work involved in sixteenth century farming. The link between the agricultural worker and his or her product was much more obvious than it may be today. The numbers employed in agriculture have been dramatically reduced in developed countries since the time of the "agricultural revolution" of the eighteenth century which was the necessary precursor of the Industrial Revolution that made Britain the first industrial nation and moved vast numbers from the countryside to the new industrial cities. As agriculture became much more efficient, so it became a minority occupation. The same process has happened in continental Europe, in the nineteenth and twentieth centuries, and in the United States in the early twentieth century. In the process agriculture became industrialized. It is easy to romanticize the situation that existed in the sixteenth century, but we believe that then the link between producer and product was stronger, that farming was by and large seen positively and that, at least with local products, it was easier for the public to judge the worth and skills of the producers. Today, we still make value judgments about food but there is perhaps a sense of alienation pervading our understanding of the relationship between product and producer. The biblical sense of the vast part of the population laboring on the land is no longer present in developed countries. The common experience of agricultural work that united much of the population no longer exists, and the cultural values associated with farming are those of an age that has disappeared. The further intensification of agriculture by the introduction of GM technology will distance even more the older cultural image of farming from the reality.

3.3.1.4 Change, Improvement, and the Sense of Stability in Sixteenth-Century English Agriculture

Technical skill as well as labor were certainly requirements for sixteenth-century agriculture. As Thomas Tusser (1524–1580), an educated farmer who farmed in Suffolk, Norfolk and Essex in the sixteenth century, put it: "... no labor no bread ... No husbandry [i.e. the craft of farming] used, how soon shall we sterve" (Tusser, 1878 [1580], 15). Knowledge and techniques, then as now, were used in farming. However, there is a strong rhetoric of improvement and change that also underscores GM technology. In the sixteenth century change was certainly on the agenda for agriculture. But there was also present the feeling that much in agriculture was unchanging. True, new agricultural products were introduced. Barnabe Gouge, the English translator of Heresbachius' work, observed:

> It is not many ages agone, since both the Peache, the Pistace, the Pine, the Cypresse, the Walnut, the Almond, the Chery, the Figge ...and a great sort of others, both Trees and Plantes, being some Perseans, some Scythians, some Armenians, some Italians, and some Frenche, all strangers and aleantes [aliens], were brought in as novelties amongst us, that

doo nowe most of them as well, and yea and some of them better, being planted amongst us in England, than yf they were at home (Heresbachius, 1577, *To the Reader*, pp. iiir–iiiv).

Barnabe Gouge also hoped that vines could be grown in England, "it is worth the tryall and the travayle to have wines [vines] of our owne though they be smaller" (Heresbachius, 1577, *To the Reader*, p. iiiv).

Moreover, the existence of a variety of techniques for a particular job implied that over time innovation had occurred. For instance, there were different types of grafting, still used today: "grafting, imbranching, the thirde inoculation or imbedding." Ploughs had been developed that suited the different soils and conditions of particular localities:

> ... and one ploughe wyll not serve in all places. Wherefore it is necessarye to have dyvers [various] maners of plowes. In Sommersetshyre, about Zelcester, the sharbeame, that in many places is called the ploughe-hedde, is four or five foote longe, and it is brode and thynne. And that is bycause the land is verye toughe and wolde soke the ploughe into the erthe, yf the sharbeame were not long, brade and thinne. In Kente they have other maner of plowes ... (Heresbachius, 1577, p. 72 r; Fitzherbert, 1882 [1534], 1).

However, there is a sense that such techniques had been around from time immemorial and were there to stay. In the didactic farming literature of the sixteenth century the systematic urge to innovate, experiment, or to praise new techniques is usually absent. To some extent all "how to" books or instruction manuals present their knowledge as timeless, but their sixteenth-century writers had a different approach to innovation from our own. One sign of this is that the work of authors over a thousand years old such as Pliny still appeared relevant. For instance, it was important to know when to plant or "set" trees: "looke that you sette them in the afternoon, in a fayre westerly winde, and in the wane of the Moone. Plinie sayth that this note is of great importance for the encrease of the tree and goodnesse of the fruite: if the tree be planted in the encrease of the Moone, it groweth to be very great: but yf it be in the wane, it wil be smaller, yet a great deale more lasting" (Heresbachius, 1577, 71 v–72 r).

Here we have an example of the *longue durée* of the French *Annales* school of history. Change in material culture and practice in this period was often very slow and the technological links to previous centuries remained unbroken. We emphasize this partly to make the contrast with the period beginning from the agricultural and industrial revolutions to the present. Constant technical innovation, of which GM technology is an example, occurs today in practice and is embedded as a social and cultural value, though it is not an unquestioned one as the debate whether the rate of innovation is a necessity or a danger to society shows. We also stress the *longue durée* in sixteenth-century agriculture because it helps to shape our romantic image of the farmer and his links to the soil.

The farming calendar was an immutable one. It was controlled by the seasons and by dates such as saints' feasts (the latter also further helping to cement the farmer into the social and religious fabric of society).

Thomas Tusser set out what a farmer had to do each month. In February, for instance, he advised:

3 Scientific and Medical Concepts of Nature in the Modern Period in Europe

Who laieth on doong er he laieth on plow
such husbandrie useth as thrift doth alow.
One month er ye spred it, so still let it stand,
er ever to plow it, ye take it in hand.
....
Go plow in the stubble, for now is the season,
for sowing of fitchis of beanes and of peason.
Sowe runcivals timelie, and all that be gray,
but sowe not the white till S. Gregories day [12th of March] (Tusser, 1878 [1580], 87).

Tusser also fixed "the farmers dailie diet" into the seasons:

When Easter comes, who knows not than,
That Veale and Bakon is the man:
And Martinmas beefe[(1)] do beare good tack,
When countrie folke doe dainties lack (Tusser, 1878 [1580], 28).[7]

The fixed links of farming to nature, to the seasons, have become less rigid as we come to the present. The immutable laws of seasonality have been subverted by simple if expensive techniques of glass or plastic protection, heating, watering, and lighting. New early and late crops also help us ignore the seasons, while the globalization of food production means that people in Britain can eat, if not taste, California strawberries in the middle of winter. It may be that GM technology will further lessen our dependence on the seasons and so further break down the old image of farming as unchanging or slow to change and working according to the fixed rhythms of nature.

3.3.1.5 The New Science and the Baconian Program: The Union of the Sciences with the Crafts

GM technology is, if nothing else, a union of science and the craft of agriculture. The idea that science should be applied was developed in the seventeenth century and is often associated with the philosophy of Francis Bacon. This view was in opposition to Aristotle who had argued that the true philosopher was concerned with universal knowledge and principles and, unlike the artisan, was not limited to practical ends:

> ... they [the first philosophers] philosophised in order to escape from ignorance, evidently they were pursuing science in order to know, and not for any utilitarian end. And this is confirmed by the facts; for it was when almost all the necessities of life and the things that make for comfort and recreation were present, that such knowledge began to be sought. Evidently we do not seek it for the sake of any other advantage; but as the man is free, we say, who exists for himself and not for another, so we pursue this as the only free science, for it alone exists for itself (Aristotle, *Metaphysics* 982^b, 21–28 in Barnes, 1984).[8]

---

[7] Beef "dry'd in the chimney as bacon and is so called because it was usual to kill the beef ...about the Feast of St. Martin, Nov. 11th."
[8] See more generally *Metaphysics* 981^{a}1–982^b 28 and *Nicomachean Ethics* 1139^b 14–1142^a 30.

Bacon refuted Aristotle's separation of natural philosophy (science) from the crafts and mechanical arts. Instead, he argued that they should be united and should mutually enrich each other:

> ... let no man look for much progress in the sciences – especially in the practical part of them – unless natural philosophy be carried on and applied to particular sciences and particular sciences be carried back again to natural philosophy .... For want of this, astronomy, optics, music, a number of mechanical arts ... altogether lack profoundness, and merely glide along the surface and variety of things (Bacon, 1860, 79).

Historians usually point out that Bacon argued that science should be "fructiferous" as well as "luciferous," that it should bear fruit or practical results as well as light or pure knowledge. What is less often remarked upon is Bacon's hope that the practical arts should be informed by luciferous experiments, by "pure science" to use our terminology:

> For the mechanic, not troubling himself with the investigation of truth confined his attention to those things which bear upon his practical work ... But then only will there be good ground of hope for the further advance of knowledge when there shall be received and gathered together into natural history [which included agriculture] a variety of experiments which are of no use in themselves but simply serve to discover causes and axioms, which I call *Experimenta lucifera*, experiments of *light*, to distinguish them from those which I call *fructifera*, experiments of fruit. Now experiments of this kind have one admirable property and condition: they never fail or miss. For since they are applied, not for the purpose of producing any particular effect, but only of discovering the natural cause of some effect, they answer the end equally well whichever way they turn out: for they settle the question (Bacon, 1860, 95, emphasis original).

Once the experiments of light "discover true causes and axioms," then the axioms "supply practice with its instruments, not one by one, but in clusters, and draw after them trains and troops of works" (Bacon, 1860, 71). In other words, pure science can produce a rich supply of applied results.

The rise of experimental philosophy or science occurred in the seventeenth century with the works of Galileo, Newton, Boyle, the philosophy of Bacon, and, it has been argued, earlier with the alchemical and natural magic traditions as well as with developments in medicine. The enthusiasm for experimentation was part of a more general movement by natural philosophers or scientists away from contemplating nature, from trying to answer Aristotle's final cause which was concerned with why something was as it was, to answering 'how' questions. Experiment and mathematics were seen as the keys to investigating how nature functioned. "How" questions were much more conducive to practical results. The investigation was to be an active and precise one, capable of being repeated and tested by others. Of course, there was disagreement over methods and over the roles of experiment and mathematics, for instance as between the philosophies of Bacon and Descartes and their followers. But there is a general consensus among historians that the investigation of nature radically changed in the seventeenth century.

It could be argued that it was only from the nineteenth century on that science actually started to produce applied results of any significance. Nevertheless, the present- day application of genetics to agriculture can be seen not only as following in the Judaeo-Christian tradition of utilizing nature but also as exemplifying the new seventeenth- century vision of science in which scientific knowledge is of a type that can produce practical results.

The positive rhetoric surrounding genetic manipulation is often concerned with progress and improvement; both were part of the rhetoric of the new science. Again, it is worth looking at what Bacon wrote, for he epitomizes the views of many of the protagonists of the new science. He was positive about progress. Unlike some of the renaissance humanists, Bacon did not believe that in the past was to be found perfect knowledge. Ancient authors and their authority, he wrote, held back progress and the acquisition of true knowledge:

> Again, men have been kept back as by a kind of enchantment from progress in the sciences by reverence for antiquity, by the authority of men accounted great in philosophy … (Bacon, 1860, 81–82).

Or, as Paracelsus, Galileo and Harvey put it, their teacher would be the book of nature and not other men's books (Pagel, 1958, 56–57).[9]

References to Pliny on the best time to plant trees would no longer do. As the links between past and present-day knowledge were being disputed and broken, so the notion of revolutionary and incommensurable knowledge as a characteristic of the new science was being articulated at this time.[10] Bacon turned down the notion of adding new knowledge onto the past. Instead of continuity, radically new fundamental theories had to be discovered:

> It is idle to expect any great advancement in science from the superinducing and engrafting of new things upon old. We must begin anew from the very foundations, unless we would revolve for ever in a circle with mean and contemptible progress (Bacon, 1860, 52).

If genetic manipulation is interpreted as revolutionary rather than in terms of continuity and development, then it falls within the concept of scientific change envisaged by the founders of the new science.

What of the rate of change and progress? Today scientific and technical progress is rapid and genetic manipulation has been developed perhaps at a faster rate scientifically and commercially that it is possible to accept and assimilate at political, ethical and social levels – at least in some countries. In Bacon's vision of the new science the rate of change in applied science, technology and the crafts would become far quicker than it had been previously. He would have recognized the picture we drew earlier of extremely slow change in agriculture, and indeed our image of the rate of change may well derive ultimately from the rhetoric of the new science:

> Nor is it only the admiration of antiquity, authority and consent, that has forced the industry of man to rest satisfied with the discoveries already made; but also an admiration for the works themselves of which the human race has long been in possession …. Again, if you …take notice of such things as … the discovery of the works of Bacchus and Ceres – that

---

[9] On Galileo see *Letter to the Grand Duchess Christina* and also The Assayer: "Philosophy is written in this grand book, the universe, which stands continually open to our gaze (Galileo, 1957)." He added that the language of the book was mathematics. See also Harvey (1628, 8), who saw himself as learning not from books and the principles of philosophers but from dissections and the fabric of nature.

[10] This is a concept which we associate with the philosophy of Thomas Kuhn and others in the modern era. See Kuhn (1964 [1962], 140–142).

is the art of preparing wine and beer and of making bread; the discovery once more of the delicacies of the table, of distillations and the like; and if you likewise bear in mind the long periods which it has taken to bring these things to their present degree of perfection (for they are all ancient except distillation), and again (as has been said of clocks) how little they owe to observations and axioms of nature [i.e., science], and how easily and obviously and as it were by casual suggestion they may have been discovered; you will easily cease from wondering, and on the contrary will pity the condition of mankind, seeing that in a course of so many ages there has been so great a dearth and barrenness of arts and inventions (Bacon, 1860, 82–83).

Innovation in scientific theories and in techniques and crafts was part of the rhetoric of the new science of the seventeenth century. GM science and technology may appear to some to be new and unexpected, but these are precisely the qualities that the new science was aiming at. Bacon proposed, and would have recognized, many of the characteristics, though not all, of the present day enterprise of science of which genetic manipulation is part. For instance, in *New Atlantis* (1627; Bacon, 1860) he envisaged scientists working in teams. However, the commercial nature of modern science would be almost unrecognizable to Bacon; scientists then often benefited from patronage by the court or the nobility, or they were men of independent means like Boyle. However, seventeenth-century chemists who sold their remedies commercially look forward to the druggists of the eighteenth century and then to the giant pharmaceutical companies of the nineteenth and twentieth centuries. What was largely absent in Bacon's time was any conception of the ethical issues that might come with scientific and technological developments.

The new science and Bacon can help us to contextualize historically the science of genetic manipulation, just as the Bible does in a different way. But has Bacon anything to say that approaches in substance to the science of the genetic manipulation of crops? On the face of it, he did in *New Atlantis,* and it is worth quoting the passage in full:

We have also large and various orchards and gardens, wherein we do not so much respect beauty, as variety of ground and soil, proper for divers [various] trees and herbs: and some very spacious, where trees and berries are set whereof we make divers kinds of drinks, besides the vineyards. In these we practise likewise all conclusions of grafting and inoculating, as well of wild trees as fruit-trees, which produceth many effects. And we make by art in the same orchards and gardens, trees and flowers to come earlier or later than their seasons; and to come up and bear more speedily than by their natural course they do. We make them also by art greater much than their nature; and their fruit greater and sweeter and of differing taste, smell, colour, and figure, from their nature. And many of them we so order, as they become of medicinal use.

We have also means to make divers plants rise by mixtures of earths without seeds; and likewise to make divers new plants, differing from the vulgar; and to make one tree or plant turn into another (Bacon, 1860, 158–159).

Many of these techniques were known about already and it is clear that Bacon approved of bypassing the seasons and altering the "natural" course. He also believed that the ideal scientists in *New Atlantis* would be able to make a new species, which is what perhaps GM science will be able to achieve. Yet Bacon was silent as to how a new species would be made: it was an example of what the new science might achieve in the distant ideal future, but it is partly the nature of the technologies involved in genetic manipulation that cause disquiet. We can also note that Bacon expressed no ethical qualms about creating new species – he would have seen it as improving humankind's life. A sign of the historical difference between the cosmos of knowledge in Bacon's time and in today's is that Christianity rather than science could be brought in to support the possibility of creating new species. Before his Fall, Adam was believed by English Protestants in the seventeenth century to have cultivated the Garden of Eden:

> God knew that Idleness corrupts the best natures, and therefore Man was imployed in that humble vocation; for though God did at first create the kinds of all Plants, yet doubtless man had and yet hath an honest Allowance to procreate a Diversity of Species by Transplantation, Ingraftings, Innoculations and other various Cultivations, which were Incestuous in other creatures (Pettus, 1674, 43–44 in Webster, 1975, 466).[11]

The new science promised to enlarge man's dominion over nature, but until the nineteenth century its practical achievements were few. In agriculture improvements came from within agriculture. Moreover, the Biblical prophesy that man was doomed to labor on the soil still appeared valid a century and a half after Bacon wrote *The New Organon*. Though, the adventitious discovery of a new plant rather than science seemed capable of wiping out the curse put upon Adam. Joseph Banks (1743–1820), who came to Tahiti in 1769 on Cook's first voyage, observed the island's bread fruit trees and wrote:

> In the article of food these happy people may almost be said to be exempt from the curse of our forefather; scarcely can it be said that they earn their bread with the sweat of their brow when their cheifest sustenance Bread fruit is procurd with no more trouble than that of climbing a tree and pulling it down. Not that the trees grow here spontaneously but, if a man should in the course of his life time plant 10 such trees, which if well done might take the labour of an hour or thereabouts, he would as completely fulfull [sic] his duty to his own as well as future generations as we natives of less temperature climates can do by [continuously toiling to sew in winter and reaping in summer] (Beaglehole, 1963, 341).

3.3.1.6 Shaping the Natural Environment

GM crops, it could be argued, are a further example of human beings shaping the natural environment. Their introduction will probably increase the trend to monoculture and uniformity. They can, therefore, be viewed as helping change and distort existing ecosystems especially with regard to habitats and plant, insect and animal variety. In addition, genetic contamination from GM plants to "natural" plant species is a feared danger. Historically, one can discuss attitudes to human changes to the environment. However, it is difficult to find historical analogues to the laboratory-based science of genetic manipulation, though a global transfer of plant genes has taken place at a "natural" level with the movement of plant species such as eucalyptus, sugar, potatoes, etc., from one continent to another in the past 500 years.

---

[11] Bacon discusses "the transmutation of plants one into another" in *Sylva Sylvarum* in *Works*, 11, pp. 506–509 and more generally pp. 506–527.

In many respects worries about man's shaping power over nature was late in coming. In the early modern period and later, changing the natural environment for human use and health was often seen as an unquestioned good, although cities and towns were viewed as foci of ill health. The early European settlers of North America, for example, unhesitatingly believed that all her natural resources were there to be exploited. Captain John Smith, one of the leaders of the colony in Virginia, wrote in 1616 of New England's wood and fisheries:

> Of woods seeing there is plenty of all sorts, if those that build ships and boates, buy wood at so greate a price, as it is in England, Spain, France, Italy and Holland ... live well by their trade ... what hazard will be here, but doe much better? And what commoditie in Europe doth more decay than wood? For the goodnesse of the ground, let us take it fertill, or barren, or as it is: seeing it is certaine it beares fruites, to nourish and feed man and beast, as well as England, and the Sea those severall sorts of fish I have related ... you may serve all Europe better and farre cheaper than can the Izeland fishers, or the Hollanders, Cape blank, or New found land ... (Captain Smith, 1616, 21–22, in Barbour, 1986, I, 337).

Smith's account, like many others, was part of the propaganda aimed at getting settlers to come over to America from England (in the same paragraph he indicates the need for more labor). Smith depicted a land of plenty and profit, where, despite his recognition of the depletion of European sources of wood, there was no acknowledgement of the need for conservation or replenishment. Such descriptions also reflected the expansionist mood of Europeans. The natural resources of old Europe were to be replenished from a new continent and from the re-discovered Far East. Moreover, there was the hope that new kinds of resources would be found, as they were with tobacco, sugar, potatoes, Indian corn, quinine, etc.

There were few initial worries about using up America's natural resources. Preservation of nature, which figures largely in the GM debate, was not in the minds of the early settlers. Indeed, they saw some aspects of nature as positively dangerous to their health. Forests and woods in hot areas were viewed as pathological, helping to create miasmatic disease-producing damp air and fogs. The mortality in the early years of Virginia and of Maryland was ascribed to the presence of surrounding woods. When they were cut down, the settlers' health, it was reported, improved (Speed, 1676, 43–44).[12]

Changing the environment, of course, has occurred throughout the course of human settlement, and GM crops fit into this pattern. The spread of GM crops and food is also linked by some commentators to power. Both state and corporate pressure has been exerted upon developing countries to accept GM crops and some developed countries have also experienced the same pressure, especially in the forum of the World Trade negotiations. The historical analogy that comes to mind is the way that colonial settlement sought to change the environment. Here, the elements of power and profit are present, though there are differences.

---

[12] See also, "In the Countries [Virginia's] minority (i.e. its first 21 years) and before they had well cleared the ground to let in ayre (which now is otherwise) many imputed the stifling of the wood to be the cause of such sicknesses" (Hammond, 1656, in Force, 1947, vol. 3, p. 10 s.p.).

Colonial changes to the environment were far more extensive either in reality or programmatically than GM crop technology has been so far. Indeed, the defenders of GM technology would argue that it will not radically change the environment. However, it is worth setting out some examples from the history of colonial settlement to show that in the past the wholesale alteration of the environment was not perceived to be a problem, and that medical theory played a role in the process.

Medical theories about the need to clear wilderness and especially woods and forests in the tropics went together with colonial settlement and with the search for profit. In the West Indies, for example, the islands' environment was transformed by sugar monoculture. Deforestation took place on a grand scale. Barbados was the first island to be cleared of trees by the later seventeenth century and many of the rest of the islands had largely lost their forests in the next century. The need of plantation owners to clear the land drove the process. There were inconveniences: "at the Barbados all the trees are destroyed, so that wanting wood to boyle their sugar, they are forc'd to send for coales from England" (Anonymous, 1676, in Watts, 1987, 186).[13]

Medical thinking supported such clearances. The environment, especially climate, was viewed up to the end of the nineteenth century as a major cause of illness. In the tropics Europeans, it was believed, were especially vulnerable when living in hot humid unventilated low-lying land. The advice was to settle on high ground. Here the air was drier and moved by breezes. James Lind (1716–1794) of scurvy fame agreed with these recommendations in his influential *An Essay on Diseases Incidental to Europeans in Hot Climates* (1768). He also believed that if settlement could not be made in healthy places then the environment should be engineered to be healthy:

> ... it is to be hoped, that as soon as Granada and the Grenadines (which have lately proved so fatal to the English planters) are cleared of woods, due attention will be given to situations so eligible for houses ... we shall then hear nothing more of fatal diseases sweeping off the inhabitants of these islands (Lind, 1768, 202–203).

Such medical thinking, which justified wholesale environmental change can also be found in the opening up of Australia in the nineteenth century. The writer, former apothecary, and amateur gold digger, William Howitt (1792–1879) wrote in 1855 of Australia:

> It is yet, with very trivial exception, one huge unreclaimed forest. Now, I do not believe that any country, under any climate in the world, can be pronounced a thoroughly healthy country, while it is in this state. The immense quantity of vegetable matter rotting on the surface of the earth, and still more of that rotting in the waters, which the visitants must drink cannot be very healthy. The choked up valleys, dense with scrub and rank grass and weeds, and the equally rank vegetation of swamps, cannot tend to health. All these evils, the axe and the plough, and the fire of settlers, will gradually and eventually remove; and when that is done, I do not believe that there will be a more healthy country on the globe (Howitt, 1855, I, 231).

---

[13] See more generally Watts, 1987, 184–187, 393–405.

Medicine in this period would have had no qualms about humans shaping the environment or indeed about the danger of monoculture to the environment. Given not only the climatological theory of disease but also the medical belief that peoples' constitutions were shaped by the constitutions of the places that they lived in, anything which made the environment more like that back home in Britain was health-giving. These medical ideas go back to the Greeks and they were genuinely believed in by British and continental European colonizers from the sixteenth to the late nineteenth centuries. However, in the present day we may, depending on our point of view, interpret the application of such beliefs to colonial environments not only as medical theories *strictu sensu* but also as colonial ideology. The new owners of the land change it, linking it more to themselves and shaping it into their own image and culture. This, it was believed, helps to make the land more healthy for the new settlers, and it also underwrites their possession and power over it. Moreover, it ties in well with the need to make profits.

In the nineteenth century such views were also linked to a racial discourse in which European agriculture was perceived to be superior and progressive, reflective of racial superiority. In India, the Presidency Surgeon of Bengal, James Ranald Martin, wrote in 1837 of the unhealthiness of uncultivated nature: "pestilential marshes" and impenetrable forests where "no wind disperses the putrid exhalations of the trees ... the soil excluded from genial and purifying warmth of the air, exhales nothing but poison and an atmosphere of death gathers over all the country." Human "industry and perseverance" can change this with the axe and fire "so that vanquished nature yields its empire to man who thus creates a country for himself" (Martin, 1837, 86–87). Martin easily slid from "vanquished nature" losing its environment to writing of Europeans creating a country for themselves to replace the one that Indians had made. The move is justified in the name of "improvement" which partly reflects a new confidence in British technological superiority in the wake of the Industrial Revolution.[14] The result of such improvement is to create a new environment that belongs to, and in terms of health, fits the English rather than its original owners.

> Agriculture must be much improved in Bengal before the European ... can be said to have created a country for himself. A Hindoo field is described by [James] Mill to be in the highest state of cultivation, where only so far changed by the plough, as to afford a scanty supply of mould for recovering the seed, while the useless and hurtful vegetation is so far from being eradicated, that when burning precedes not, the grasses and sterils which have bid defiance to the plough, cover a large portion of the surface (Martin, 1837, 87).

Such lack of improvement was typically seen by the British as the result of Indian "laziness" and their antipathy and blindness to European notions of progress: "The same author [Mill] concludes that 'everything which savours of ingenuity, even the most natural results of common observation and good sense, are foreign to the agriculture of the Hindoos'" (Martin, 1837, 87).

---

[14] On technology and empire see Headrick (1981, 1988) and Adas, 1989.

Racist discourse is no longer prevalent amongst medical writers. However, one can ask whether the discourse of scientific progress which surrounds GM technology also implies that some countries are more advanced than others; that, to use the terminology of the nineteenth century, some are higher than others in the scale of civilization. Other questions that come to mind from reading material such as Martin's are: to whom does the environment that GM technology produces belong, whom does it "fit," and are such questions still relevant today? We can note that today the medical benefits of changing the environment through GM technology are not expressed as positively as they were when colonization was reshaping nature. This is partly because the climatological theory of disease had been replaced after the laboratory-based bacteriological revolution of the later nineteenth century, and partly because of the development in the nineteenth century of the neutral stance (in an ideal world) of statistically based clinical trials. In medicine the link between places and people has been attenuated. What is now important are bacteria, viruses, and genes that may have preferred locations but which can in principle be found in any laboratory in the world. Laboratory-produced crops may also have lessened for some, but not all, the sense of crops "belonging" to a place, and perhaps it is no accident that the bio-medical developments that loosened the ties between climates, places and diseases are echoed by GM crops since the laboratory is central for both.

Paradoxically, earlier colonial settlement and its reshaping of the environment also began to dilute the sense of place in agriculture. European exploration, trade, and colonization started a wave of globalization. From the sixteenth century onwards, plants, animals and human beings were transported between and across continents.[15] The early English settlers, for instance, looked to see if English plants would grow and if they and their animals would thrive in North America, though they were also quick to see the potential of indigenous products like tobacco. In one description of Southern Virginia, Indian corn (and also rice) was promoted, but so was English wheat:

> But lest our palats ... dislike whatever is not native to our owne Country [England] and wheat is justly esteemed more proper this happy soyle, though at first too rich to receive it, after it hath contributed to your wealth by diminution of its owne richnesse, in three or four crops of Rice, Flax, Indian Corne, or Rapeseed, will receive the English wheat with a gratefull retribution of thirty for one increase (Edward, 1650, III, 12).[16]

Globalization took place within a localist framework: in the example above the soil had to be altered to grow wheat. Conversely, English bodies, in the view of medical and lay writers, had to be "seasoned" to the constitution of places like Virginia by not working hard in the first year and by gradual acclimatization (see Kupperman, 1984). In the process of recognising the local limits to life and growth, the colonial globalizing enterprise attempted to attenuate them. We leave it to others to elaborate further on the global and localist dimensions of GM technology.

---

[15] For a Darwinian discussion of this see Crosby, 1986.

[16] See also Wood (1634) on the need to lessen the richness of the soil before planting wheat.

3.3.1.7 Conclusion: Improvement, Population and Agriculture

In some ways the previous section has been related only very generally to the spread across the world of GM crops. However, as we come to the nineteenth century, the historical parallels and roots to GM technology become closer, though differences remain apparent.

In this section we want to bring out the links between GM crops and the theory of evolution. Competition for food and selective breeding were two formative aspects of Charles Darwin's (1809–1882) theory of evolution by natural selection. The ability of GM crops to alleviate world hunger is advertised as one of their benefits. And, the analogy is often made between GM crop technology and traditional selective breeding of plants and animals, especially by those who see genetic manipulation as a continuation with the past.

Thomas Malthus (1766–1834) in his *Essay on the Principle of Population* (1798) famously feared that the rise in population would outstrip the increase in the food supply. Hence a perpetual struggle for existence was inevitable. The environment now became the arena for this struggle, and the more populous it was the greater the struggle:

> Whatever was the original number of British Emigrants that increased so fast in the North American Colonies; let us ask, why does not an equal number produce an equal increase, in the same time, in Great Britain? The great and obvious cause to be assigned is, the want of room and food, or, in other words, misery (Malthus, 1798, 109).

In his autobiography Darwin wrote that having been convinced that variation occurred in nature he then enquired into "domesticated productions" "by conversation with skilfull breeders and gardeners and by extensive reading … But how selection could be applied to organisms living in a state of nature remained for some time a mystery to me." Then:

> In October 1838, that is, fifteen months after I had begun my systematic enquiry, I happened to read for amusement 'Malthus on *Population*', and being well prepared to appreciate the struggle for existence which everywhere goes on from long continued observations of the habits of animals and plants, it at once struck me that under these circumstances favourable variations would tend to be preserved, and unfavourable ones to be destroyed. The results of this would be the formation of new species. Here, then, I had at last got a theory by which to work (Darwin, 2002, 72).

In Chapter 3, "Struggle for Existence," of *The Origin of Species* (1859) Darwin gave Malthus a prominent place in his argument: "Hence, as more individuals are produced than can possible survive, there must in every case be a struggle for existence …It is the doctrine of Malthus applied with manifold force to the whole animal and vegetable kingdoms" (Darwin, 1974 [1859], 117). Food production and selection of plants and animals thus became enmeshed into the most powerful biological theory of the last two centuries. The science of genetics, when it developed after Darwin's theory, played a crucial role in the creation of GM crop technology. But, it is, we think, no accident that this technology is the material expression of two major constituents of Darwinian evolution: food production and selection.

Paradoxically, GM crop technology is a child of evolution yet it is also claimed to be a way of controlling and of bypassing evolution. The technology thus continues the tradition of human beings having dominion over nature and shaping it. It is also an example of the Baconian and new science's hope that practical results should issue from pure science.

Finally, one could note how in Chapter 3 of *The Origin of Species* Darwin stressed the interdependence of plants and animals with each other. The science of ecology, and with it campaigns for conservation, were emerging at this time. For instance, George Marsh's (1801–1882) *Man and Nature* (1864) challenged the wholesale destruction of the environment for profit, and noted that in the long run it produced more loss than gain. Such views also shape some present day perceptions of the environment and inform part of the GM debate. But that, we think, is another story.

3.3.2 Medical Intervention to Correct the Deficiencies of Nature

3.3.2.1 Introduction

In the preceding case study, we saw that nature was, by and large, positively valued and thus appeals to it guided judgment and behavior. Such appeals were warranted, because nature involved a unity of plants, animals, and humans, the goods of which were intimately bound up with each other. Utility, improvement, and progress in agriculture became positive values in the context of this unity. Altering nature, which was achieved late in the period and only on a modest scale, was acceptable if it served these goals and thus enlarged man's dominion over nature. The proper exercise of that dominion, however, was understood in terms of the unity of nature. There were justified limits on our use of nature.

A concept of nature as normative plays a major role both in the philosophy of medicine of Dr. John Gregory and in the science of man and morals of David Hume (1711–1776) of the late eighteenth-century Scottish Enlightenment. As in the case of foods, it was thought at the time that the normative dimensions of nature were directly observable, following the discipline of Baconian scientific method. Disease was thought to involve either an over-response or under-response of nature to changes in anatomy or physiology and the purpose of medicine was to correct these deficiencies of nature by restoring the body to it normal conditions. The capacities of medicine to accomplish this purpose were thought to be limited and expandable only with cautious attention to the increased iatrogenic risks of efforts to improve medical care.

Gregory provides the following conception of nature:

> We shall consider what is meant by *nature* – It is sometimes meant by the word nature the general system of laws by which God governs the world at other times it is used to express the particular system of laws by which any part of the world is governed at other times to express that unknown principle in the animal oeconomy by which the Body of itself may thro' off Diseases (Gregory, 1769, 2).

He does not treat nature as impersonal, but personifies nature, referring to nature with the feminine personal pronoun. Nature, Gregory holds, nurtures her creatures, though slowly. He has this to say about nature, in the second of these two senses, in his *Comparative View of the State and Faculties of Man, with those of the Animal World* (1772a) (most of which he presented to meetings of the Aberdeen Philosophical Society, which he helped to found along with his cousin Thomas Reid and other literati of that Northeast city):

> Nature brings all her works to perfection by a gradual process. Man, the last and most perfect of her works below, arrives at this by a very slow process (Gregory, 1772a, 55–56).

During the seventeenth and eighteenth centuries, in the period roughly from Bacon to Hume, there developed in British philosophy of science and medicine the view that the normative dimensions of nature were directly observable and that the normative itself existed in nature and was directly observable. As can be seen in the above passages from Gregory, far from being understood as value-neutral, nature was understood to be value-laden and also to include realities that were intrinsically normative: standards for correct structure (anatomy) and function (physiology) were written into nature itself, constitutive of it. Because nature was understood to be observable by science and medicine (using the methods of Baconian science) the normative dimensions of nature and the normative itself were understood to be both real and observable. On such a view, the normative did not somehow supervene on nature or the real (as we now, to our loss, tend to think). Instead, the normative was understood to be something real and therefore as something that helped to constitute nature. Insofar as the normative was constitutive of or found in nature, the normative was directly observable. Put another way, there developed in this period of British philosophy of science and medicine a robust realism of the normative in nature, a robust moral realism.

In medicine this moral realism of the observable normative was developed especially in the work of John Gregory, a Scottish physician, medical educator, and ethicist (Gregory, 1772b; McCullough, 1998a, b), and in the moral psychology and philosophy of David Hume (1739). Gregory held that nature is good at doing things that are good for human beings, maintaining health and correcting its loss, in particular. Health is an example of a normative dimension of nature that is directly observable. Hume provided an account of the intrinsically normative in nature, in particular the principle of sympathy that is a constitutive causal principle of human nature and by its very nature and function normative. Both Gregory and Hume were thoroughgoing Baconians, and so we start our consideration of seventeenth- and eighteenth-century British views of the observable normative in nature with a brief review of Bacon's experience-based method and its influence on medical science.

3.3.2.2 Baconian Experience-Based Method and Its Influence on Medical Science

Baconian scientific method provided the intellectual foundations for this moral realism and realism of the normative. The prevailing spirit of the time included a three-pronged attack on then-current science and medicine. The first prong involved a frontal assault on dogmatism, i.e., uncritical adherence to authorities, especially adherence to authorities that would not or, worse, could not be corrected by appeal to observation and experiment. The second prong assailed the "metaphysical" character of old ways of thinking. It is crucial to be clear what this critique involved: an objection to *speculative* metaphysics that was beyond the ken and therefore correction of observation and experiment. This critique did not involve rejection of metaphysics *per se*. It cannot be emphasized enough that this critique did *not* involve a rejection of a metaphysics of causes and their effects, i.e., that there really existed causes in nature. Speculative metaphysics appealed to the Cartesian method of introspection, the examination of ideas without any reference to experience, to determine which ideas were clear and distinct and therefore true (i.e., which had as their extension separately existing *res*). Speculation that was undisciplined by experience (speculation disciplined by experience is acceptable and involves hypothesis formation and testing) was rejected and along with it *a priori* metaphysics and all other forms of *a priori* reasoning. The third prong of the critique rejected system-building as often disconnected from reality, despite the intricacy and evident beauty of classification systems. Systematic nosologies of disease, for example, spun out in their own momentum ran the risk of being mere words, nominal, and not descriptive of something real.

Bacon's philosophy of science and medicine provided an alternative to dogmatism, speculative metaphysics, and nominal systems. Bacon proposed to ground science and medicine in experience, properly understood.

> There remains simple experience; which, if taken as it comes is called accident; if sought for, experiment. But this kind of experience is no better than a broom without its band, as the saying is; – a mere groping, as of men in the dark, that feel all around them for the chance of finding their way; when they had much better wait for daylight, or light a candle, then go. But the true method of experience on the contrary first lights the candle, and then by means of the candle shows the way; commencing as it does with experience duly ordered and digested, not bungling or erratic, and from it educing axioms, and from established axioms again new experiments; even as it was not without divine order and method that the divine word operated on the created mass (Bacon, 1875b, 81).

The true investigator of nature, who aims to produce results that we can rely on in subsequent investigations and in interventions in human lives intended to "relieve man's estate," goes about gathering experience in a disciplined, not serendipitous, fashion, by forming and testing hypotheses about the order of nature, an order given to nature by the creator god of deism.

By following the method of experience, the investigator of nature submits to the discipline of identifying discoverable reality, especially the principles of things, the real constitutive causal forces or components in them. In this way, the investigator of nature establishes claims that range along a continuum of "certainty" – a better word choice now would be 'reliability'. This continuum ranges from the "certainly true," i.e., highly unlikely to be refuted by further experience, at one end through the "doubtful whether true," to "certainly not true" (Bacon, 1875b, 260). Accumulating knowledge (justified, i.e., well founded, and therefore reliable belief) is understood by Bacon to be a rigorous process.

> ... if in any statement there be anything doubtful or questionable, I would by no means have it suppressed or passed in silence, but plainly and perspicuously set down by way of note or admonition. For I want this primary history to be compiled with a most religious care, as if every particular were stated upon oath; seeing that it is the book of God's works, and (so far as the majesty of heaven may be compared with the humbleness of earthly things) a kind of second Scripture (Bacon, 1875c, 261).

Bacon's influence on medical science was profound (and still is, in the form of evidence-based medicine, which Baconian medical scientists anticipated more than two centuries ago). Thomas Sydenham (1624–1689) followed Bacon and understood nature to be "the totality of events according to a strict and impersonal causal nexus" that was ordained by a creator god (King, 1970, 117). Sydenham insisted that physicians have rationales for their remedies – and so he rejects "empirics" who offered remedies but no rationales about their supposed efficacy – and such rationales were to be developed on the basis of experience, the careful observation of disease and how its course is altered by various medical interventions. Hermann Boerhaave (1668–1738), the founder of the modern science of human physiology whose teaching drew students from across Europe to the medical school at Leiden, becomes the great champion of Baconian method as the basis for physiologically based medicine.

Lester King points out that during this period explanation in medicine was understood in general as "the process whereby some particular observation or phenomenon is referred to something else" (King, 1970, 185). The something else may be a "hypothetical entity," a concept, or a principle or constitutive causal component of a thing (King, 1970, 185). Baconian method provides explanations of the third type: referral to real, constitutive causes in things that generate subsequent events or things. These real, constitutive causal principles are the origin of "motion:" regular, ordered change in things that can be observed following Bacon's prescriptions for natural and laboratory experiments. This law-governed change cannot be explained in terms of the old, static metaphysics of substantial forms, but can be explained in terms of a dynamic metaphysics such as Leibniz's (McCullough, 1996). Baconian method rejects speculative metaphysics and, far from rejecting metaphysics altogether, calls for and presumes a dynamic metaphysics of (causal) principles.

In summary, Baconian method in science generally and in medical science and clinical practice in particular promoted a critical attitude toward any proposition of fact. Claims about what the facts are should not be taken on authority but tested against experience. Baconian method is open to anyone willing to submit to its rigorous discipline of observation, hypothesis formation, hypothesis testing, classifying hypotheses for the level of "certainty" or reliability, and using the collection of tested and therefore more reliable hypotheses as basis for further investigation of nature and clinical intervention in the care of patients. Scientific and medical knowledge and clinical practice are therefore based on a secular method that is not the exclusive province of scientists and physicians. Explanation proceeds by reference to experientially observed principles, or, more precisely, from observing the steady, repeated effects of principles. Gregory and Hume both admit that they do not have observation directly of principles but do hold that they have sufficient grounds in experience to hold that such principles exist. Gregory and Hume are thus both causal realists, a crucial ingredient in their metaphysics and therefore in their moral realism. Finally, Baconian method is compatible with deism. Bacon was committed to deism – the existence of a creator god who is the author of the causally ordered nature that we can observe using proper method – and Gregory followed Bacon in this. Indeed, Gregory goes out of his way to argue that being a physician did not require one to be an atheist. Instead, Gregory appeals to the argument from design and experience of physicians for the existence of a creator god (McCullough, 1998a). Hume, famously – or, more commonly then, infamously – was emphatically not a deist. Baconian method can be reliably applied in a science of human nature, a science of man, and a science of morals, to all of which Hume subscribed and contributed.

3.3.2.3 Gregory on Medicine's Role in Correcting Nature's Deficiencies

Gregory took the view that the main purpose of medicine was to correct nature's deficiencies. Nature sometimes under-reacted to disease and injury and sometimes over-reacted to them. Return to normal function was the goal. Normal structure and function were thus normative: they set the standard for judging the success of medical treatment. Normal structure and function were also observable, which mean that the natural as both normal and normative was observable.

Gregory's views on the observable normative in nature have their roots not only in Baconian method but in the views of such Continental figures as Friedrich Hoffmann (1660–1742) and Georg Stahl (1660–1734) on the human body, not as a simple body-machine, but as a complex of mind and body. Hoffmann held that the "life of man consists in the uninterrupted communion of mind and body" (Hoffmann, 1971, 11). Hoffmann went on to claim that there is a "vital principle" in humans, that provides a causal source not only of mechanistic bodily functions but also of "thinking and reasoning" (Hoffmann, 1971, 12–13). For Hoffmann, the mind possessed an animal spirit that originated purposive movement – of the body and of the mind.

On such an account each human being has a nature and that nature is the same in each of us, a dynamic nature originating in the *vis viva* (living force). With enough persistent, disciplined observation the investigator of human nature can identify with high reliability the principles of regular activity or habit that constitute human nature as an active force or animal spirit. For Gregory this investigation leads to the identification of two constitutive principles in human nature: reason, "a weak principle" (Gregory, 1772a, 15); and the social principle, surely the stronger principle that "unites them [human beings] into societies, and attaches them to one another by sympathy and affection. This principle is the source of the most heartfelt pleasure which we can ever taste" (Gregory, 1772a, 86).

Gregory's views on the primacy of the social or moral principle in human nature are not unique; they reflect widely held views of Scottish-Enlightenment figures (McCullough, 1998a). Gregory takes his account of the moral principle from Hume (about which principle more in the next section). It is worth noting that the discovery of the social principle helped to solve a major political problem in Scotland after unification of 1707 and failed rebellion/civil war four decades later: how to have a nation or people with a distinct national identity in the absence of their having any longer their own, independent state. The normative dimensions of the social principle include both morality and politics, with consequences felt to the present day in the devolution of power to Scotland and its new – more accurately, restored – Parliament.

Gregory understood nature to bring her animal works to perfection by a faster gradual process than she does in the case of humans. Animals, for example, have a built-in capacity to heal from their illnesses and wounds – what he calls the "animal above." This capacity works better in animals than it does in humans; animals usually recover and do so quickly, unless they are overwhelmed by their disease or injury in which case they quickly die. Here Gregory is reporting the results of Baconian observation of nature doing a good thing, perfecting animal works by preserving their health and granting them quick deaths. Both the perfection and health are observable normatives.

In the case of humans, nature most of the time responds to disease and injury effectively. However, sometimes it over-responds and at other times under-responds. One of the purposes of medicine is to correct these inappropriate responses, so that nature can do its good work of restoring health.

> These efforts [to restore proper structure and function] however are sometimes very irregular and not at all uniform and sometime so very violent as entirely to destroy the machine and at other times insufficient to do any service from this weakness. Here art must be called in to regulate, to restrain [when it violently over-responds] and assist nature [when it weakly under-responds], according as she shall stand in need of either. When nature makes no efforts toward her own relief, physicians must be guided by experience and practice (Gregory, 1768, 61–63).

Medicine aims to correct nature's deficiencies, in order to restore normal function.

Gregory also holds that there is a natural language of pain, which is especially pertinent for adults knowing when to respond to the needs of their children.

> The cries of a Child are the voice of Nature supplicating relief. It can express its wants by no other language (Gregory, 1772a, 40).

The natural language of pain does crucial work, because it activates our response to children from sympathy (about which more in the next section). Our subsequent attention to them results in their protection and the meeting of needs that otherwise would go unmet, to the harm of the child.

In his views of nature and human nature and in his concepts of health and disease, Gregory combines what we would call a purely descriptive statistical concept of species-typical anatomy and physiology and what we would call a value-laden account of anatomy and physiology. In Gregory's philosophy of medicine, health becomes an observable norm of the structure and function of human anatomy and physiology. The norm in question is partly descriptive of the range within which most human anatomy and physiology falls. The norm in question is also partly normative in that this species-typical anatomy and physiology result in death not occurring too early in life (mortality by age 16 was then 60%) and such that we have a decent functional status throughout life, especially in our advanced years. In taking these two views he follows Bacon's prescription that the purpose of medicine is to relieve man's estate. Diseases constitute an observable departure from this observable normative.

The capacities of medicine to relieve man's estate, i.e., reduce mortality, morbidity, and lost functional status, Gregory reminds his reader regularly, are limited. Those limits should be pushed out in clinical research always attentive to the risks of doing so. Medicine takes it moral authority from its capacity to assist nature when it is too weak and to lessen its force when it is too violent, always appreciative of medicine's limits to do so. Medicine's value comes from its attentively and carefully supplementing the already intrinsically valuable work of nature's corrective processes. Thus, Gregory's concepts of health and disease and of the moral authority of medicine are based on a powerful, discoverable normative concept of the normal in nature.

The moral authority accorded medicine thus understood is not paternalistic. The ability to observe a normative nature in general and to observe the descriptive-normative health and disease in particular provide sufficient warrant for clinical judgment about patients' clinical needs and action to relieve them of the pain, distress, and suffering of injury, disease, and impending death. This warrant includes support for Gregory's views on patients who refuse needed intervention: "No people would be stupid or obstinate as to leave a dislocated arm to nature" (Gregory, 1762, 162^r). However, he also holds that treatment cannot be forced on those who refuse it, even when necessary to save life. They have the right to go out of this world in any way of their own choosing, Gregory says, which then included self-dosing and over-dosing with laudanum (McCullough, 1998a).

3.3.2.4 Hume on the Observable Normative

David Hume wrote *A Treatise of Human Nature* with the crucial subtitle, *Being an Attempt to Introduce the Experimental Method of Reasoning into Moral Subjects* (Hume, 1739). In doing so, Hume wrote what he called a "science of man," which is self-consciously Baconian language. As a scientist of man, Hume held commitments typical of the Baconian experiential metaphysics of activity that had by the middle of the eighteenth century firmly dislodged the older, speculative, static metaphysics of substantial forms. Hume held that there exists something called human nature that is active, not static. We can discover the constitutive operating, causal, physiologic principles of human nature by following the Baconian method of experience, not *a priori* metaphysics and its speculation. The mistake of the older metaphysics was therefore not to do metaphysics but to do it in the wrong way. This is apparent, for example, in Hume's account of the distinction in the Treatise between impressions and ideas, based on direct and repeated observation of them. When it comes to a science of morals, Hume identifies the core moral principle of sympathy – a real, constitutive, causal force in human nature that binds us socially.

Before Hume, there was a considerable body of literature on sympathy that sought to address and explain such phenomena as a room of previous gaiety being plunged into gloom when an evidently very sad person entered and circulated among the crowd, actors who were able to communicate emotions such as fear from the stage to an audience (stage plays were a major form of entertainment to a populace with a high rate of illiteracy), and also physical phenomena such as a wound healing when a dressing was removed from it, taken to another room, and soaked in a medicated solution, followed by wound healing. Until Hume, there had been no scientifically satisfactory solution to this problem of affective action-at-a-distance, although there had been progress in explaining action-at-a-distance between bodily organs via nervous sympathy.

Hume's account of sympathy should be read within his larger account of human social and political origins. Hume denies that humans were ever once in a "state of nature," in which they existed as isolated, independent, self-sufficient social atoms, a concept that must, he said, be regarded as a "mere fiction" (Hume, 1739, 317). Human nature is inherently social and actions that sunder the social connection within families or within the larger society, which he calls "ingratitude," constitute the most "horrid and unnatural" crimes (Hume, 1739, 300). Hume treats understanding (or reason) and the affections (or the social principle) as "inseparable" components of human nature (Hume 1739, 317), evoking the metaphysical category of distinct inseparables.

Morals cannot, he argued, be derived from reason because morals "excite passions, and produce or prevent actions. Reason of itself is utterly impotent in this particular" (Hume, 1739, 317). Sympathy is the principle of our moral physiology that generates morals.

> No quality of human nature is more remarkable, both in itself and in its consequences, than that propensity we have to sympathize with others, and to receive by communication their inclinations and sentiments, however different from or even contrary to our own (Hume, 1739, 206).

Sympathy "communicates" or acts at a distance, a view that Hume models on the inter-organic account of sympathy that had previously been developed by others (McCullough, 1998a). Sympathy causes us to "enter so deep into the opinions and affections of others, whenever we discover them" (Hume, 1739, 208). Sympathy, as a physiologic principle of human nature, is the same in all of us, although it does need to be developed and properly regulated.

Hume's great scientific contribution was to explain sympathy by reference to the mechanism of the double-relation of impressions and ideas or sympathy. When one, for example, sees another human being in pain, one has an impression – an immediate, physical sensation of normative visual images against the eye. One then forms, by abstraction from the particulars of that other person's pain, the idea of oneself being in pain. This idea, in turn, leads naturally and automatically to the impression of that same pain in oneself, a physical sensation activating the nerves. When sympathy functions properly – when it is not deformed by self-interest or ingratitude – then this double-relation of impressions and ideas functions automatically, just as sympathetic changes in the stomach in response to irritation of the heart function automatically. Hume is writing a neurophysiology of sympathy. As a result of the causal workings of sympathy, one has the same impression – or direct, vivid experience – of pain that the individual observed has.

This experience of pain moves one to act to relieve that other individual promptly, again unless one has willfully deformed one's principle of sympathy by self-interest or ingratitude. Sympathy also moves us prospectively, to act to prevent pain, distress, and suffering in others. The way that sympathy for another's pain or distress works is by creating, as a result of the double-relation of impressions and ideas, an unease that we can resolve only by acting for the other in need.

In Hume's *Treatise* sympathy is a real, constitutive, active, moral, mental physiologic principle of human nature that explains morals. The mind is an active force or active power and sympathy is one of its main constitutive principles. Sympathy explains the social principle: the natural, evident, built-in, observable, and fundamental other-regardingness of human beings that Scottish moral and political philosophers took to be an empirically established fact.

Hume in his account of sympathy in his *Treatise* thus provides an account of the observable principle of human nature, sympathy, which is intrinsically normative. Baconian method thus allows for the direct observation in nature of the normative. The result is a robust realism of the normative in nature, which Gregory takes up whole.

3.3.2.5 Implications for Accounts of the Normativity of Nature

The Normativity of Nature and Our Sympathetic Response

Gregory's account of health and disease draws on observable norms in nature interpreted through the sympathetic response. This account has implications for current debates about whether the concepts of health and disease should be regarded as value-neutral or value-laden. Gregory comes down on the side of the latter view and the reasons he does so helps us to see why this is the more reasonable conceptualization to adopt. Recall that Gregory holds that the interventions of medicine are needed when nature under- or over-responds to disease or injury. Gregory provides his reader with no account of how great a magnitude of departure from normal function is required to warrant a clinical judgment that nature is under- or over-responding. He does indicate that the physician diagnoses under- or over-response on the basis of symptoms, changes that patients experience and that they can observe and report or that the physician can observe and report. These include such changes as the distress or fright that a patient can experience from heart palpitations (an over-response) or from a depressed spirit (an under-response). Gregory is not explicit about this, but the following seems a reasonable reconstruction of his thinking. Judgments about what should count as disease states are based on departures from the observed, species-typical (as we now say) normal structure and/or function. These are judged abnormal on symptomatic grounds that themselves count as such in virtue of being mediated through the physician's sympathetic response to the patient.

Such a view encounters challenges when it comes to detection of signs of disease, i.e., changes in structure or function that are not experienced by patients. Gregory does not address this problem directly, but the following response to it can be constructed reliably in Gregorian and Humean terms. Findings of the physician that count as signs of disease or injury gain this clinical status in virtue of a judgment by the physician that the findings fall outside the normal range. Here 'normal' means a state in which human beings can function well enough, which surely is normative in nature. Signs of disease or injury can also invoke the sympathetic response in its anticipatory, preventive aspect. This is because it is sometimes the case that signs are predictors of disease or injury, the prospect of which provokes a sympathetic response in the physician, thus inciting the physician to act to prevent disease or injury. A sympathy-based account of value-laden concepts of health and disease would therefore appear to carry from symptoms to signs of disease and injury.

This eighteenth-century, Gregorian account of the value-laden nature of concepts of disease and health has implications for how the normativity of nature should be understood. First, nature's healing capacities or powers are valued because they contribute to a state that humans value, health, and, when properly assisted or regulated by medicine, these powers can be harnessed to relieve man's estate by reducing frightful rates of mortality and morbidity. The normativity of nature in such an account is instrumental: nature's healing powers can operate to restore health unaided by medicine and so we value them. Second, in the clinical setting nature also has this instrumentally valuable dimension when it directly or prospectively activates a sympathetic response. The normativity of nature in the clinical setting is not simply a function of nature (as it is when nature acts unaided by medicine); that normativity is also a function of properly functioning human judgment. To use an older language of predication, the normativity of nature is not present in it as a constitutive elements of it, but is said of nature in virtue of features of nature that provoke a sympathetic response. Put in Gregory's own terms, in the clinical setting the normativity of nature exists in response to a natural language of pain. Third, the normativity of nature varies not according to features of nature considered in and of themselves, but according to the strength of the sympathetic response to those features. Not all normatively abnormal features of nature are abnormal to the same degree; we can and do set priorities. Fourth, the normativity of nature is not value-neutral in the sense of disvalue being attributed to states simply in virtue of their departure from a species-typical or some other statistical norm. The observation of a human hand with six fingers is not enough, by itself, to make polydactyly a pathological anatomical formation of that hand. Value-neutral accounts of the abnormal in nature that are meant to generate obligations on someone's part to correct such conditions cannot be generated from the Gregorian account of health and the healing powers of nature. In other words, a simple observation, unmediated by the sympathetic response, of a departure from the observed normal does not count as a normatively important departure. A person with six fingers who suffers shame or embarrassment from this abnormality does evoke the sympathetic response and so this anatomic irregularity becomes a clinical deformity meriting intervention to correct it, since nature does not self-correct this problem.

The Normativity of Animal and Human Nature

Current debates about the acceptability of altering nature frequently appeal to nature as normative and the (sometimes plain) observability of the normative in nature, the "natural." It would therefore appear that these current debates owe a considerable, unacknowledged debt to the Baconian tradition of nature as normative in such Scottish Enlightenment figures as Gregory.

This intellectual debt presents interesting intellectual challenges to those who would claim now that nature as we find it is normative and that attempts to alter it can be judged acceptable or unacceptable on the basis of this observed normative or normality. One challenge is whether such claims can be separated from a dominant strand in the Baconian tradition, namely deism. Bacon's method is explicitly deist, reflecting Bacon's theological commitments. Gregory is also a deist in his account, explicitly so in his lectures on medical ethics (McCullough, 1998a, b). The view that there are normative components of nature appears to rely very much on an account of the origin, external to nature, of its normativity, namely, in the creative act of the deists' god. In this strand of the Baconian tradition, deism is an essential intellectual foundation for claims that there is an observable normative in nature. The normativity of the natural presumes the existence and creative action of the deists' god.

By sharp contrast, Hume is explicitly not a deist. Indeed, he took himself conclusively to have shown that the argument from design – the cornerstone of a Baconian, experiential, deist account of the existence of a creator god – failed miserably. Hume appears to have taken the world as he found it, as a good Baconian should (and long before Wittgenstein thought to do so). In the course of his observations, Hume appears to have taken it as a fact of nature as we find it that it includes realities that are intrinsically normative. These can be observed like any other reality or aspect of nature by following the experimental method of Bacon.

It is important to underscore the (sharply) limited scope of the Humean account of the normativity of nature. It appears that only animal and human natures include as a constitutive principle the social principle in humans and its counterpart in non-human animals. Entities that lack this principle in their make-up do not possess and therefore do not exhibit an intrinsically normative nature. Thus, the Humean normativity of nature does not extend to include vegetation, rocks, lakes and oceans, and all of the rest of the larger non-human and non-animal environment. This limited scope makes Humean normativity of nature quite distinct from the universal vitalism of a Leibniz, who held that monads – conscious (to different degrees) centers of activity – are constitutive principles of all creatures, from the lowliest and loneliest dust mote out at the edge of the galaxy all the way up to humans and, above them still, the angels. (God is not a monad but the creator of them.) On Leibnizian grounds one could hold that all of nature is normative, but on Humean grounds one could not, for the very good Humean reason that the proposal that monads exist is an unrestrained flight of fancy of speculative metaphysics. Hume would likely also add that the normative is epitomized by the social principle, the other-regardingness that generates the ties that bind: of family, clan, and nation.

These ties are natural and naturally normative. The regard that we show to the larger non-animal and non-human nature we elect to show; such regard is artificial, not natural. It is, to be sure, normative, but of a much more unstable and unreliable sort than the naturally normative. The protection of forests perhaps can be built on such a slender reed, but not a nation or people.

The Unity of the Sciences

Finally, it would appear that only the human and animal sciences touch upon the normativity of nature. The physical sciences do not. We can get from neurophysiology a science of man with a robust realism of the normative and moral realism. We can get from physics a robust realism of physical causes but no realism of the normative or moral realism. In this respect, some sciences are privileged in explicating the normativity of nature, while others play no role at all in undertaking this task.

3.3.2.6 Conclusion

By way of summary, it would therefore appear that, within the period and in the cases under consideration here, it was considered intellectually possible and respectable to hold on secular scientific and philosophical grounds that nature was real, was ordered and stable, that this order and stability originated in principles or real, constitutive causal forces in the world, that some of these principles such as self-correcting animal and human physiology had normative dimensions (nature accomplishes things that are good for animals and human beings by preventing early mortality and responding effectively most of the time to disease and injury), that some of these principles are intrinsically normative, such as sympathy, and that the normative dimensions of principles and normative principles themselves are observable.

One could hold such views both as a deist, i.e., on the view that we could discover from the study of nature and its evident design that there is a creator God. "Nature's God" is the author of all of reality, including nature itself. Gregory himself embraces deism. However, one could also hold such views and as an anti-deist or atheist, because the intellectual authority of such views did not originate in belief in nature's God. Instead, belief in nature's God originated in and was thus a function of such views. With Hume, for example, one could reject the argument from design but embrace these views of nature. To be sure, Hume was famous or infamous (depending on your views on the matter) for his irreligiosity. Yet Scots of many religious stripes and none at all embraced his concept of sympathy and the social principle as fundamental truths about human nature and Scottish national character. It would therefore appear that a secular account of the normativity of the natural can be constructed without deism or at least on neutral grounds. Whether an account based on scientific observation of the normative in nature can succeed two centuries later – when a fact-value distinction is to be taken seriously – becomes a central methodologic question in debates about altering nature.

3.3.3 Dynamic Nature: The Ever-Changing Historical Conceptions of Aging

3.3.3.1 Introduction

Our case study of Gregory's views on nature reveal an appeal to nature as a source of cautious, disciplined alteration of nature, especially in response to nature's deficiencies in the form of diseases and injuries. Gregory was an avowed enemy of enthusiasm, i.e., adoption of an hypothesis or practice in the absence of an evidence-based evaluation of it. He thus can be read as embracing disciplined, melioristic alteration of nature. Our third case study presents a complex mix of both positive valorization of nature (e.g., following natural law to achieve healthy longevity) as well as negative valorization (e.g., aging as a disease to be altered by some advocates of anti-aging medicine.).

Aging has always been a curious and troubling phenomenon, made especially disturbing by its apparent inevitability. Throughout the history of science, aging has been considered inherent in all living beings except for one-celled organisms, which simply divide in two and thereby seem to cheat both aging and death (Cole, 1992, 177). For humans though, the intractability of aging challenges our unquenchable hopes for improved health and longevity and our pervasive dreams of immortality. The rise of "anti-aging medicine" (Cole and Thompson, 2001–2002, 9–14) and the emergence of new scientific possibilities for age-retardation and/or life-extension (Aubrey et al., 2002, 452–462) in the 1990s are both testaments to these wishes. Private biotechnology companies, hucksters, and entrepreneurs fed off aging baby boomers and the peculiarly American fear of decline, frailty, and old age (Hall, 2003; Gullette, 2004). In the early twenty-first century, these developments are gaining momentum and will pose difficult moral, social, economic, and political challenges well into the future (President's Council on Bioethics: Staff Working Paper, 2003, 21–30; Juengst et al., 2003).

How do concepts of nature affect these challenges? Historically, this question has taken the following form: Is aging natural (or normal) and universal? Or is it pathological (a disease process)? Scholars and physicians on both sides of the issue have used scientific answers in search of moral guidance on questions of life extension and age retardation. Generally speaking, those who argue that aging is "natural" mean that it is a normal (not pathological) and universal process. The moral guidance (usually implicit) embedded in this view derives from the ancient Stoics, for whom "nature" is an orderly, unified, and ultimately benevolent force in individuals and in the cosmos. The task of the aging individual is to bring herself into alignment with the orderly span of human life.

On the other hand, those who consider aging a pathological process are also operating from the (implicit) assumption that "nature" is benevolent, along with its corollary – that pathology/disease are evil deviations. This corollary grounds their moral justification for ameliorating intervention into the biology of human senescence. Although the natural/pathological distinction pervades Western thought about aging, it has proven inadequate to the scientific and normative the challenges posed by the complexities of human aging in its irreducibly historical circumstances.

From a legal and statistical view, no one has died of old age since 1951, when "old age" was deleted from standardized federal and state lists of official causes of death. This would seem to settle the question: old age is not a disease but rather a natural process. But, of course, this deletion does not answer our question; it simply reflects a decision to look for more specific causes of death. This relationship between aging and illness, between a normal state and disease remains vexing to both the scientific community and the general population. The prevailing view among biogerontologists has been articulated by Leonard Hayflick (2002, 416–421), who defines aging (very carefully) as "the normal biological processes that are collectively the single greatest risk factor for the pathologies of old age" (Hayflick, 2002). Yet there are those who argue that aging itself is a disease process (Murphy, 1987, 237–255; Caplan, 1992). This section undertakes an historical examination of concepts of aging, its causes, and modalities of improvement. We will be interested in whether there has ever been (or can be) a scientific consensus or definitive explanation of the nature of aging that might guide us in making difficult policy choices about age-retardation and/or life extension.

3.3.3.2 Ancient Greece and Rome

Any developed historical examination of concepts of aging in western civilization must begin with the ancient Greek and Roman thinkers, whose views remained dominant well into the eighteenth century. But even as classical causal models of aging held sway, Medieval and Early Modern physicians and scientists challenged classical resignation to the inevitability of aging. This difference in sensibility is best captured by Gerald Gruman, who applies the categories of "apologism" and "meliorism" to the history of ideas about aging and prolongevity. According to Gruman, most Greek and Roman thinkers on the subject of aging were apologists, "condemn[ing] any attempt by human action basically to alter earthly conditions" (Gruman, 1966, 6). Gruman does acknowledge that ancient thinkers put forward modest aspirations toward longevity and the extension of life, but in our view, he trivializes ancient notions of meliorism by dismissing their moral and cosmological contexts. Certainly, with regard to extension of the physical lifespan, the ancients could be considered apologists. However, by no means should this reflect a pervasive cultural orientation. The ancients were most definitely interested in notions of improvement, only less quantitatively than the lifespan meliorists who have more pronouncedly populated history.

In his discussion of the history of ideas about aging, Richard L. Grant elaborates on this cautious, more qualitative meliorism. Hippocratic theory (c. fourth century BCE), one of the earliest attempts at a rational causal explanation, describes aging as the gradual loss of a finite amount of innate heat with which all people are born (Grant, 1963). And, while Hippocratic theorists believed that "[t]his innate heat could be fortified and replenished by various means," they maintained that "it could never be completely restored to a given previous level; the total reserve of innate body heat continuously diminished" (Grant, 1966, 449). Aristotle, one of the ancient world's most important theorists of aging, echoes this assessment of aging. In works like *On Longevity, On Life and Death*, and *On Youth and Age*, he posits his own biological theories in an attempt to explain the processes affecting humans throughout their lives. According to Aristotle, people age as their bodies become increasingly dry and cold, losing the vital elements of moisture and heat (Gruman, 1966, 15). And, like the Hippocratic physicians, Aristotle believed this to be inevitable: aging and mortality were inherent and unavoidable elements of human nature that might be moderately delayed, but could never be wholly subverted (Gruman, 1966, 16).

Cicero (c. first century BCE) goes even further than Aristotle and the Hippocratic physicians, asserting that not only do humans inevitably age, but that aging, properly understood, becomes a virtuous process. In fact, according to Cicero, the only problematic aspect of aging is the pervasive negative societal attitude toward it. In his landmark *De senectute*, as cited by Gruman, Cicero posits rebuttals to the four chief complaints people lodge against to old age:

> To the complaint by the aged that they are excluded from the important work of the world, Cicero replied that courageous old men can find a way to make themselves useful in various advisory, intellectual, and administrative functions. To the charge that senescence undermines physical strength, Cicero answered that bodily development counts for little as compared with the cultivation of mind and character. To the complaint that aging prevents the enjoyment of sensual pleasures, Cicero replied that such a loss is good riddance, because it allows the aged to concentrate on the promotion of reason and virtue. Finally, to the charge that old age brings with it increased anxiety about death, Cicero answered along Platonist lines that death should be considered a blessing, because it frees the immoral soul from its bodily prison on this imperfect earth. Even if the soul were not immortal, he adds, it is desirable that the duration of life be limited just as a play is limited in length (Gruman, 1966, 14–15).

Far from seeking longer life, Cicero advocated that individuals fully accept, and even embrace, the naturally ordained stages of life.

Finally, Galen's writings continue this classical, apologist trend. In his book on hygiene, and its place in the maintenance of human health, Galen (second century CE) adopts a Hippocratic and Aristotelian foundation for his own theories. For Galen, aging is caused by the loss of both the innate heat and the innate moisture with which people are born. Accordingly, the moment a person is created, the male sperm causes a distinct drying effect on the female material, initiating the long, inevitable, and natural sequence of aging. Thus, just as soon as humans are born they begin to die. However, that is not to preclude any room for improvement in the inevitable aging of human beings. Within the context of natural and inevitable aging, Galen leaves room for the benefits of hygienic measures to "moderate, but not ... alter, the inexorable development of the constitutional imbalance of old age" (Gruman, 1966, 16). According to Carole Haber, "[T]hose able to maintain their vital energy, whether through diet, moderation, or simple good luck, were apt to experience a healthy and vital old age" (Haber, 2001, 9). Those who were either not so diligent or so fortunate would reach old age sooner. Thus, for Galen, as well as the rest of the Classical thinkers, aging is an inevitable and, therefore, natural life process that can be slowed, though never stopped. Further, thinkers like Cicero even go so far as to question why anyone would want to prevent aging. As a *natural* phenomenon, Cicero believes the process to be valuable and to serve an important, even necessary, and inviolable function in human society.

3.3.3.3 Medieval and Early Modern Europe

From antiquity until the thirteenth century, Western conceptions of aging underwent little change. Per the historical medieval synthesis, Christian nuances were relatively smoothly integrated with classical sensibilities. Aging remained a natural, inherent part of the human condition, as determined by God after the Fall of humans in the Garden of Eden (Cole, 1992, 9). However, at the same time, the mortality of humans' earthly forms seemed less troublesome in the face of the spiritual immortality promised by virtue of Christian religious faith. For many Medieval and Renaissance Christians physical aging remained natural and normal, with this earthly existence occupying only an infinitesimal part of human life. Pathological theory remained centered upon the four humors and the four qualities, and old age still resulted from a loss of vital moisture, heat, and/or energy, leaving one increasingly susceptible to illness and ultimately death. However, within this paradigm, there occurred a gradual shift in beliefs about the limits of the lifespan. For some pre-Enlightenment theorists of aging, like Roger Bacon (1214–1294) and the alchemists, death no longer remained a foregone conclusion, and attempts to subvert old age and death seemed not only possible, but even achievable. Thus, gradually, meliorism became an increasingly predominant philosophical perspective. According to Gruman, the meliorists maintained that "human effort can and should be applied to improving the world" (Gruman, 1966, 10). For the radically meliorist alchemists, e.g., Roger Bacon and the like, who formulated "the first systematic prolongevitism to appear in Western civilization," improving the world included attempts to rid the world of all disease, old age, and death (Gruman, 1966, 49).

Born early in the thirteenth century, Roger Bacon may have been the earliest of the important pre-Enlightenment aging theoreticians. Publishing such works as *On Retardation of Old Age, the Cure of Old Age, and the Preservation of Youth*, Bacon remained obsessed with aging for much of his scholarly career (Freeman, 1965). And while his conception of the process of aging, like that of most pre-Enlightenment theorists, is now considered mostly unoriginal and derivative of that handed down by the ancients, his hybrid approach to prolongevity is a rather unique and noteworthy contribution to the historical development of ideas about aging (Gruman, 1966, 64). Adopting the hygienist strategy set forth by Galen before him, Bacon set himself apart by the end to which he thought this approach useful. As quoted by Grant, Bacon elaborated his beliefs in *Of the Wonderful Power of Art and Nature*:

> The Possibility of Prolongation of Life is confirmed by this, that Man is *naturally* immortal, that is, not able to dye: And even after he had sinned [original sin], he could live near

3 Scientific and Medical Concepts of Nature in the Modern Period in Europe

a Thousand Years [the antediluvian span], afterwards by little and little the Length of his Life was abbreviated. Therefore, it must needs be, that this Abbreviation is Accidental; therefore it might be either wholly repaired, or at least in part. But if we would but make Enquiry into the accidental Cause of this Corruption, we should find, it neither was from Heaven nor from ought but want of a Regiment of Health.... Now the Remedy against every Mans proper Corruption is, if every Man from his Youth would exercise a complete Regiment, which consists in these things, Meat and Drink, Sleep and Watching, Motion and Rest, Evacuation and Retention, Air, the Passions of the Mind. [The six so-called "Non-Naturals," a term coined by Galen to indicate those things outside of the body, the proper use of which was beneficial to the body.] For if a Man would observe this Regiment from his nativity, he might live as long as his *Nature* assumed his parent would permit, and might be led to the utmost Term of *Nature*, lapsed from Original Righteousness (Grant, 1963, 456; see also Burns, 1976, 202–211).

Thus, for Roger Bacon, nature seemed both mutable and normative, and hygiene would play a significant role in extending the life of people who seriously followed such a regimen. However, reliance on the benevolence of Nature alone did not conclude Bacon's attempts to improve the aging of humankind; he also maintained that alchemy might play a role in this process. According to Gruman, Bacon, in his *Opus majus*, asserted that alchemy might further prolong the lives of people: "[S]ilver and purest gold ... [are] thought by scientists to be able to remove the corruptions of the human body to such an extent that it would prolong life for many ages" (Gruman, 1966, 49). Via alchemy and heightened attention to hygiene, Roger Bacon sought to extend the longevity of people, increasing the lifespan to 150 years or more (Gruman, 1966, 64). Thus, whether Bacon believed this fixed lifespan to be some natural limit to be transcended or that humans had been mired for centuries in an artificially shortened existence, he believed hygienics and alchemy might together improve, or at least live up to, the nature that governed human life.

Other early Modern theorists of aging maintained less extreme hopes for their forays into the improvement of the aging process. Gruman classifies the Latin alchemists, including Bacon, as "radical" prolongevitists, or "thinkers ... so optimistic that they foresaw a decisive solution to the problems of death and old age, ... [and] aimed at the attainment of virtual immortality and eternal youth" (Gruman, 1966, 7). Some even believed that they might be able to gain control of the Aristotelian fifth element, that ethereal substance which, unlike the other four, was harmonious and would yield eternal life (Gruman, 1966, 16). Other early Modern thinkers, like Luigi Cornaro (1467–1565) and the non-alchemical hygienists, are considered more "moderate," in the scope of their sought-after improvement (Gruman, 1966, 8). Late in his life, Cornaro published a book of essays entitled *Sure and Certain Methods of Attaining a Long and Healthy Life with Means of Correcting a Bad Constitution* (1562), in which he extolled the virtues of exercise and temperance in leading a life that made full use of our naturally determined and defined human lifespan (Freeman, 1938, 328). His theory of aging and his hygienic regimen were both formulated in the traditional Renaissance fashion, with the appropriation of established classical thought for a slightly different purpose. He modified Cicero's arguments against prolongevity to support his own prolongevity stance and then "simplified and popularized" the Galenic regimen of hygiene, making unproven claims of its unrivaled effectiveness (Gruman, 1966, 68). According to Gruman, Cornaro set forth four primary reasons, contra Cicero's, in support of the "desirability of longevity:" first, he believed that a long life was worthwhile, so that one might continually improve upon his/her relationship with God; second, he believed old age was "the most beautiful period of life;" third, Cornaro believed that as much time as possible was necessary for men to achieve perfection "in learning and virtue;" and finally, he asserted that only a sufficiently long life would guarantee a quick, painless, and "natural" death (Gruman, 1966, 69–70). In many of these arguments, Cornaro's seems a departure from most, especially radical, prolongevitist thought. He still sought after improvement, but for different reasons and with different goals in mind than those proposed by Bacon and the alchemists. According to Haber, what set Cornaro apart was his maintenance of a positive perspective regarding aging: "Cornaro did not see old age itself as an enemy to be vanquished. Instead, if good health were maintained, old age was seen as a friend with whom he could share contentment" (Haber, 2001, 10). The balance struck by Cornaro between apologism and meliorism set him apart from most thinkers in the history of prolongevity, and especially from the more scientifically-inclined thinkers to come during the Enlightenment.

3.3.3.4 Origins of the Scientific Method

The sixteenth and seventeenth centuries witnessed a sea change in the scope and perspective of Western science and rationality (Debus, 1978; Porter, 1997, ch. 9). Not surprisingly, this advance affected conceptions of aging and prolongevity as drastically as it did other intellectual circles. Faith in the possibility of scientific progress replaced the stagnant primitivism that had dominated the intellectual landscape of aging for so many centuries. According to Gruman, prior to the scientific revolution, people could only look to the past for examples of what was possible (Gruman, 1966, 75). Even the Alchemists, with their quasi-scientific prolongevitist ventures, looked more to the past to justify their work than they did to the future, with hope and faith in progress (Gruman, 1966, 63). Now, in the late sixteenth and early seventeenth centuries, with thinkers like Rene Descartes (1596–1650) and Francis Bacon, conceptions of prolongevity and aging began to evolve again, and no longer would the classical inheritance be so stifling (Gruman, 1966, 75). Further, with the gradual turn of Enlightenment thinkers away from millennia-long constraints of what they increasingly believed to be Christian dogmatism, many thinkers established longevity as the progressive, rational, and somewhat anti-Christian victory over death (Gruman, 1966, 77). Descartes's primary contributions to this discussion of aging and prolongevity involve a similar paradigm shift. In his *Discourse on the Method of Rightly Conducting the Reason, and Seeking Truth in the Sciences* (1619), Descartes, as quoted by Gruman, exemplifies his belief in the potential of progress and justifies his support for the cause of prolongevity:

> ... [A]ll at present known in it [medicine] is almost nothing in comparison of what remains to be discovered ... we could free ourselves from an infinity of maladies of body as well as of mind, and perhaps also even from the debility of age, if we had sufficiently ample knowledge of their causes, and of all the remedies provided for us by *nature* (Gruman, 1966, 78, emphasis added).

For Descartes, the whole of scientific and medical knowledge of the body and aging was not to be found via study of the theories of the Ancients, but in the future, as the result of experimentation and discovery, that might use *natural* remedies to subvert *natural* aging. Descartes advocated for the employ of mathematics and speculation by humans in their search for the knowledge necessary to secure dominion over the whims of nature (Gruman, 1966, 77). His revolutionary reconceptualization of the body as a machine, rather than a container of Hippocratic vital spirit or Aristotelian humors only further validated the realistic possibility of such grand ambitions (Gruman, 1966, 79). In Descartes, the prolongevity movement found a melioristic ally sharing more in common with the alchemists than the more pervasive and influential hygienists. In such a system, based upon the credibility of scientific investigation, even radical prolongevity seemed a likely possibility. Rather than continuing to cling to the classical, hygienist view that humans already lived out their naturally-defined lifespans or the Alchemist notion that nature would actually allow for much longer life, had it not been accidentally abbreviated by human behavior, the early Moderns like Descartes began to believe that, via science, nature might be mutable. In other words, given the progress promised by science, for the first time in human history, nature seemed truly and realistically conquerable, and the existential meaning provided by nature and its limits might no longer be so consequential. Of course such declarations trade in irresponsible hyperbole, but with Descartes and the newly planted seeds of the soon-coming Enlightenment, such a characterization of nature as mutable, like that by which we operate today, did not appear to be far off.

Building upon the work begun by Descartes, Francis Bacon even further invigorated the early Enlightenment trend toward prolongevity via this progressive new science. In fact, according to Gruman, Bacon was "the very personification of the idea" of meliorism and sought, like Descartes, through science, "to increase human well-being by gaining a greater command over *nature*" (Gruman, 1966, 80, emphasis added.). And, in his *Advancement of Learning* (1605), Bacon made clear his commitment to improving the process of aging, heralding "prolongevity as the 'most noble' goal of medicine" and even going so far as to criticize physicians of his day for not pledging the same: "… [T]he lengthening of the thread of life itself, and the postponement for a time of that death which gradually steals on by *natural* dissolution and the decay of age, is a subject which no physician has handled in proportion to its dignity" (Gruman, 1966, 80, 81, emphasis added). Bacon then went on to elaborate exactly why he believed the concept to be so central, offering four similarly volatile justifications for his rather outspoken stance:

> First, that, to date, all works on the subject have been unsound; Aristotle's contribution was only of slight value, while more modern writers (apparently, the alchemists and iatrochemists) were vain and superstitious. Second, that naïve efforts to preserve natural warmth and moisture do more harm than good. Third, and perhaps the most apt, that prolongevity is a long and complex undertaking:
>
> … men should cease from trifling, nor be so credulous as to imagine that so great a work as this of delaying and turning back the course of nature can be effected by a morning draught or by the use of some precious drug; by potable gold, or essence of pearls, or suchlike toys.

And fourth, that it is necessary to distinguish between the regimen for health and that for longevity, for that which exhilarates the body and spirit is not necessarily conducive to long life (Gruman, 1966, 81, quoting Bacon, 1860–1864, 40–41).

Thus, not only did Bacon allege that the members of the contemporary medical establishment were not spending nearly enough time and interest on such issues, but he also called into question much of the work that had been done on longevity by thinkers in the past. Like Descartes before him, Francis Bacon has proved invaluable part of the history of ideas regarding the improvement of the aging process, not because of any revolutionary scientific discovery, but instead because of the faith he had in science to do this work. This displacement of faith from religion to science truly marked a transition in perspective regarding the sanctity of the natural. What had previously seemed permanent and perhaps even sacred began to appear, at least to some, like yet another superable frontier for man to conquest.

3.3.3.5 The Enlightenment

Building upon the landmark contributions of founding fathers like Descartes and Francis Bacon, the Enlightenment was soon fully underway. Thinkers like Benjamin Franklin demonstrated the culmination of this evolution, wholly trimming from their perspectives any primitivist or religious tendencies, at least as far as science and progress were concerned: "Primitivism no longer carried any weight; … Franklin looked entirely to the future. And the dogma of the fall of man … no longer influenced Franklin, for his religious views were founded on reason rather than revelation" (Gruman, 1966, 83). Similar views were incorporated into the theories of thinkers like Antoine-Nicolas de Condorcet and William Godwin, whose thoughts, according to Gruman, "mark the culmination of eighteenth-century prolongevitism" (Gruman, 1966, 85). Nonetheless, however anti-primitivist such thinkers claim to be, their ideas about aging and its improvement demonstrate the incorporation of and development upon the thoughts of those who came before them. Condorcet's was a sort of hybrid approach to improvement, in which he made room for the hygiene of old amidst his devout passion for scientific progress. Fundamental to his conception of prolongevity were three main strands: "the improvement of the environment," via public health measures and the like; "the inheritance of acquired characteristics," by which Condorcet believed each generation's proclivity to longer life would increase given adequate support (for the same reasons Roger Bacon alleged the lifespan had diminished); and "the advancement of medical science" (Gruman, 1966, 88). The culmination of these three facets, as cited by Gruman, would ultimately satisfy, according to Condorcet, what can be only considered a highly idealistic vision of longer human life:

Would it be absurd then to suppose that this perfection of the human species might be capable of indefinite progress; that the day will come when death will be due only to extraordinary accidents or to the decay of the vital forces, and that ultimately, the average span between birth and decay will have no assignable value? Certainly man will not become immortal, but will not the interval between the first breath that he draws and the time when in the *natural* course of events, without disease or accident, he expires, increase indefinitely? (Gruman, 1966, 87, emphasis added).

Although his ideas resulted primarily from a novel conception of scientific progress, Condorcet's vision strongly resembled that of Roger Bacon and the Latin alchemists, who themselves sought a virtual fountain of youth, albeit via alchemical enhancement. Nevertheless, for Condorcet, nature had come to represent merely a boundable hurdle on the path to pure scientific progress.

William Godwin, an Enlightenment thinker and contemporary of Condorcet's, took this rapidly developing notion of rationality and reason even further, into relatively new, and as yet uncharted, prolongevitism territory. According to Godwin, the end of improvement of the aging process would find its fulfillment, not in scientific advancement, as proposed by Condorcet and prominent Enlightenment predecessors, but in the increased power of the human mind: "These authors [Francis Bacon, Franklin, Condorcet, etc.] ... have inclined to rest their hopes, rather upon the growing power of art, than, as is here done, upon the immediate and unavoidable operation of an improved intellect" (Gruman, 1966, 85). Thus, Godwin maintained that with increasing omnipotence, which humans would undoubtedly achieve with the steady advance of progress, the matter of the body might be sufficiently controlled to ultimately allow such people immortality. However radical Godwin's psychosomatic hypothesis might seem, it too betrays historical roots. In fact, upon further analysis, Godwin's theory strongly resembles that of Cornaro and the hygienists, and perhaps even the ancients. Inherent in Godwin's theory is the notion "right thinking and right living:"

> It followed that right thinking and right living should increase longevity, and the recipe for immortality is "cheerfulness, clearness of conception and benevolence." Moreover, the belief in prolongevity is a factor making for these desirable qualities; we become sick, and we die, partly because we expect such a fate and consent to it; but if we had faith in prolongevity, our more sanguine temper would prolong our lives (Gruman, 1966, 86).

Like Condorcet, Godwin advances the prolongevity movement, accommodating it to fit the rubric of the new science. However, neither can fully escape the legendary aspects that have dominated dreams of added longevity for the history of the movement. Further, neither moves significantly beyond the long-lived theories and ideas of how to achieve such immortality, whether relative or absolute. Thus, however anti-primitivist their beliefs, such thinkers' Enlightenment sensibilities seem merely more revised and/or complicated routes to the same end.

3.3.3.6 Nineteenth and Early Twentieth Centuries

As the Enlightenment mentality progressed and persisted, the scope of scientific inquiry into aging became proportionately limited to science. Thomas R. Cole suggests that the increasing utility of science soon marginalized many other longer-lived disciplines:

> Since the early nineteenth century, when elite French physicians in Paris established a clinical basis for geriatric medicine, the scientific study of aging had increasingly freed

itself from the influence of speculative philosophy, ancient medical theory, and theology. Scientists discarded older theories based on the exhaustion of some vital element (e.g., heat, moisture, energy) and focused their attention on progressively narrower empirically observable changes taking place in organs, tissues, and cells. In the late nineteenth century, experimental methodology began influencing the biomedical study of aging (Cole, 1992, 197–198).

Freeman corroborates this statement, asserting that "[t]he change from speculative philosophy to physiology is complete in [Jean] Charcot (1825–1893)" (Freeman, 1938, 331). Coincident with this evolution, "Charcot also acknowledged the limits of scientific inquiry. 'It does not seek to find out the essence or the *why* of things.... It remembers that beyond a certain point, *nature*, as Bacon says, becomes deaf to our questions and no longer gives an answer' " (Cole, 1992, 197, emphasis added.). In other words, according to Charcot, the progress of science would increasingly limit any normative meaning provided by nature. Charcot's contemporary Claude Bernard emphasizes a similar point, "warn[ing] scientists against wasting their time thinking about the origin, destiny, meaning, or purpose of living things. 'If our feeling constantly puts the question *why*, our reason shows us that only the question *how* is within our range' " (Cole, 1992, 198, emphasis original). This turn away from the metaphysical, to focus solely on the physical, ushered in an era of callous distaste both for aging as well as for those afflicted by it. Unlike Cicero and Luigi Cornaro, the prolongevitists of the late nineteenth and early twentieth centuries levied "unambivalent hostility toward weakness and illness in old age" and "declared infirm old age ... an unacceptable condition" (Cole, 1992, 175).

Through his work, Charles Asbury Stephens (b. 1844) exemplified this period's tendency to revulsion at the thought of old age:

> For Stephens old age was as horrifying as death. ... [He] could not abide by physical decline. He found old age a condition of "grossness, coarseness, and ugliness," and described the aging body as "a sad, strange mixture of foulness and putrefaction in which the sweeter, purer, etheric flame of life struggles and smolders (Cole, 1992, 176–177).

For Stephens, the possibility of immortality was a necessity, if people were to escape the detestable stage of old age. In fact, he even went so far as to assert that not only was immortality possible, but that "aging constituted a pathological process whose removal would lead to deathless life" (Cole, 1992, 177). Thus, he proceeded to his biological studies seeking a sort of hygiene for individual cells. While the hygienists of the past had focused on right living for the human body, Stephens suggested that perhaps a paradigm shift was necessary to achieve the desired end of an increased lifespan. Ultimately, "[h]e concluded that, under proper conditions of nutrition and stimulation, the cell was a potentially deathless unit" (Cole, 1992, 177). Through his microscopic approach to prolongevity, Stephens appears somewhat visionary. Still, his theory seems heavily dependent upon the roots of the hygienic movement, which bear their own historical legacy. Just how original or effective his proposed regimen would prove remained to be determined.

Elie Metchnikoff (1845–1916), another prominent prolongevitist of this era, maintained similar fidelity to the Enlightenment ideals mentioned above, building his theory "around the Enlightenment tradition of rational obedience to benevolent natural law" and pledging his devotion in "a new faith ... in the all-powerfulness of science" (Cole, 1992, 187). However, unlike his contemporary Stephens, Metchnikoff did not hold immortality as the ultimate goal of life and science. Rather more moderate in his views, Metchnikoff sought after " 'orthobiosis' – a completely fulfilled life cycle, regulated by reason and knowledge" (Cole, 1992, 187). Accordingly, Metchnikoff set out to improve the latter part of life, old age, both in duration and in quality, seeing life extension merely in terms of added years as foolish and fruitless. Rather, far from pursuing longevity and immortality at all costs, Metchnikoff even demonstrated an understanding of the potential value of death: "He [even] hypothesized that at the end of a completely fulfilled life, an instinct for death would replace the desire to live" (Cole, 1992, 188). Thus, in Metchnikoff and his prolongevity theories, there lingers a distinctively pre-Enlightenment, perhaps Cornaro-esque, value for the sanctity of life, not necessarily above death, only before it.

Remnants of the thought of Luigi Cornaro exist in other thinkers of the time as well. In his work *Old Age Deferred* (1891), Arnold Lorand reincarnates the age-old notion of hygienics as a supplement to his scientific theories about aging and its deterrence. In his studies, Lorand belonged to the school that maintained that old age resulted from the "degeneration of the ductless (endocrine) glands" (Cole, 1992, 184). Thus, Lorand maintained, the forestalling of old age, at least until age 90 or 100, required obeisance to the " 'Twelve Commandments' for a 'green old age:' "

> These included plenty of open air, sunshine, exercise, and deep breathing; a carefully regulated diet; daily baths and bowel movements, assisted by purgatives if necessary; porous cotton underwear, loose clothing, a light hat, and low shoes; early rising and retiring; six to eight hours of sleep in a dark, quiet room with an open window; one complete day's rest each week; avoidance of unpleasant emotions, discussions, or activities; careful sexual relations within marriage; and temperate use of alcohol, tobacco, coffee, and tea. If functioning of the ductless glands was weakened by age or disease, Lorand recommended a glandular transplant or injections, "but only under the strict supervision of medical men." (Cole, 1992, 184–185)

For Lorand, although couched in the paradigm of Enlightenment rationality, the legacy and wisdom of the hygienists lived on. Further, this verifies that even while attempting to establish a new science upon which to reorganize our conceptions of life, the theories of the past remain all but impossible to evade, however antiquated.

3.3.3.7 Contemporary

In the last third of the twentieth century, geriatrics became established as a medical subspecialty, and longitudinal physiological studies were initiated by the U.S. National Institutes of Health. The ancient controversy over whether aging is "natural" or "normal" (i.e., universal) and healthy or whether it is natural and pathological took on new colorations. Leading figures – e.g. Robert Butler and T. Franklin Williams – in geriatrics and in the National Institute on Aging (est. 1975) firmly distinguished between "normal" aging and disease processes. Diseases of old age – not aging itself – became the primary target of clinical and scientific improvement efforts. Sharon Curtin captured this view in the title of her book *Nobody Ever Died of Old Age* (Curtin, 1973).

By the 1980s however, mainstream scientists and geriatricians acknowledged that the relationship between aging and disease was not a dichotomy but existed on a continuum. Some normal (i.e., universal) biological processes are harmful and clinically debilitating, thereby meriting classification as disease. Normal changes in the eye, for example, will cause cataract formation and blindness if they proceed far enough. Changes in the immune system and in the mechanical properties of the lungs sooner or later lead to pneumonia (Rowe, 1985, 200; Johnson, 1985).

By the 1990s, a group of researchers and scientists declared war on the processes of aging in itself. While philosophical justification for the view of aging as a disease can be found in the work of Arthur Caplan (1981), the driving forces behind the "anti-aging" movement are: fears of aging baby boomers, financial incentives of clinicians, drug companies, and the biotech industry; and advances in molecular science and medicine (Hall, 2003). In 1993, anti-aging medicine was established as a full-fledged and licensed (if controversial) medical specialty. Ronald Klatz, founder of the American Academy of Anti-Aging Medicine, offers the following definition of anti-aging medicine:

> Anti-aging medicine is a medical specialty founded on the application of advanced scientific and medical technologies for the early detection, prevention, treatment, and reversal of age-related dysfunction, disorders, and diseases. It is a healthcare model promoting innovative science and research to prolong the healthy lifespan in humans. As such, anti-aging medicine is based on principles of sound and responsible medical care that are consistent with those applied in other preventive health specialties (Klatz, 2001, 59).

Mainstream gerontology and geriatrics has vigorously attacked anti-aging medicine as a form of hucksterism with no sound scientific basis (Butler, 2001, 63–65). But serious scientists and pharmaceutical and biotechnology companies insist that an anti-aging pill will be available within 15–20 years. For our purposes, it is important to note that classification of aging as "normal" or "pathological" no longer holds scientifically; even if it did, the dichotomy itself no longer provides any clear guidance in an era when enhancement (rather than treatment) drives a good deal of scientific research and clinical care. The meliorist struggle against frailty, disease, and death goes on under both banners. However, the more radical quest to extend – even to double – the human life span will continue to wrestle with the normative implications of the apparently "natural" limit of about 120 years (Callahan, 1995, 21–27).

3.3.3.8 Conclusion

As we have seen, ambiguity over whether aging is a natural process or a disease cannot be resolved on scientific grounds. Definitions of aging are always historically and socially constructed. A complete understanding of any particular definition requires that one ask: who created it? What moral and religious values are in play? What economic and social interests are at stake? (A task that is beyond the old-fashioned "history of ideas" approach taken in this chapter). In the current environment, people on both sides of the question have used their answers to justify the search for life extension and age-retardation. Pathologizing aging is a seductive way to increase public support for research at life extension, as well as an effective way that advocates of anti-aging medicine argue that their therapies constitute treatment rather than enhancement. At the same time however, mainstream biogerontology calls for research into "normal" aging and the basic biological mechanisms of senescence in order to garner support for its own meliorist agenda.

Beginning with the Stoics, historical attempts to find a singular moral norm in "nature" haved been faced with a persistent, transhistorical problem: nature is both a source of beauty and goodness *and* a source of destruction and suffering. Although Reason told the Stoics that the Nature was "good beyond improvement," observation and experience reveal that it is also "evil beyond remedy" (Hays, 2003). Contemporary efforts at improving aging might take heed from this persistent paradox.

3.4 Implications of Appeals to Nature in the Ethical Assessment of and Policy Concerning Biotechnologies

The complexities and challenges of appeals to nature as normative, illustrated in our three historical care studies, should be responsibly managed. We therefore close this chapter by turning to the task of drawing some general lessons from our three case studies, especially with a view toward identifying major concepts and lines of reasoning that should be considered in contemporary assessments of biotechnology that appeal to nature, the natural, and the unnatural. We consider the scope of 'nature', nature as normative, constructing historically informed appeals to nature as normative (to correct current ahistorical approaches), altering nature and evolution, the two contrasting historical themes of caution in altering nature and enthusiasm for doing so, the persistence of historical conceptions of nature that often goes unrecognized in contemporary discourses, historical resistance to altering nature, and policy implications.

3.4.1 The Scope of 'Nature'

'Nature' is understood in these three case studies to have the following scope and features. First, nature is understood to be all that exists, including both organic and inorganic entities. Second, human beings exist in and are part of nature. Third, human beings can create nature, e.g., through agriculture. Fourth, human beings can detect and attempt to correct deficiencies in nature, e.g., through medical care. Fifth, human beings can alter nature for the better, e.g., by means of anti-aging regimens.

The scope of nature is also concerned with whether we should understand nature as an undifferentiated whole or nature as differentiated. Organic nature is seen as undifferentiated, because plants, animals, and humans share constitutions and characteristics. On this account, it would appear that inorganic entities do not share constitutions and characteristics with organic entities. Appeals to nature as normative make more sense for organic nature than they do for nature as all that exists, because the latter includes both organic and inorganic entities.

3.4.2 Nature as Normative

Nature as normative means that something about nature itself sets or requires standards to which human judgment, decision making, and behavior should conform or against which they can be confidently evaluated. Various appeals are made in the three case studies to nature, both as the basis or justification of judgments about and as guides to creating, correcting, and altering nature.

Sometimes nature is understood as a norm to which human judgment and behavior should conform. On this account, a concept of nature becomes a moral norm, against which human interventions into nature are judged as permissible or impermissible. Opponents of anti-aging interventions often make such an appeal. In a sense, Gregory appeals to nature in this way, as the basis of a concept of normal anatomy and physiology that it is the purpose of medicine to preserve and restore.

However, sometimes nature is in need of improvement. On this account, the present state of nature is not accepted as morally normative in the sense that efforts to improve nature are prohibited. The Baconian injunction to "relieve man's estate" implicitly adopts this conception of nature. There are at least three different ways in which nature can be thought of as needing improvement: (a) nature functions well but not well enough for human purposes or desires; (b) nature is deficient or defective and needs mild to moderate correction to get it back to "normal;" (c) nature is pathological, harmful to human beings, and needs major correction to make its constraints morally acceptable.

Still, at other times, nature is provided both positive and negative normative meanings in the science and medicine described throughout these case studies. It is not the purpose of this chapter to advocate for a particular conception of nature as normative or not. Rather the case studies are intended as descriptive accounts of some of the various ways in which nature has been conceived over the course of scientific and medical history, and especially in the last several centuries, since Bacon and the beginnings of the Scientific Revolution.

In all three of the above-mentioned approaches to concepts of nature, the question arises: to what sources are appeals made in articulating a particular conception of nature as normative for human judgment and behavior? We should be aware, in any account of nature as normative or as ethically irrelevant, of the meaning of nature that is being invoked and the implications this has for determinations of what is natural in the sense of normal and what is natural in the sense of normative (i.e., creates standards to which our judgment and behavior should conform). This has particular relevance for contemporary thinking about the normative authority of a concept such as nature in the context of policy development and implementation.

3.4.3 Constructing Historically Informed Appeals to Nature as Normative in the Assessment of Biotechnologies

Our three case studies and the general reflections drawn from them about nature as normative suggest the lines along which historically informed appeals to nature as normative should be constructed for use in the moral assessment of biotechnologies. First, such an appeal should make clear the meaning of 'nature', i.e., the scope that the term is meant to encompass. What nature includes should be clearly specified; otherwise appeals to nature put themselves, unnecessarily, at risk of equivocation. Such specification may range from all finite reality, both physical and biological, to only biological entities. Whether nature includes human beings, their behavior, and their cultural and physical artifacts should be made clear.

Second, accounts of natural structure and function should draw, as they did in the past, on the best available science. Accounts of nature that go beyond science should be labeled for what they are, speculation. The role of speculation about nature in ethical inquiry into biotechnology needs to be acknowledged and defended against the quite reasonable Baconian charge that it is worthless.

Third, the historical examples sometimes treat nature as having not just functions, i.e., physiology, but also purposes or ends toward which physiology tends, usually as perfection of nature. This way of thinking about nature (and natural law) would appear to require deism as its intellectual foundation: there is a creator God that designed nature and ordained its purposes, its tendency toward its own perfection. On such a deistic account, nature is dynamic and has a direction, its own perfection as a (poor) reflection of the perfection of the Creator.

In the past century or so, modern biology has abandoned deism and, with it, reference to such purposes or ends as required for adequate explanations of nature. The sciences of biology are now undertaken, for the most part, without appeals to teleology. The focus is on structure and function of organisms, very small to very large. Current structure and function are understood to be the products of centuries of evolutionary processes. The phrase, "natural selection," is unfortunate in its connotation of a "selector" who makes selections for various purposes, e.g., to confer advantage in reproduction. Disciplined scientific thinking in biology, medicine, and biotechnology, in our view, treats teleological explanations of nature are unnecessary. In contemporary biology nature is dynamic, to be sure, but purposeless.

This way of thinking does not rule out thinking of nature as purposive, provided that nature itself is not understood to be the origin of its purposiveness. As explained in Chapter 1 of this volume, many of the world's religions do indeed view nature as purposive, with its purposes originating in God. Thus, theological appeals to nature as purposive certainly make sense, provided that we remain clear that the origin of nature's purposes is outside of nature. Nature can even be interpreted as exhibiting these God-given purposes. Appeals to nature as intrinsically purposive, however, such as those made explicitly by Leon Kass and implicitly by the President's Council at the very least face a burden of proof – in our judgment, a very steep burden of proof.

Fourth, the basis for establishing nature as normative must be made clear and defended. In particular, clear warrant must be provided for the transition from a descriptive observation of something as normal (within acceptable deviations from observed normal range) or abnormal (outside such deviations) to a prescriptive judgment of something as normal (ethically acceptable) or abnormal (ethically unacceptable). 'Nature as normative' means that nature itself provides a reference point, benchmark, standard, or set of criteria for the moral assessment of biotechnologies.

Nature can be valorized positively or negatively. There appear to be two main senses in which nature is thus valorized and taken to be ethically normative: (a) nature provides a positive norm to which human judgment and behavior should conform; and (b) nature needs improvement, it could be better. Within this second sense of nature as normative, there are three sub-positions: (a) nature functions well but not always well enough for human purposes or desires; (b) nature is usually deficient or defective and requires ongoing human intervention to be corrected; and (c) nature is pathological or harmful to human beings.

In the first of these two ways of valorizing nature, the normativity of nature is intrinsic to it. Appeals to nature result in the objective identification of moral norms for the assessment of human interventions in nature. We admit that there is something immensely attractive to such an appeal; ethical judgments will be solidly grounded and transcultural, highly resistant to the perils of moral relativism. In addition, on this account of nature as normative, there is no naturalistic fallacy: one can confidently reason from observed norms to moral judgments based on and implementing them.

In the second of these two ways of valorizing nature, either the normativity of nature is partly a function of nature but also partly of human purposes or desires (a and b) or the normativity of nature is solely a function of human purposes or desires (c). The normativity of nature is either partly or wholly constructed. Any adequate construction requires that the norm-other-than-nature be clearly identified and justified by argument. This way of valorizing nature lacks the immense appeal of the first way; argument is needed and the problem of moral relativism must be confronted head on. On this account of nature as normative the naturalistic fallacy does indeed become a fallacy and must therefore be avoided.

3.4.4 Altering Nature and Evolution

There are particular concerns when altering nature alters evolution, especially in the long-term and with potentially unimaginable and/or unmanageable effects. We saw above that Bacon expected that the experimental method could be used to make new

3 Scientific and Medical Concepts of Nature in the Modern Period in Europe   191

species and also shape the natural environment. We would now say that we can expect to make new species and alter the course of evolution. Bacon would have such science and technology guided by the imperative to relieve man's estate – to reduce mortality, morbidity, and impaired functional status. The difference that exists in contemporary conversations of this sort, as opposed to those that occurred in Bacon's time, is captured by the infinitely greater scientific, biomedical, and technological power we wield today. Pervading our understanding of the scientific capacities of our historical predecessors is an appreciation of the limits of their abilities to truly alter nature in any grand or long-lasting way. The same cannot be maintained for biomedical science today, which, given its increasingly microscopic bent, is now or will soon be able to effect changes that will be long-lasting, or even permanent, perhaps irrevocably altering the nature of nature as we currently understand it. Such prospects cannot have been thought truly possible in past centuries, during earlier stages of scientific understanding and development. For this reason, it is all the more important that we heed some degree of the apologist notion of caution and attempt to curb science's and commerce's historical outstripping of ethical, political, and societal conversation and thought. The potentially irrevocable and/or frightening implications of our current scientific abilities and capacities make such thought and conversation an essential component of contemporary attempts to alter or improve nature.

The concept of the organic unity of nature bears special significance to conversations regarding evolutionary altering, as shared constitutions and characteristics play a major role in assessing what is natural and, therefore, in determining the normative implications of that naturalness. Scientific and technological attempts to alter the course of evolution may result in the creation of species that do not share their constitutions and characteristics with the rest of plants, animals, and humans. This would especially appear to be the case for cross-species fertilization and also for human-machine hybrids. The entities that resulted would be radically new species if they had qualitatively different constitutions and characteristics. To the extent that such differences disrupted the organic unity of nature, entities with such differences would be unnatural or would force a thoroughgoing reconception of nature and the natural. Thus, determinations of their value would have to derive from sources other than nature or conceptions of nature would have to be expanded to prevent discontinuity and the disruption such a scenario would entail.

3.4.5 *Two Contrasting Themes*

Among the many themes presented in these case studies, two extreme and contrasting themes emerge from these case studies as significant. The first is that human beings should exercise caution in attempts to alter nature. In part this caution arises from the recognition of the always-limited capacity of science and technology to alter nature. This caution also arises from the recognition that attempts to alter nature can and do result in making things worse – either for human beings or for nature (in the sense of not including human beings). Here we might expect of science and medicine that they provide incrementally more effective tools for managing nature, to make it function normally more often and better – attentive always to the risks of efforts to make our management tools more powerful. One way to characterize this first group is as apologists or disciplined meliorisists, as exemplified by the Ancients, the Hygienists, and Gregory.

The second theme is an enthusiasm for altering nature. This enthusiasm originates in a more ambitious understanding of the capacities of science and technology to alter nature. This enthusiasm needn't assume that such capacities are unlimited; only that they are not as limited as we might think at any one moment and are, instead, considerable. This enthusiasm need not deny risks to altering nature, although it sometimes does. It does need to assume that these risks are in the majority, if not almost all, of cases manageable. This second group may be characterized as enthusiasts, exemplified by many anti-aging advocates. Members of this camp typically assume that, through science and medicine, we can control and direct nature largely to ends or goals of our own choosing.

3.4.6 Persistent Importance of Historical Conceptions of Nature

Many of the various historical conceptions of nature we have discussed in these three case studies endure in contemporary thought on the subject. Old values reappear in new contexts. On Bernard I. Cohen's (1985) account of the introduction of new ideas into the history of science and thought generally, it is often the case that such ideas are appropriated in contemporary debates without recognition of their historical origins. Awareness of those historical origins, which would include awareness of historical alternatives, provides a critical perspective often lacking in the appeals to nature in contemporary bioethical discourse.

3.4.7 Historical Difficulties in Altering Nature

Claims that nature should be corrected or even altered have been difficult to accept in the past. Why has such resistance occurred? What are its implications for contemporary bioethical debates about altering nature?

One source of resistance is experiential. It comes from the hard-won recognition that human ability to alter nature – to correct its deficiencies and to improve its capacities – is always limited and often weaker than nature itself. Gregory emphasizes the limits of medicine's capacities repeatedly, echoing the Hippocratic concern that nature can sometimes overmaster us, especially in the end-stages of disease and injury. Gregory also warns against the perils to patients of forgetting these limits and succumbing to enthusiasms – a problem that still plagues us, e.g., in the overselling of what we should expect from the "revolution" and "miracles" or molecular medicine and biotechnology.

Another source of resistance is conceptual. One such conceptual source is the view that organic nature sets its own standards of normality and therefore of what we should seek. Another is the view that correcting and improving nature's capacities can be reliably understood as part of treating and preventing disease, whereas enhancement cannot. The former, because they "relieve man's estate" have higher priority than the latter in the competition for scarce resources.

Yet another source of resistance can be attributed to the religious and theological dimensions and implications of altering nature, as discussed in the three case studies. Science and medicine in the modern period, especially since the late eighteenth century, have understood themselves to be secular. That is, science and medicine can do their work of developing reliable accounts of nature and of interventions to correct or alter nature without appeal to transcendent reality, deity or deities, or sacred texts and revealed traditions. This methodologic commitment does not require that science and medicine need be intrinsically hostile to religious beliefs and traditions and the moral theologies in which they are expressed. Instead, in the science and medicine of this period, 'nature' means all that exists and is observable. Supernature – that which, by definition, cannot be observed but is thought or believed to exist – is not an ontological category in modern science and medicine. This leaves open how we should understand the relationships between nature, thus understood, and supernature. In particular, 'nature' in science and medicine names a collection of observable organic and inorganic entities that exhibit function but not ultimate purpose or telos. Appeals to the perfection of nature as the justification for its alteration thus necessarily appeal to a telos or purpose of nature outside of nature itself. In particular, scientific and medical concepts of nature are indifferent to claims that a creator God has made nature and ordered it to perfection. Nature is surely understood to be dynamic in our three cases studies and in the modern period – anatomy, as it were, is supplanted by physiology as the major focus of scientific and medical concern – but that dynamism can be adequately understood and managed without positing some purpose outside of or beyond nature toward which nature tends. Nature is not understood to be dynamic in a particular direction.

3.4.8 Policy Implications

As described above, the value of this discussion, and especially the case studies, for policy development and implementation lies in the broad range of historical perspectives they provide. Historical conceptions of nature, its alterability, and the attendant moral implications continue to weigh heavily upon us today, both as explicit philosophical and/or religious talking points and as implicit, constitutive parts of the moral, historical, and cultural legacy that frames any contemporary conversation of this sort. Neglecting to attend sufficiently to such matters of historical worth would be to ignore a fundamental part of that which has informed and continues to inform these issues.

At the same time, however, appeals to historical religious and/or philosophical traditions cannot be considered sufficient as an end for contemporary conversation regarding policy development and implementation. This seems an inherent weakness in such declarations as those made by the President's Council, as described above. Fundamental to the argument(s) put forth by the President's Council is an acceptance of the normative moral standing that nature necessarily possesses in contemporary American culture and society. Of course the shards of such beliefs remain implicit in much of our cultural understanding and identity (MacIntyre, 1984). However, as we have attempted to describe both in the case studies as well as in this summary, concluding section, an apologistic acceptance of nature as normative can no longer be taken as a given upon which to base theoretical statements that are to guide discussions of biomedical technological policy discussions. Surely these strands/shards must be part of the conversation, but they can not stand alone, as they are intended by the President's Council, without an appreciation of their historical and conceptual complexities, which require a much more rigorous defense of their applicability to the pluralistic and secular nature of our contemporary scientific establishment. Awareness and understanding of these historical legacies are necessary as a part of any such conversation of the normativity of nature today, but they are not sufficient and cannot be relied upon to do the work that requires the reconciliation of a number of equally powerful and competing cultural value claims.

This cautionary note, we believe, will become especially compelling as we enter an era of biotechnology in which we do not simply study or manipulate nature but explore the minimum genetic material required for life and even undertake to create life forms "from scratch," as it were. Bacon, perhaps at his most ambitious, we saw above, wrote of the power to make "divers new plants, differing from the vulgar; and to make one tree of plant turn into another" (Bacon, 1875a, 159). It would, we think, not surprise Bacon that we are on the threshold of making quite "divers" new life forms. Our historically informed, first response to the ethical challenges of this remarkable biotechnology should be to turn to the concepts of nature as normative that we have inherited from the history of modern European and North American science. We should not assume either that such advances are altogether new and unprecedented or that somehow they outstrip our moral sensibilities and intellectual resources to assess and manage them responsibly.

References

Adas, M. (1989). *Machines as the Measure of Men, Science, Technology and Ideologies of Western Dominance*. Ithaca, NY: Cornell University Press.
Aubrey, D.N.J. et al. (2002). "Time to Talk SENS: Critiquing the Immutability of Human Aging," *Annals of the New York Academy of Sciences* 959, 452–462.
Bacon, F. (1860–1864). *The Works of Francis Bacon*, J. Spedding, R.L. Ellis, and D.D Heath (eds.), 15 vols. Boston, MA: Brown & Taggart.

3 Scientific and Medical Concepts of Nature in the Modern Period in Europe

Bacon, F. (1875a). "The New Atlantis," in J. Spedding, R.L. Ellis, and D.D. Heath (eds.), *The Works of Francis Bacon*, Vol. III. London: Longmans, Cumpers.
Bacon, F. (1875b). "The New Organon," in J. Spedding, R.L. Ellis, and D.D. Heath (eds.), *The Works of Francis Bacon*, Vol. IV. London: Longmans, Cumpers, pp. 39–248.
Bacon, F. (1875c). "Preparative toward a natural and experimental history," in J. Spedding, R.L. Ellis, and D.D. Heath (eds.), *The Works of Francis Bacon*, Vol. IV. London: Longmans, Cumpers, pp. 249–263.
Barbour, P.L. (ed.) (1986). *The Complete Works of Captain John Smith (1580–1631)*, 3 vols. Chapel Hill, NC: University of North Carolina Press.
Barnes, J. (ed.) (1984). *The Complete Works of Aristotle*, 2 vols., Bollingen Series LXXI. Princeton, NJ: Princeton University Press.
Beaglehole, J.C. (ed.) (1963). *The Endeavour Journal of Joseph Banks 1768–1771*, 2 vols. Sydney, Australia: Angus & Robertson.
Boorde, A. (1870 [1542]). *A Compendyous Regyment or A Dyetary of Helth* (1st ed., London: Wyer, 1542), this ed., E.J. Furnivall (ed.), Early English Text Society, Extra Series 10, London.
Bullein, W. (1558). *The Government of Health*. London: Ihon Day.
Burns, C.R. (1976). "The Nonnaturals: A Paradox in the Western Concept of Health," *The Journal of Medicine and Philosophy* 1, 202–211.
Butler, R. (2001–2002). "Is There an 'Anti-Aging' Medicine?," *Generations* 25 (Winter), 63–65.
Callahan, D. (1995). "Aging and the Life Cycle: A Moral Norm?," in C. Daniel et al. (eds.), *A World Growing Old: The Coming Health Care Challenges*. Washington, DC: Georgetown University Press, pp. 21–27.
Caplan, A.L. (1981). "The Unnaturalness of Aging: A Sickness unto Death?," in Caplan et al. (eds.), *Concepts of Health and Disease*. Reading, MA: Addison Wesley, pp. 331–345.
Caplan, A.L. (1992). "Is Aging a Disease?," in A.L. Caplan (ed.), *If I Were a Rich Man Could I Buy a Pancreas? And Other Essays on the Ethics of Health Care*. Bloomington, IN: Indiana University Press, pp. 195–209.
Cogan, T. (1612). *The Haven of Health*, Epistle Dedicatory. London: W. W. Norton.
Cohen, B.I. (1985). *Revolution in Science*. Cambridge, MA: The Belknap Press of Harvard University Press.
Cole, T.R. (1992). *The Journey of Life: A Cultural History of Aging in America*. Cambridge, MA: Cambridge University Press.
Cole, T.R. and Thompson, B.A. (eds.) (2001–2002). "Anti-Aging: Are You for It or Against It?," *Generations* 25 (Winter), 9–149.
Crosby, A.W. (1986). *Ecological Imperialism. The Biological Expansion of Europe 900–1900*. Cambridge: Cambridge University Press.
Curtin, S. (1973). *Nobody Ever Died of Old Age*. Boston, MA: Little Brown.
Darwin, C. (1974 [1859]). *The Origin of Species by Means of Natural Selection* (1st ed. 1859), this ed., Harmondsworth, England: Penguin Books.
Darwin, C. (2002). *Autobiographies*, M. Neve and S. Messenger (eds). London: Penguin Books.
Daston, L. and Park, K. (2001). *Wonders and the Order of Nature 1150–1750*. New York: Zone Books.
Debus, A. (1978). *Man and Nature in the Renaissance*. New York: Cambridge University Press.
Edward W. (1650). *Virginia: More especially the South Part Thereof, Richly and Truly Valued*. London in Force, *Tracts*, vol III.
Fitzherbert, A. (1882 [1534]). *The Book of Husbandry*, in W. Skeat (ed). London: Trübner for the English Dialect Society.
Force, P. (ed.). (1947). *Tracts and Other Papers Relating Principally to the Origin, Settlement and Progress of the Colonies in North America*, 3 vols. (1st ed. 1844, reprinted New York: Peter Smith).
Freeman, J.T. (1938). "The History of Geriatrics," *Annals of Medical History* 10, 328.
Freeman, J.T. (1965). "Medical Perspectives in Aging (12–19th Century)," *The Gerontologist* 5, part 2 (March), 1–24.

Galen (1916). *On the Natural Faculties*, Translated with introduction by Arthur J.B. London: William Heinemann/Cambridge, MA: Harvard University Press.
Galen (1968). *De Usu Partium (Galen on the Usefulness of the Parts of the Body)*, in M. Tallmadge (trans. and ed.), May. Ithaca, NY: Cornell University Press.
Galileo (1957). *Discoveries and Opinions of Galileo*. Translated with introduction and notes by Stillman D. New York: Anchor Books.
Glacken, C. (1967). *Traces on the Rhodian Shore*. Berkeley, CA: University of California Press.
Grant, R.L. (1963). "Concepts of Aging: An Historical Review," *Perspectives in Biology and Medicine* (Summer), 443–478.
Gregory, J. (1762). "Dr Gregory – Whether the Art of Medicine, as It Has Been Usually Practised, Has Contributed to the Advantage of Mankind," July 12, 1761, Question 59 for the Aberdeen Philosophical Society. Aberdeen, Scotland: Aberdeen University Library MS. 37. Reprinted in L.B. McCullough (ed.), *John Gregory's Writings on Medical Ethics and Philosophy of Medicine*. Dordrecht, The Netherlands: Kluwer, 1994, pp. 59–66.
Gregory, J. (1768). "The Practical Course Delivered by Dr. Gregory at Edinburgh 1768/9," Royal College of Surgeons of Edinburgh. Bethesda, MD: National Library of Medicine MS. B7.
Gregory, J. (1769). " 'Lectures on Medicine' of John Gregory. One of Several Extant Student-Note Versions of Gregory's Medical Ethics Lectures." Edinburgh, Scotland: Royal College of Physicians and Surgeons MS. D27. Reprinted in L.B. McCullough (ed.), *John Gregory's Writings on Medical Ethics and Philosophy of Medicine*. Dordrecht, The Netherlands: Kluwer, 1997, pp. 85–87.
Gregory, J. (1772a). *A Comparative View of the State and Faculties of Man with Those of the Animal World*, 5th ed. London: J. Dodsley.
Gregory, J. (1772b). "Lectures on the Duties and Qualifications of a Physician," in W. Strahan and T. Cadell (eds.), London. Reprinted in L.B. McCullough (ed.), *John Gregory's Writings on Medical Ethics and Philosophy of Medicine*. Dordrecht, The Netherlands: Kluwer, 1997, pp. 161–245.
Gruman, G.J. (1966). "A History of Ideas About the Prolongation of Life: The Evolution of Prolongevity Hypotheses to 1800," *Transactions of the American Philosophical Society* 56, part 9, 6.
Gullette, M.M. (2004). *Aged by Culture*. Chicago, IL: University of Chicago Press.
Haber, C. (2001). "Anti-Aging: Why Now? A Historical Framework for Understanding the Contemporary Enthusiasm," *Generations* 25(Winter), 9.
Hall, S.S. (2003). *Merchants of Immortality*. Boston, MA: Houghton Mifflin.
Hammond, J. (1656). "Leah and Rachel or, the Two Fruitfull Sisters Virginia and Maryland," in P. Force (ed), *Tracts and Other Papers Relating Principally to the Origin, Settlement and Progress of the Colonies in North America*, 3 vols (1st ed. 1844, reprinted New York: Peter Smith, 1947).
Hansard (1999). House of Commons Debates for 21 Oct, 1999. London: United Kingdom Parliament.
Harvey, W. (1628). *Exercitatio Anatomica de Motu Cordis et Sanguinis in Animalibus*. Frankfurt, Germany: W. Fitzer.
Hayflick L. (2002). "Anarchy in Gerontological Terminology," *The Gerontologist* 42, 416–421.
Hays, G. (trans.) (2003). *The Meditations of Marcus Aurelius*. New York: Modern Library, xiv.
Headrick, D.R. (1981). *The Tools of Empire*. New York: Oxford University Press.
Headrick, D.R. (1988). *The Tentacles of Progress. Technology Transfer in the Age of Imperialism 1850–1940*. New York: Oxford University Press.
Heresbachius, C. (1577). *Foure Bookes of Husbandry*, Collected by M. Conradus Heresbachius, newly Englished, and increased, by Barnarbe Googe, London.
Hoffmann, F. (1971). *Fundamenta Medicinae*, in L.S. King (trans.). London: MacDonald.
Howitt, W. (1855). *Land, Labour and Gold or, Two Years in Victoria*, 2 vols. London: Longman, Brown, Green & Longman.
Hume, D. (2000 [1739]). *A Treatise of Human Nature*, in D.F. Norton and M.J. Norton (eds.). Oxford: Oxford University Press.
Johnson, H.A. (ed.) (1985). *Relations Between Normal Aging and Disease*. New York: Raven.

Jonas, H. (2001 [1966]). *The Phenomenon of Life: Toward a Philosophical Biology*. Evanston, IL: Northwestern University Press.

Juengst, E., Binstock, R., Mehlman, M., Post, S., and Whitehouse, P. (2003). "Biogerontology: 'Anti-aging Medicine,' and the Challenges of Human Enhancement," *Hastings Center Report* 33(4), 21–30.

Kass, L. (1977). "The Wisdom of Repugnance," *The New Republic* (June 2), 17–26.

Kass, L. (1985). *Toward a More Natural Science*. New York: Free Press.

King, L.S. (1970). *The Road to Medical Enlightenment, 1650–1695*. London: MacDonald/New York: American Elsevier.

Klatz, R. (2001). "Anti-Aging Medicine: Resounding, Independent Support for Expansion of an Innovative Medical Specialty," *Generations* 25(Winter), 59.

Kuhn, T. (1964 [1962]). *The Structure of Scientific Revolutions*. Chicago, IL: University of Chicago Press.

Kupperman, K. (1984). "Fear of Hot Climates in the Anglo-American Colonial Experience," *William and Mary Quarterly* 41, 213–240.

Lind, J. (1768). *An Essay on Diseases Incidental to Europeans in Hot Climates*. London: Becket & De Hondt.

MacIntyre A. (1984). *After Virtue: A Study in Moral Theory*. Notre Dame, IN: Notre Dame University Press.

Malthus, T.R. (1798). *An Essay on the Principle of Population*. London: John Murray.

Marsh, G.P. (1964 [1864]). *Man and Nature*. Cambridge, MA: Harvard University Press.

Martin, J.R. (1837). *Notes on the Medical Topography of Calcutta*. Calcutta: G.H. Huttmann, Bengal Military Orphan Press.

McCullough, L.B. (1996). *Leibniz on Individuals and Individuation: The Persistence of Premodern Ideas in Modern Philosophy*. Dordrecht, The Netherlands: Kluwer.

McCullough, L.B. (1998a). *John Gregory and the Invention of Professional Medical Ethics and the Profession of Medicine*. Dordrecht, The Netherlands: Kluwer.

McCullough, L.B. (ed.) (1998b). *John Gregory's Writings on Medical Ethics and Philosophy of Medicine*. Dordrecht, The Netherlands: Kluwer.

Mouffet, T. (1655). *Healths Improvement: Or, Rules Comprizing and Discovering the Nature, Method and Manner of Preparing All Sorts of Food used in this Nation: Corrected and Enlarged by Christopher Bennett*. London: Newcomb & Thomson.

Murphy, T.F. (1987). "Cure for Aging?," *Journal of Medicine and Philosophy* 11, 237–255.

National Bioethics Advisory Commission (1997). *Cloning Human Beings*. Washington, DC: U.S. Government Printing Office. Available at http://www.georgetown.edu/research/nrcbl/nbac/pubs.html, accessed April 3, 2005.

National Bioethics Advisory Commission (1999). *Ethical Issues in Human Stem Cell Research*. Washington, DC: U.S. Government Printing Office. Available at http://www.georgetown.edu/research/nrcbl/nbac/pubs.html, accessed April 3, 2005.

Pagel, W. (1958). *Paracelsus*. Basel, Switzerland: Karger.

Pettus, J. (1674). *Volaties from the History of Adam and Eve*. London.

Porter, R. (1997). *The Greatest Benefit to Humanity*. New York: W. W. Norton.

President's Commission for the Study of Ethical Problems in Medicine and Biomedical and Behavioral Research (1982). *Splicing Life: A Report on the Social and Ethical Issues of Genetic Engineering with Human Beings*. Washington, DC: U.S. Government Printing Office.

President's Council on Bioethics (2003). *Beyond Therapy: Bioethics and the Pursuit of Happiness*. Washington, DC: U.S. Government Printing Office. Available at http://www.bioethics.gov/reports/beyondtherapy/index.html, accessed April 3, 2005.

President's Council on Bioethics (2004). *Reproduction and Responsibility*. Washington, DC: U.S. Government Printing Office. Available at http://www.bioethics.gov/reports/reproductionandresponsibility/index.html, accessed April 3, 2005.

President's Council on Bioethics: Staff Working Paper (2003). "Age-Retardation: Scientific Possibilities and Moral Challenges." Available at http://www.bioethics.gov/background/age_retardation.html.

Rowe, J. (1985). "Interaction of Aging and Disease," in C.M. Gaitz and T. Samorajski (eds.), *Aging 2000*, Vol. 1. New York: Springer.

Smith, Captain J. (1616). *A Description of New England*. London: H. Lownes, for R. Clerke.

Speed, J. (1676). *The Theatre of the Empire of Great Britaine*. London: Basset & Chiswell.

The Times (2007). "Egg Mastermind Duped Shoppers with 500 million Free-Range Fakes," March 19. London.

Thomas, K. (1984). *Man and the Natural World*. London: Penguin Books.

Tryon, T. (1683). *The Way to Health, Long Life and Happiness*. London: Andrew Sowle.

Tusser, T. (1878 [1580]). *Five Hundred Pointes of Good Husbandrie* (1580 ed.), in W. Payne and S. Herrtage (eds.). London: Trübner, for the English Dialect Society.

Venner, T. (1628 [1620]). *Via Recta ad Vitam Longam: Or, a Plaine Philosophical Demonstration of the Nature, Faculties and Effects of all such Things as ...Make for the Preservation of Health*. London: E. Griffin for R. Moore.

Watts, D. (1987). *The West Indies: Patterns of Development, Culture and Environmental Change Since 1492*. Cambridge: Cambridge University Press.

Wear, A. (2000). *Knowledge and Practice in English Medicine, 1550–1680*. Cambridge: Cambridge University Press.

Webster, C. (1975). *The Great Instauration*. London: Duckworth.

Wood, W. (1634). *New England's Prospect*. London: Tho. Cotes.

Chapter 4
Ethical Challenges of Patenting "Nature": Legal and Economic Accounts of Altered Nature as Property

Mary Anderlik Majumder, Margaret M. Byrne, Elias Bongmba, Leslie S. Rothenberg, and Nancy Neveloff Dubler

4.1 Introduction

Contemporary American society is increasingly focused on business and profit rather than social justice and the common good. As this direction gained ascendancy in the three decades before the 21st century the idea of asserting a monopoly over some idea or substance derived from nature seemed more and more justified and commonplace. So, for example, patenting the changes wrought in otherwise found compounds, substances and genetic material began to seem less and less like violation or mere copying of "nature" and more in balance with the government and social policies that surround general scientific development. A new boundary began to emerge from the mélange of judicial opinions, legislation and business practice: nature itself cannot be patented, but processes that alter nature and the products of those processes can be patented. The biotechnology industry, ever aware of its dramatic position and possibly controversial role, has attempted to reassure the public that it does not want to usurp the role of creator, claiming authorship of nature. However, when biotechnology companies augment, refashion or reconfigure products of nature, they assert that their investment should be recognized and recompensed. Controversy persists nonetheless, and hence it is worthwhile to consider how ethical concepts of nature might place limits on efforts to alter nature and make it property.

Honing the content for this sort of chapter – that seeks merely to confront all of creation in all of its fullness and mystery – presented a daunting task of definition and delineation. Even tentative decisions regarding the outline risked excluding huge segments of the world's populations, ignoring serious religious texts and practices, and de-emphasizing important schools of economic theory. From a universe of theoretical constructs and practical government and individual entrepreneurial behaviors, how could we, as authors, pick and choose among the fullness and breadth of examples? Our ultimate choices were dictated by our own individual interests and passions, and by our hopes that these examples presented generalizable material that would be inherently interesting and might provide templates for considering like issues in the future. We thus considered one primary example and different contrasts as a way of providing useful material and complementary elaborations. It is a given in bioethics that when ethical principles fail to produce a clear and definitive solution, one must turn to examine the process. If substantive justice fails to emerge, then one must consider procedural issues.

Our first decision was to focus our work on the legal and economic structures that govern the appropriation of nature for private ends, as private property. In the biotechnology arena, patenting of nature has emerged as a locus of conflict, hence an analysis of contemporary Western patenting regimes seemed a good starting point and organizational anchor for our chapter. This then led us to the consideration of DNA patenting as a primary case study of the issues arising at the intersection of biotechnology, legal and economic concepts of nature as property, and religion. But that turned out to be a bit dry and narrow. We then supplemented the discussion of DNA patenting with discussion of other patent controversies that help to illuminate core concerns about the commodification of nature and injustice in the distribution of the benefits of the bio-pharmaceutical enterprise.

Lest this be charged as Western-centric and self-interested, we then added a focus on the ways in which two very different religious worldviews entered the discussion of nature, property, and biotechnology. At one end of the spectrum, we considered Protestant Christianity, which has often been identified with contemporary Western property and patenting regimes. At the other end of the spectrum, we considered African and other "indigenous" religions, religions that are playing a significant role in challenging those regimes.

As you read through this chapter, consider if we have answered the question: In economics and law, what does it take to treat nature like property? When has the law resisted treating nature as property? Can we pinpoint what we find troubling about patents? Are there standpoints from which Western ideas about nature as property appear bizarre or strange? Consider this last question reframed in the following way: Does an examination of other cultural perspectives highlight how strange the notion of nature as property actually is?

One major theme emerges in these discussions: the more human ingenuity and productive labor have been exerted to reach the ultimate product, that is, the more distant the product from its "found" state in nature, the more it seems amenable to the claims of property. This altered state of nature then acquires a moral and legal status distinct from its origins. These sorts of claims are particularly compatible with American notions of productive labor creating the basis for individual profit. In a society consumed with the idea of the common good and with a clear fiduciary relationship to nature, these claims of property would seem less compelling. But America has not, for most of its history, been such a nation. Resources are there not to be conserved but to be spent to produce wealth, and wealth is under no moral imperative to be shared. Thus altering nature for profit is a comfortable political perspective. There are those who argue for conservation and preservation of nature in its most common sense, that is forests and sea shores and unique lands. Their voices are, at present, muted, and have been for some decades. The idea of capturing nature for profit is the contemporary "Robber Baron" who does not build railroads, but rather directs a biotechnology company, develops drugs, or patents genes and genetic tests. It may be too extreme to state that this idea of nature and the natural as the frontier of the profitable enterprise may be the only widely shared notion about nature in this society – but it also may not be too extreme.

Later in this chapter we state: "There are many ways in which property arrangements may prove dysfunctional. While the Lockean justification for private property rests on the assumption that common property (or nature absent a means of private appropriation) will be underdeveloped, the phrase the 'tragedy of the commons' has been used to describe the overuse and degradation of common property. Some now argue that a multiplication of intellectual property rights is likely to lead to a 'tragedy of the anticommons' as the many private owners of pieces of the genome and other research tools hinder each others' efforts." In the conclusion we will return to this discussion and detail some recent events that may demonstrate a less aggressive stance toward nature and the natural.

Ideas about "nature" function in a variety of ways in law and economics. In understanding how nature comes to be conceptualized as property, other concepts come into play, such as, *state of nature*, *natural law*, and *laws of nature*. In the first instance, different approaches to law and economics rest on differing assumptions about human nature. However, the relationships between particular accounts of human nature and particular positions on legal or economic questions can be quite complex. Humans may be described in terms that emphasize individuality, a tendency to form and identify with groups, or a broad interdependence that cannot be confined to groups or even to the species. Accounts also vary in portraying humans as governed by rational calculation, the passions, or a transcendent purpose. Introducing one more layer of complexity, views of human nature within a society, the structure of that society's social institutions and its laws, and actual patterns of behavior are in a dynamic relationship (Hirschman, 1977). Observations of behavior may generate an account of human nature that, embraced as timeless and universal truth, reinforces and so strengthens certain existing tendencies. On the other hand, an account that offers a radical departure from the established account may achieve influence and alter institutions and behavior in significant ways.[1]

In the West, the cluster of cultural developments commonly referred to as "the Enlightenment" brought with it the notion that lawmakers should favor a realistic, and perhaps even pessimistic, view of human nature. For example, Vico wrote: "'Philosophy considers man as he ought to be and is therefore useful only to the very few who want to live in Plato's Republic and do not throw themselves into the dregs of Romulus. Legislation considers man as he is and attempts to put him to good uses in human society'" (Hirschman, 1977, 14). In classical economics and "policy science" the dominant account of human nature is individualistic, stressing self-interest as the key to understanding human motivation. Rational calculation is accepted as the basis for understanding behavior and for fashioning laws and economic policies to control or guide behavior (Bewley, 2003; Fehr and Rockenbach, 2003). Adherents often express discomfort with symbolic, religious

---

[1] See Ball (2000). The bioethics debates discussed by Ball concerned ART. Ball argues that attention to Rousseau's critique of earlier views of human nature would have improved those debates.

or moral discourse. For example, on the question of patenting of higher life forms, a recent editorial in the journal *Nature Biotechnology* urged that any legislation "steer clear of moral and ethical definitions" and instead "stick to rational and scientific benchmarks" (Editorial, 2003, 341). A quite different understanding of human nature and its relation to law is expressed in a document published by the Pontifical Academy for Life on the ethics of biomedical research. The Academy asserts that science must transcend the material, "intuiting in the corporeal dimension of the individual the expression of a greater spiritual good." International legislation should ensure a unified and uniform system of regulation "based on the values inscribed in the nature of the human person" (Pontifical Academy for Life, 2003).

Recent empirical research in the fields of economics and sociology points to the importance of cooperation, altruism, and perceptions of fairness in human behavior (Bewley, 2003; Fehr and Rockenbach, 2003). A defense of a more other-regarding, value-oriented view of human nature is not necessarily tied to a belief that motivation is irrational, passion-driven or supernatural. As one set of researchers put it, "[i]f people have altruistic aims, altruistic behavior is a rational means by which to achieve their proximate goals" (Fehr and Gächter, 2003).

The state of nature, a description of human existence in the absence of government, has figured prominently in Western philosophy, especially the philosophy of the Enlightenment. Although the state of nature has usually been portrayed as a state existing in the past, historical accuracy has not been the goal. Rather, the state of nature is a heuristic device. It has functioned as a counter-example to the civil society familiar to the philosopher and his contemporaries. It has also been invoked as a guide or basis for the reform of that society. Although the notion may seem a bit antique, there are recent parallels, for example, the "original position" in the theory of justice developed by John Rawls (1971).

The concept of natural law was pivotal in the work of many pre-Enlightenment and Enlightenment philosophers and jurists. It virtually disappeared from the mainstream with the rise of positivism, but it remains influential even today within the Roman Catholic tradition and underwrites much of international human rights law.[2] Natural law reasoning rests on a complex of related beliefs: that the universe is lawful, that the principles or rules that govern its operation can be discerned or intuited by humans, that these have moral force, and that they provide a template for civil law. Natural laws have "moral force" because they were created by God (or originated in his intellect and not his creative freedom), the source of all goodness in the world. A commitment to natural law is usually coupled with the view that the principles and rules so derived are true for all times and places, i.e., transhistorical and trans-cultural.

---

[2] For example, "God favours and supports the efforts of human reason, and enables the human being to recognize so many 'seeds of truth' present in reality" (Pontifical Academy for Life, 2003).

Finally, the term "laws of nature" is used as a label for a domain of fundamental knowledge about the world that should be treated as given to all rather than as susceptible to appropriation as private property. A similar term, "natural phenomena," referring to things as they are apart from human intervention or improvement, may be used in a similar way. The justification for marking off such a domain from individual property claims may borrow from natural law, deontological or utilitarian ethics, or principles of economics. The concern is shared by many critics of current approaches to intellectual property, especially patents.

4.2 Patenting in Biotechnology

The trend over time has been to expand the reach of laws that confer property rights with respect to inventions – new, potentially useful "things-in-concept" – by means of patents. In this section, we begin by reviewing the statutory foundations for the US patent system and the current legal standards for patenting. It should become apparent that, in a sense, the system aims at the alteration of nature. It does this by taking progress as its object, and by making change to the found, given or known a condition for obtaining patent rights. In many cases, the change will consist of a direct alteration of physical nature; biotechnology patents will almost always involve changes to organisms. As we shall see, although "laws of nature" and "natural phenomena" are excluded from patentability, the concern motivating these exclusions is not preservation of nature in its pristine state, but rather preservation of the general freedom to use whatever is assigned to these categories.

4.2.1 *Background on Patenting in Biotechnology: Legal Foundations and Controversies*

4.2.1.1 Patent Law from the US Constitution Forward

In the United States, Article I, Section 8 of the Constitution is the basis for patents and other forms of intellectual property. This provision empowers Congress to "promote the Progress of Science and useful Arts, by securing for limited Times to Authors and Inventors the exclusive Right to their respective Writings and Discoveries." In 1790, Congress enacted the first patent statute under this constitutional grant of authority. While many key concepts can be traced back to the original statute, suggesting continuity over time, there have been dramatic swings in attitudes toward patenting in the courts, and in the US Supreme Court in particular, reflecting both "institutional" concerns and broader social concerns (Chisum, 2002, §1–7). The courts have sometimes led the way, developing doctrines that were later incorporated in legislation. In other cases, amendments to the patent statute have been used to override judge-made law.

For roughly the 1st century the patent system was in existence, all appeals in patent infringement suits went before the US Supreme Court. By the mid-1800s, patent litigation was consuming a significant amount of the Court's time. Some of the justices expressed skepticism about the value of an expansive approach to patenting. For example, Justice Bradley remarked that " 'It was never the object of [the patent] laws to grant a monopoly for every trifling device, every shadow of a shade of an idea, which would naturally and spontaneously occur to any skilled mechanic or operator in the ordinary progress of manufactures' " (*Atlantic Works v. Brady*, 1882).

A shift from hostility to receptiveness in the 1890s followed a general improvement in economic conditions and the enactment of a new federal law creating regional courts of appeals, a development that dramatically decreased the number of patent cases before the US Supreme Court. The pro-patent policy continued until the 1930s. At that time, the US Supreme Court once again used its decisions to rein in the patent system. First, the Court expanded the patent misuse doctrine, which rendered a patent unenforceable if the patent holder tried to extend the scope of the patent through tying agreements and other improper practices. Second, the Court tightened the standards concerning claims, requiring greater clarity and restricting the breadth of claims. Third, the Court raised the bar for "invention," emphasizing a creative leap as an aspect of the process of invention necessary to secure patent protection.

In 1952, Congress enacted the patent statute that remains in effect today. To a large extent, Congress simply re-arranged the provisions of the prior statute and codified case law. At the same time, the statute was more liberal than the body of case law emanating from the courts in the two decades immediately preceding its enactment. The section on conditions for patentability, 35 USC. 103(a), includes a statement that patentability "shall not be negatived by the manner in which the invention was made," putting an end to scrutiny of the process of invention. In 1982, Congress created a special appellate court, the Court of Appeals for the Federal Circuit, with exclusive jurisdiction over cases arising in whole or in part under the patent laws. As might have been expected, this structure has resulted in greater consistency in the case law, and a more positive legal climate for seekers and holders of patents. The pro-patent posture seems to be in line with the mood in Congress. For example, in 1988 Congress enacted the Patent Misuse Reform Act, sharply restricting the application of the patent misuse doctrine.

It is important to emphasize some of the limits of patents. First, the rights conferred by patents are negative. Specifically, the patent holder has the right to exclude others from making, using, or selling the claimed invention for a limited period of time. This is not equivalent to a right to make, use, or sell the invention, as these activities may be prohibited by other law or limited by the rights of others, including patent rights. (Also, unless required by other law, the patent holder has no mandate to exploit the invention or allow anyone else to do so.) In addition, patent rights are limited in time. The term of a US patent is currently 20 years from the date the application was filed.

4.2.1.2 Current Standards Related to Patent Issuance and Infringement

The act of invention is a necessary but not sufficient step to secure patent protection under US law. The requirements for issuance of a patent that can withstand challenge in court include patent-eligible subject matter, novelty, nonobviousness, utility, enabling disclosure, and clear claiming. The current patent statute addresses the first requirement, eligible subject matter, in the section that lays the foundation for patent rights: "Whoever invents or discovers any new and useful process, machine, manufacture, or composition of matter, or any new and useful improvement thereof, may obtain a patent therefor, subject to the conditions and requirements of this title..." (35USC. §101). As the patent law has been interpreted by the courts, only three categories have been consistently recognized as ineligible for patenting: laws of nature, natural phenomena, and abstract ideas. At least since 1952, courts have been very liberal in finding patent-eligible subject matter. In *Diamond v. Chakrabarty*, the US Supreme Court cited a committee report that accompanied the 1952 act for the proposition that Congress intended statutory subject matter to "'include anything under the sun that is made by man'" (447 US 303, 309, 1980). The court determined that the bacterium at issue in *Chakrabarty* had "markedly different characteristics from any found in nature," and hence was "not nature's handiwork" but Ananda Chakrabarty's (447 US 303, 310, 1980).

In most cases, then, the first requirement of any concern will be novelty. "If an invention is not new, then the invention is not patentable. That ends the inquiry" (Chisum, 2002, §5.01). However, the applicant is not required to show superiority over the existing stock of inventions in the public domain and the patent record. In this context, "novelty" signifies departure from the prior art. A related question is whether the invention would be obvious to one with ordinary skill in the art. It may be that a search of the record turns up no reference to a particular process or composition of matter, but any peer of the inventor could arrive at the same result unaided by her disclosure. The application of the nonobviousness requirement to certain patents in the field of genomics has emerged as a significant area of controversy. The same is true of the utility requirement. In general, the utility requirement is met if the invention passes three tests. "First, it must be operable and capable of use. It must operate to perform the functions and secure the result intended. Second, it must operate to achieve some minimum human purpose. Third, it must achieve a human purpose that is not illegal, immoral or contrary to public policy" (Chisum, 2002, §4.01). Still, given the patent-friendly atmosphere in the appellate courts and the silence of the patent statute on morality, a credible assertion that an invention has even a single legitimate use will probably pass muster.

Patents may be more or less aggressive depending upon the type and scope of the claims they encompass. In general, process patents are more limited in coverage than composition of matter or other "product" patents, and the distinction has assumed significance for many who object to patents on organisms on religious grounds. A process patent stakes out rights to a certain way of making or doing something; a product patent stakes out rights in the thing itself and in any activity that makes or makes use of the thing.[3] The scope of patent claims and the adequacy of disclosure are the focus of much patent litigation. Although the Patent Office may scrutinize claims before issuing a patent, the patent holder's rights are not settled upon issuance of the patent, as claims remain subject to legal challenge. One study found that in 46% of disputes over the validity of biotechnology patents in US courts, the patent is invalidated (OECD, 2002, 85, n.12).[4]

By way of example, a patent issued to the University of Rochester in 2000 included a claim for a method of treating pain and inflammation by inhibiting the Cox-2 enzyme and a test to screen potential drug candidates targeting the enzyme. Inhibition of the Cox-2 enzyme is the mechanism for several commercially successful drugs, including Celebrex and Vioxx. The university filed a lawsuit against Pharmacia and Pfizer, the makers of Celebrex, as part of an enforcement strategy it expected to generate billions of dollars in royalties. The university's press release announcing issuance of the patent was subtitled "Historic drug patent is likely to be the most lucrative in US history" (University of Rochester Medical Center, 2000). It was not to be. The patent was invalidated. A judge found that the written description in the patent application was not clear and comprehensive enough to enable others to duplicate the work. In a statement, the university president suggested that the academic research enterprise would be served by a less demanding standard – a contestable proposition.[5] In any event, uncertainty over the ultimate outcome of any legal challenge limits the value of patents.

The liberal approach toward issuance of patents in the US, at least in recent years, is matched by a conservative approach toward exemptions from liability for patent infringement. Most relevant here are the medical exemption and the research (or experimental use) exemption. The medical exemption was created through an amendment to the patent statute in 1996, as a response to the flood of patents on medical methods that followed a decision upholding such a patent in *Ex parte Scherer*.[6] It is limited to medical and surgical procedures, and it expressly excludes

---

[3] A product patent (also known in Europe as a "substance *per se*" patent) is "without limitation to any particular process of purification or isolation and without any limitation as to its intended use" (OECD, 2002, 28). The inventor of a new process or use for a thing subject to a product patent may patent her invention, but she and her licensees cannot exploit the invention without permission from the holder of the product patent. India and some other developing countries have refused to issue product patents, but under the Agreement on Trade Related Aspects of Intellectual Property Rights this will no longer be an option. See Lanjouw (1998). (Lanjouw concludes that the change will have little effect on the poor and suffering in India because they lack the resources to purchase pharmaceuticals under the current patent regime, and so even if substantial price increases materialize this will not affect their access.)

[4] The percentage is not out of line with the figure for all US patent disputes.

[5] The statement of Thomas H. Jackson read in part: "'While courts are comfortable with narrow patents, there is widespread interest among research universities in ensuring that our broader, more basic research work is likewise protected by the nation's patent laws'" (Pollack, 2003). Interestingly, the work at the University of Rochester was supported by a grant from the National Institutes of Health. The University recently lost its appeal, and its attempt to get the case heard by the US Supreme Court failed, meaning the loss is definitive. *University of Rochester v. G.D. Searle and Co.*, 375 F.3d 1303 (Fed. Cir. 2004), *cert. denied*, 125 S.Ct. 629 (2004).

[6] 103 USPQ 107 (Board of Patent Appeals and Interferences, 1954).

any infringing use of a patented machine, manufacture, or composition of matter, or the practice of a process covered by a biotechnology patent. Also, the exemption does not extend to commercial activities that are (a) directly related to the development or other use of a machine, manufacture, or composition of matter or the provision of pharmacy or clinical laboratory services (other than clinical laboratory services provided in a physician's office), and (b) regulated under the Federal Food, Drug, and Cosmetic Act, the Public Health Service Act, or the Clinical Laboratories Improvement Act.[7] So, for example, the exemption would not permit a physician or hospital to perform a patented diagnostic test or compound a patented drug without authorization.

The general research exemption has been created by courts rather than Congress. The opinion in the recent case of *Madey v. Duke University*[8] emphasizes that the exemption is "very narrow and strictly limited," confined to activities performed "'for amusement, to satisfy idle curiosity, or for strictly philosophical inquiry.'"[9] The court states that even conduct with no commercial implications whatsoever, engaged in by a non-profit entity, would fall outside the exemption if it furthered the alleged infringer's legitimate business objectives. These objectives would include, for a university, "educating and enlightening students and faculty."[10] At the same time, in ruling as it did, the court in *Madey* took notice of the increasing commercialization of university-conducted research. A footnote contains the wry observation that "Duke...like other major research institutions of higher learning, is not shy in pursuing an aggressive patent licensing program from which it derives a not insubstantial revenue stream."[11] The judge-made exemption is complemented by a very narrow statutory exemption for the use of a patented invention "solely for uses reasonably related to the development and submission of information" for approval by the US Food and Drug Administration.[12] This provision was enacted into law in response to a case in which a company conducting research preparatory to manufacture of a generic drug attempted to defend its activity as non-infringing based on the judge-made research exemption, and lost.[13]

Other Western patenting regimes draw the lines somewhat differently. In European law, medical practice and research are more broadly shielded from the potentially restrictive effects of the intellectual property system. For example, Article 52 of the European Patent Convention excludes surgical, therapeutic, and

---

[7] 35 USC § 287(c).

[8] 307 F.3d 1351 (Fed. Cir. 2002), *cert. denied*, 123 S. Ct. 2639 (2003).

[9] Ibid. at 1362 (quoting *Embrex, Inc. v. Service Engineering Corp.*, 216 F.3d 1343, 1349 (Fed. Cir. 2000)).

[10] Ibid.

[11] Ibid. at 1363, n. 7.

[12] 35 USC. § 271(e). The product cannot, however, be placed on the market during the life of the patent without infringing.

[13] *Roche Prods., Inc. v. Bolar Pharm. Co.*, 733 F.2d 858 (Fed. Cir. 1984). This chain of events is discussed in, e.g., Madey, 307 F.3d at 1355.

diagnostic methods from patentability.[14] Are the Europeans likely paying a price for this policy in terms of diminished innovation? After our discussion of the controversy over standards for patenting of DNA, we will turn to the underlying economic considerations.

4.2.1.3 The Controversy over DNA Patenting

Although the patenting of "DNA" or "genes" has been at the center of controversy in recent years, the precedents for these kinds of patents in the US date back to the turn of the century. In 1912, a purified protein derived from a naturally occurring protein was found patentable.[15] There are rulings suggesting that so long as the claim covers a substance that is isolated and purified, and other requirements are met, a patent will be granted (Goldstein and Golod, 2002).[16] The US Patent and Trademark Office has in fact issued many patents related to genes or parts of genes. In 1991, the Court of Appeals for the Federal Circuit upheld a patent that included a claim on a DNA sequence coding for a known therapeutic agent.[17] Further, the volume of activity is escalating. As of August 2002, the GENESEQ database showed 18,174 patent applications filed worldwide with claims covering human DNA sequences (Thomas et al., 2002). Between 1976 and 2002, the US Patent and Trademark Office issued 9,456 patents with claims using the term "nucleic acid," with the vast majority (8,334) falling between 1996 and 2002; an estimated 1,500 of the patents issued in 2001 covered human genetic material (OECD, 2002, 34).

A decision at the National Institutes of Health (NIH) to seek patents on gene fragments identified in the course of the Human Genome Project was an important turning point. In 1991, the NIH filed patent applications on several hundred gene fragments sequenced by scientist J. Craig Venter, then at NIH.[18] These fragments of

---

[14] Products for use in surgery, therapy, and diagnosis can be patented.
[15] *Parke-Davis and Co. v. H.K. Mulford and Co.*, 196 F. 496 (2nd Cir. 1912).
[16] The 1912 case was *Parke-Davis and Co. v. H.K. Mulford and Co.*, 196 F. 496 (2nd Cir. 1912).
[17] *Amgen, Inc. v. Chugai Pharmaceutical Co.*, 927 F.2d 1200 (Fed. Cir. 1991).
[18] Venter gained greater fame, or notoriety, through his involvement with several for-profit ventures. In 1998, Venter helped found Celera Genomics, a private company, to compete directly with the Human Genome Project. Celera employed Venter's "whole-genome shotgun sequencing strategy." Using information generated by the public initiative, the private sector activity could be characterized as parasitism or "cream-skimming." Yet private funding enabled Venter to develop a strategy that Project insiders, comfortable with the systematic approach to sequencing, would have left to the sidelines. And, while keeping up their criticism of the Venter strategy as slapdash, the leaders of the Human Genome Project accelerated their investment and modified their strategy, sacrificing a certain measure of accuracy in order to provide a "rough draft" ahead of schedule. The private sector activity was indeed parasitic on the public sector effort, and its existence and success can scarcely be used as an argument against public science. At the same time, private firms displayed the qualities of agility and innovativeness so often heralded by defenders of private enterprise and they prodded their public counterparts to shift course in a manner that might be described by some as beneficial.

4 Ethical Challenges of Patenting "Nature": Legal and Economic Accounts

DNA snipped out of the genome and described by their sequences of base pairs are known as "expressed sequence tags" or ESTs.[19] The NIH action served as a catalyst for criticisms of DNA-related patents. Some argued that at least certain categories of DNA patents were antithetical to the norms of science, scientific progress, and broader goals of social welfare, or should not exist on technical, legal grounds (e.g., owing to the exclusion for natural phenomena, or failure of tests of novelty, nonobviousness, and utility), while others asserted that these patents violated the laws of God or nature or constituted an affront to human dignity.

From the viewpoint of some NIH administrators, filing the patent applications was a way of keeping options open (Cook-Deegan, 1995). After all, the filing of the patent applications was consistent with the pro-patent policy reflected in statutes such as the Bayh-Dole Act.[20] If successful, NIH would end up with a carrot that could be dangled in front of industry – the promise of a monopoly in pursuing further work and a basis for additional patents for achievements further along in the stream of development. The step would also address a doomsday scenario – a foreign firm or government applies for a patent on the fruits of the Human Genome Project and NIH draws fire from Congress. On the other hand, the effort to remove information generated by the project from the public domain was in tension with the rhetoric used to justify the large public investment in the project (Eisenberg and Nelson, 2002). It also created friction with foreign governments, inclined to view the project as an international collaborative venture to reveal the common heritage of humankind. Nor was the scientific community aligned behind NIH. Many scientists saw the NIH move as premature. James Watson, then head of the US initiative, strongly objected to patent applications on expressed sequence tags on the grounds that automated sequencing machines " 'could be run by monkeys' " (Cook-Deegan, 1995, 311).[21] Science and industry groups feared that patents on expressed sequence tags would hinder research aimed at actually understanding the molecular mechanisms of disease; the latter were particularly concerned about new expenses in the already costly process of drug development (Eisenberg and Nelson, 2002).[22]

In the continuing debate concerning DNA patenting, the same arguments appear repeatedly. The differences are both ideological and factual. Unfortunately, competing

---

[19] According to the glossary to the OECD Report, an expressed sequence tag is "a small part of the active part of a gene which can be used to fish the rest of the gene out of the chromosome" (OECD, 2002, 88).

[20] See notes 30–31 infra and accompanying text.

[21] The authors thank Ted Peters for the reminder that Watson cited the applications as a reason for his resignation, and that Francis Collins, the current head of the National Human Genome Research Institute, also opposes patents on raw genomic data. Collins has taken steps to encourage placement of information in the public domain, for example, requiring recipients of large-scale sequencing grants to release raw genomic sequence information to the public promptly (Marshall, 1997). However, Bayh-Dole makes it difficult for NHGRI to announce an outright prohibition on patenting (Rai and Eisenberg, 2003, 307).

[22] Perhaps owing to these concerns, a pharmaceutical company, Merck, financed a competing sequencing effort to put expressed sequence tags in the public domain.

assertions about the relationship between patents and scientific progress and social benefit are difficult to test.[23] Some of the points of contention are not peculiar to patents on DNA. For example, although it is clear that unaltered nature cannot be appropriated by means of a US patent, it appears that even small alterations requiring little ingenuity, effort or investment will be sufficient to take a natural substance out of the category of "natural phenomena" and so make patenting possible, assuming other requirements are met. Patents have been issued on purified vitamins and viruses, among other substances (Demaine and Fellmeth, 2003). At the same time, even critics of the current patent regime have reservations about reliance on a reinvigorated nature-linked limitation as a general solution for the shortcomings of the patent system (Rai and Eisenberg, 2003).

What, if anything, is distinctive about DNA viewed from within the standard law and policy framework? There are concerns that DNA sequences are so basic that patents including claims on stretches of the genome may be difficult to "invent around," a standard response in other areas where a patent holder's unreasonable behavior threatens to shut down further research and refinement of an invention (OECD, 2002, 11). At present, for the most part, discovery of a gene provides a therapeutic target only, meaning information to guide the search for a therapy, rather than a possible therapy. Traditional concerns about tangibility and utility (in the robust sense of a direct contribution to society) surface when patents are sought on discoveries so remote from any true applications.

Traditional concerns about preserving or enhancing the "knowledge commons" are also implicated. Much of what we are learning about genomics seems to fall within the categories of knowledge of the natural world (at least with genes, the "set" for study is quite limited and quickly exhausted), foundational ideas, and increasingly, everyday ideas.[24] What is perhaps most valuable about DNA is its informational content.[25] An esoteric but important discussion among scholars of patent law centers on a radical *extension* of patents to what is in essence information (Barton, 2002; Eisenberg, 2002). Professor John Barton of Stanford Law School lists all of the following as almost certainly patent-eligible under the law as currently interpreted: use of a particular genetic characteristic to infer a particular phenotypic characteristic; use of a particular genetic characteristic in deciding whether to administer a particular drug; use of a particular receptor as a drug target; a technique of statistical analysis for evaluating clinical data; use of a particular equation in prediction; and a process for comparing gene sequences to look for homologies. He concludes: "[F]or the first time, the patent system is effectively controlling the use of natural information, and taking out of the public domain

---

[23] Contrast the negative assessment of DNA patents in relation to social goals in Andrews (2002) with the positive assessment in Goldstein and Golod (2002), and Bendekgey and Hamlet-Cox (2002).

[24] See Section 4.2.3 of this Chapter for discussion of these concepts.

[25] Incidentally, for this reason some commentators expect that copyrights will soon be as significant for genomics firms as patents. See, e.g., OECD (2003, at 21, 60).

information that is there for anyone to measure. We are now issuing patents with claims analogous to claims on the use of blood pressure to evaluate health as distinguished from the more traditional claims on the use of a specific device to measure blood pressure" (Barton, 2002, 1343).

In January 2001, the US Patent and Trademark Office issued guidelines with a more demanding standard of utility for DNA-related patents than in the past. Applications must now disclose a "specific, substantial and credible" use (US Patent and Trademark Office, 2001). However, utility may still be established by computer-based analysis showing a claimed sequence to be homologous to genetic sequences of accepted utility. The guidelines stress that a patent on a gene covers the "isolated and purified" gene but does not cover the gene "as it occurs in nature" and that patents "do not confer ownership of genes, genetic information, or sequences." These guidelines have yet to be tested in the courts.

The controversy over patenting of DNA is not unique to the US. As biotechnology industries concentrate on the potential of biological materials for new therapies, diagnostic tests, and vaccines, the number of patents has increased, both in the United States and in Europe. When ruling on challenges to the viability and infringement of these patents, courts in the UK are being pulled between the competing goals of protecting an industry in a competitive global market and enforcing moral principles, while trying to reinterpret patent and property law. In considering the patenting of DNA, courts must not only wrestle with the new classification of genetic material but, in addition, the applicability of long-standing patent law precedent. As one commentator notes, "bemused English lawyers don't know whether to treat DNA as tangible property or intellectual property or human tissue or information" (Foster, 2003).

Patents in the UK may be filed through procedures designated by the European Patent Convention (EPC), created in 1973. The EPC provides an overarching framework for patent law in countries that are members of the European Union. The European Patent Office was established by the EPC to grant patents and received its first applications in 1978. Now, all European Union members and some additional contracting states can file for patents with the European Patent Office, a process which creates parallel patents in each of the member countries (Nuffield Council on Bioethics, 2002).

The EPC codifies the moral concerns inherent in the granting of patents. Not specific to biotechnology, the guidelines prohibit the granting of any patents for "inventions the publication or exploitation of which would be contrary to 'ordre public' or morality" (European Patent Convention, Article 53(a)). There is no equivalent of this provision in US patent law, although a similar qualification exists in the Agreement on Trade Related Aspects of Intellectual Property Rights (TRIPS) (OECD, 2002, 45).[26] No further definitions or clarifications are given, leading the UK's Nuffield Council on Bioethics to suggest that the European Patent Office look

---

[26] There is also no US parallel to a European law that permits any person to address a challenge to the issuance of a patent to the European Patent Office, or go to court to challenge a patent.

to the European Group on Ethics to assist in producing guidelines which clarify these principles (Nuffield Council on Bioethics, 2002, 15).

In 1998, the European Parliament and the European Council adopted the final draft of Directive 98/44/EC concerning biotechnology (Directive), intended to harmonize law on a particular subset of patent issues across member countries. This document was the result of a ten-year contentious political debate on the patentability of "biotechnological inventions" and represented a compromise between the European Commission and the European Parliament on the extent of legal protection for biotechnological inventions. The central disputes were between members of the biotech industry who were looking for protection for expensive research and development costs and the special interest groups that opposed patenting life (Jenkins, 2001).

The Directive has been implemented by the UK and is incorporated into the European Patent Office's guidelines for patent examination as a "supplementary means of interpretation."[27] The first draft of the Directive contained no reference to morality (Jenkins, 2001), but the implemented draft adds that patents will not be awarded to inventions "where their commercial exploitation would be contrary to ordre public or morality," (Directive, Article 6.1) closely paralleling the EPC. The Directive specifies further exclusions, namely, human cloning processes, germ line modifications, commercial uses of human embryos, and processes for genetic modification of animals which may cause suffering and do not provide substantial medical benefit (Directive, Article 6.2).[28]

A central tenet of the EPC is that discoveries are not patentable (EPC, Article 52(2)(a)). The guidelines for the European Patent Office provide the following explanation of Article 52: "If a new property of a known material or article is found out, that is mere discovery and unpatentable because discovery as such has no technical effect and is therefore not an invention... If however that property is put to practical use then this constitutes an invention which may be patentable... To find a previously unrecognized substance occurring in nature is also mere discovery and therefore unpatentable. However, if a substance found in nature can be shown to produce a technical effect it may be patentable" (European Patent Office Guidelines, Part C, Chapter IV, Paragraph 2.3).

Much of the debate surrounding DNA patents centers on this argument of whether DNA, in any form, already exists in nature, or whether the process of isolating and sequencing the genes constitutes practical use or producing a technical effect. The Biotechnological Inventions Directive strives to resolve this issue by creating a dichotomy between discovery and isolation. It asserts that "the human body, at the various stages of its formation and development, and the simple discovery

---

[27] Implementing Regulations to the Convention on the Grant of European Patents: Part II Chapter VI Rule 23b

[28] For an example of the implications, see, e.g., the statement from the UK Patent Office on inventions involving human embryonic stems cells at http://www.ipo.gov.uk/patent/p-decisionmaking/p-law/p-law-notice/p-law-notice-stemcells.htm.

of one of its elements, including the sequence of partial sequence of a gene, cannot constitute patentable inventions" (EPC, Article 5.1). It contrasts this with the patentability of "an element isolated from the human body or otherwise produced by means of a technical process, including the sequence or partial sequence of a gene ... even if the structure of that element is identical to that of a natural element" (EPC, Article 5.2).

The difficulty with the Directive's approach to resolving this potentially difficult ethical issue, is that no gene can be sequenced without first being isolated from the body, rendering the distinction virtually useless (Schertenleib, 2003, 124, 127). The use of language from the EPC in the Directive and the incorporation of the Directive into the rules of the European Patent Office brings courts no closer to guidance on the ethical issues involved. In the opinion of one critic, the Directive "does little more than state what is understood to be the current position of the European Patent Office in relation to the exclusions from patentability under the EPC for immoral inventions" (Jenkins, 2001, 11). Though the Directive represents a compromise between two opposing viewpoints, "the tone of the rules is very much in favour of recognising intellectual property rights in genetic products" (Foster, 2003, 31).

UK case law has mainly centered on the patentability of DNA from the perspective of whether the product or process being patented fits into the requirements that inventions are novel, involve an inventive step, and have industrial applications (Directive, Article 3). The seminal cases involving DNA patenting focus primarily on the technical, not the ethical implications of validating or revoking patents.

In *Genentech v. Wellcome* (1989) the Court of Appeals revoked Genentech's patents for tPA and cDNA vectors, upholding a lower court's ruling, but on the grounds that DNA sequencing was not an inventive step and cannot be patented. A wide patent for the Hepatitis C virus was upheld by the Court of Appeals in *Chiron v. Murex* (1996). The court found that the claims covered inventions, not discoveries because the process used novel molecular cloning technologies.

Biogen Inc. v. Medeva (1997) was heard in the House of Lords, the UK's highest court, in 1996. The patent for a recombinant DNA molecule was held to be invalid as it was too broad. In contrast, the patent court in the *Kirin* case (*Kirin-Amgen Inc. v. Roche Diagnostics GmbH*, 2002) upheld the validity of a patent for the coding sequence of erythropoietin and found that Roche Laboratories had infringed on that patent. *Kirin* provides the leading UK example of biotechnology patent infringement.

All of the above cases were decided on grounds other than ethical or moral. Article 53(a) of the EPC has not been the basis of many opinions concerning the patenting of DNA. Perhaps the most explicit discussion of its applicability is found in the *Howard Florey/Relaxin* case, heard by the European Patent Office Opposition Division in 1994. In upholding the patent for a protein and its associated DNA sequence, the court rejected the argument that the patent offends morality by patenting human life. "DNA is not 'life' but a chemical substance which carries genetic information and can be used as an intermediate in the production of proteins which may be medically useful. The patenting of a single human gene has nothing to do with the patenting of a human life" (*Howard Florey/Relaxin*, 1995, 541, 551).

The *Florey* decision was written between the first and final drafts of the Directive and acknowledged the raging debate at the time, where the European Parliament had voted to prohibit patents on genes and the Council of Ministers was in favor of the patents. It concluded that a moratorium would be inappropriate and that the European Patent Office was not the forum to be deciding intricate ethical issues. "Only in those very limited cases in which there appears to be an overwhelming consensus that the exploitation or publication of an invention would be immoral may an invention be excluded from patentability under Article 53(a)" (*Howard Florey/Relaxin*, 1995, 541, 553).

In the influential Harvard oncogenic mouse decision, involving a mouse engineered to develop tumors to serve as a research tool, Article 53(a) was interpreted to require the balancing of the "suffering of animals and possible risks to the environment on one hand and the invention's usefulness to mankind on the other" (*HARVARD/Oncomouse*, 1990, 501). Again in an Appeals Court of the European Patent Office, the patentability of a transgenic mammal was sent back to the Examination Court to determine whether it survived this analysis of morality under the EPC.

Thus, the UK courts have not squarely confronted the ethics of patenting DNA. The difficult ethical and moral questions have been skillfully avoided by confining legal and public debates to the more technical aspects of patent law and concentrating on the definitions of the processes used and product made. There seems to be little common ground between the positions of the biotech firms, which maintain that an industry will collapse if they are not allowed to protect the results of their work, and staunch opponents who view these patents as a way to profit from human body parts. "The rhetorical landscape is littered with slippery slopes and dubious parallels between gene sales and prostitution" (Foster, 2003, 31).

Suggestions on how to resolve the inconsistencies in the application of the law include the separation of pure product patents from those with substantial qualitative modifications (Schertenleib, 2003, 137) or allow patents only for use claims (Jacobs, 2001, 505). Either of these approaches would require significant movement in public and legal thought. Until the ethical considerations are resolved, UK patent courts may be reluctant to take on issues that need to be settled elsewhere.

4.2.2 Economic Foundations and Controversies

4.2.2.1 Economic Considerations and Biotechnology Patents

The innovation and development of biotechnology has been an important driver in improving standards of living in the United States and throughout the world. However, in societies with competitive economic environments, as in the United States, the economic system might fail to allocate resources optimally to invention for several reasons (Arrow, 1962; Clement, 2003). First, with research and development of any new technology, there is great uncertainty; it is almost always unknown whether the time and money allocated to innovation will pay off with a successful project. The uncertainty of success diminishes the incentives to innovate. Second is

4 Ethical Challenges of Patenting "Nature": Legal and Economic Accounts

a problem that economists call inappropriability. That is, the private benefit of the invention to the inventor may be much less than the total social benefit of the invention. Thus, although the total social benefit of the invention may outweigh the costs of its development, the inventor herself may only receive some small part of the total benefit. From inventor's perspective, the private benefits may not outweigh the cost of invention. Third, a related obstacle is indivisibility or nonrivalry. Many inventions involve a substantial upfront cost of development, yet after that investment has been made, the new technology or item can be reproduced at much lower cost. In societies with free markets and no barriers to entry, products are priced at their marginal cost (i.e., the cost of producing one unit). Thus, the initial expenditures for research and development are not recouped by the inventor as competitive imitators will push the price of the product to the marginal cost. Again, the inventor will have no incentive to undertake research which leads to the new invention.

Historically, societies have encouraged research in a variety of ways (Glennerster and Kremer, 2000). The most prominent incentive has been patent protection that grants inventors monopolies over goods that are produced from their ideas. (Other means of encouraging innovation will be discussed later.) Under traditional economic views, the rationale for granting patents is to prevent competitors from producing and selling a newly invented product or process for some period of time. This allows the inventor to have a (temporary) monopoly on the new invention; with a monopoly, the inventor can charge a higher than marginal cost price (and will charge the price that maximizes profits) and be the sole vendor of this invention. If all goes well, profit from sales will, over a sufficient period of time, repay the innovator's research and development costs. The presence of patent protection thus may be one of the primary factors determining the willingness of potential inventors, including pharmaceutical manufacturers, to invest in the development of new technologies (Scherer, 2002; Schweitzer, 1997). Without the guarantee of a temporary monopoly – during which time inventors can recoup their investments in a new product or process – economic incentives for innovation may be weak.

With the above-described situation, the newly developed product will be priced monopolistically and will be unavailable to other inventors, but consumers are still better off than they would be if the innovation had not been discovered. The benefit to consumers who purchase the new product can be measured as the difference between what they would be willing to pay and what they have to pay for the product. In the case of life-saving medical products, for example, consumers' willingness to pay may be very high indeed, leading to large consumer surpluses. In economic terms, individuals who do not purchase the product do not benefit by its discovery – relative to if no product were there – but neither are they harmed in any tangible way.[29] Thus, both inventor-developers and consumers gain from the new product. Overall, society as a whole benefits from the existence of a new product.

---

[29] One might argue however, that individuals who are unable to afford a new drug for example are harmed psychologically by their inability to obtain a potential cure. Whether considerations of this effect are important enough so that we should prevent an expensive drug from coming into existence is, clearly, a decision that is outside the realm of classical economics.

The preceding view of the incentives for innovation has long been held by economists as justifying and necessitating the patent system. Recent theoretical and empirical work has questioned this view, and controversy as to the role of patents on innovation continues.

Until recently, much of the theoretical economics literature had assumed an unambiguous relationship between the strength of patent protection and the rate of innovation (Kamien and Schwartz, 1974; Gilbert and Shapiro, 1990; Klemperer, 1990; Waterson, 1990). However, even "[f]rom an economic standpoint, patents are not an unmixed blessing. Patent rights motivate private firms to invest in research, but they also introduce significant inefficiencies that may inhibit future research" (Eisenberg and Nelson, 2002, 1394). There are a couple of potential reasons for this. First, many of the earlier models are based on the assumption that the patent award does not affect costs for subsequent researchers to pursue innovation (Lerner, 2002). However, recent theoretical work relaxing this assumption shows that when there are sequential innovations, the nature of the protection offered the initial innovator does affect the incentives and costs of subsequent researchers (Bessen and Maskin, 2000; Scotchmer and Green, 1990). Thus, "patents on essential materials and processes may require researchers to seek licenses before they proceed, which can impose significant transaction costs" (Eisenberg and Nelson, 2002). When patented technologies are precursors to an innovation, each patent holder will seek to extract royalties or other considerations from downstream innovators through material transfer agreements. The demands of upstream patent holders could accumulate to a level at which the next innovative steps are seriously impeded. In addition, the costs of screening and maintaining patent protection can be substantial, at least 2–5% of a research laboratory costs (Murashige, 2002). Thus, patents on products or processes which are essential to subsequent innovative efforts can add large transaction costs to the costs of research and development (Murashige, 2002; Scherer, 2002; Thomas et al., 2002). If this is the case, the social welfare effects of patents are unclear.

Second, worse than increasing costs alone, patents may enable patent holders to block research by competitors. This is especially problematic where the discoveries are ground-breaking: "Patents on fundamental discoveries that open up new research areas are typically broader than patents on incremental technological advances in established fields, because the principal constraint on the scope of patent claims is the prior state of knowledge in the relevant field." Yet it is precisely these patents that give rise to the greatest concern "in light of the great value, demonstrated time and again in the history of science and technology, of having many independent minds at work trying to advance a field" (Eisenberg and Nelson, 2002, 1394–1395). By way of contrast, patent protection does appear to play an important role in the generation of new products and processes in certain industries, such as the pharmaceutical industry.

In the US, economic concerns have shaped recent public policy around patenting of the results of publicly funded research, and these policies have in turn had significant economic consequences. Prior to 1980, federal agencies conducting or funding research generally kept any resulting inventions, although specific policies varied and case-by-case exceptions were possible. Concerns developed that inventions were languishing on the shelf; certainly, universities had little incentive to find outlets for inventions owned by the government, and the lack of a uniform policy was a source of frustration in the commercial sector. In 1980, Congress addressed these concerns with two landmark pieces of legislation, the Stevenson-Wydler Technology Innovation Act (P.L. 96–480, October 21, 1980) and the Bayh-Dole Act (formally, the Patent and Trademark Act Amendments of 1980) (P.L. 96–517, December 12, 1980). Stevenson-Wydler made the promotion of commercial innovation part of the mission of federal agencies and mandated that agencies take specific steps in fulfillment of this mission, such as establishing an office in each laboratory to identify technologies with commercial potential and transfer them to US industry. Bayh-Dole also had the stated intent of promoting commercial innovation, in this case by empowering universities, not-for-profit corporations, and small businesses to commercialize (e.g., patent) inventions made with federal funds (USGAO, 2002, 5–7).

In a 2002 survey of federal agencies, the US General Accounting Office found that NIH officials had become more selective over time, arriving at a policy that patents on biomedical technologies should be sought only where they would facilitate the availability of the technology to the public for preventive, diagnostic, therapeutic, or research use or other commercial use, and also a policy that patenting should be reserved for inventions requiring a substantial additional investment. Further, NIH officials said that it was NIH policy "to pursue nonexclusive or co-exclusive licenses whenever possible" (USGAO, 2002, 55–56).[30] However, Bayh-Dole does not require others to adopt similar policies. In a recent article, Arti Rai and Rebecca Eisenberg argue that Bayh-Dole should be modified to empower funding agencies to make these kinds of determinations for their contractors and grantees, as well as for their intramural research programs (Rai and Eisenberg, 2003).

The large public investment in genetic research matters in the larger debate because it bears on a fundamental justification for government-created and-enforced monopolies.

The argument that patents leave future users…no worse off…rests on the assumption that, but for the efforts of the patent holder (or others similarly motivated by the prospect of getting a patent), the invention would not have been made. If, instead, the invention would, in all likelihood, have soon become freely available in the public domain through the efforts of others working in the field, patents are much harder to justify (Rai and Eisenberg, 2003, 1384). Some government projects, such as the Human

---

[30] These policies are affirmed in recent "best practices recommendations" to those involved with technology transfer. Regarding patents, "when significant further research and development investment is not required, such as with many research material and research tool technologies, best practices dictate that patent protection rarely should be sought. Regarding licensing, "[w]henever possible, non-exclusive licensing should be pursued as a best pratice." Patent claims to gene sequences are offered as an example; exclusive licensing might be appropriate for development of a specific type of therapy if combined with concurrent non-exclusive licensing for diagnostic testing or research concerning gene expression (USDHHS, 2004, 67747–67748). An earlier document contained guidelines for NIH funding recipients specifically focused on sharing of research tools (USDHHS, 1999).

Genome Project, aim to put information in the public domain. Federal agencies also have the legal authority to restrict the rights of holders of patents that result from research they fund. While the Bayh-Dole Act encourages commercialization, it also contains a "march in" provision that could be invoked given sufficient public pressure. This provision gives the funding agency the right to require licensing upon reasonable terms, or to grant such a license itself, if it finds this action is "necessary to alleviate health or safety needs which are not reasonably satisfied" by the patent holder or its licensees. However, NIH has consistently adopted a narrow interpretation of this provision that precludes intervention based on considerations of price or affordability.[31]

It should be noted that empirical studies have added to the debate as well, although not without controversy as they have questioned the traditional view that strong patent protection encourages and is essential for innovation. Lerner (2002) evaluates the effect of 177 patent policy changes on rates of innovation across 60 countries and a 150 year time period. Similar to previous work on single patent policy reforms (Sakakibara and Branstetter, 2001; Kortum and Lerner, 1998; Hall and Ziedonis, 2001; Lanjouw, 1998; Scherer and Weisburst, 1995), Lerner found evidence that policies which strengthen patent protection do not increase innovation. In addition, surveys by Levin et al. (1987) of industrial research and development laboratory managers reported that patents were the second least effective means for capturing the benefits of new products, and the first least effective for capturing the benefits of processes. Another survey by Cohen et al. (2000) also showed that patents were ranked the second least effective means of procuring the benefits of innovation.

A number of alternatives to patents for promoting research and innovation have been suggested and discussed in the literature. These include government funded research (such as through the National Science Foundation, the National Institutes of Health, and the National Aeronautics and Space Administration), prizes for innovations, patent buyouts, and, for pharmaceuticals, guaranteed purchase commitments (Glennerster and Kremer, 2000; Baker, 2004). Each of these options has advantages and disadvantages, often depending on the characteristics of the innovating industry and the nature of the research and development projects. Government funded research provides assured resources to cover the cost of innovation, but may not provide strong enough incentives for completion of innovation development, and may not select appropriate research topics. Prizes do provide incentives for completion of research and development, but of course can be used only when it is possible to describe the desired invention ahead of time. Patent buyouts, and guaranteed

---

[31] This interpretation has been stated in a series of position papers issued in response to petitions, e.g., Elias A. Zerhouni, March-in Position Paper in the Case of Norvir, September 17, 2004, and Elias A. Zerhouni, March-in Position Paper in the Case of Xalatan, August 2, 2004, available at http://ott.od.nih.gov/NewsPages/news.html. This interpretation is consistent with the views of the authors of the Bayh-Dole Act. For example, in connection with the Norvir case, Senatory Bayh stated that the march-in provision was included in the Act to address concerns that companies might license technologies purely to suppress them, and hence should be triggered only where the patent holder is making no effort to bring a product to market. See Statement of Senator Birch Bayh to the National Institutes of Health, May 25, 2004, available at http://www.essentialinventions.org/drug/nih05252004/birchbayh.pdf.

purchase commitments, allow the research community to decide which projects to pursue and would only be compensated when discoveries were made. However, it would seem difficult to secure such a commitment, and would – as with prizes – only be applicable for discoveries which can be described in advance.

Given the arguments for and against patents, some authors assert – and we agree – that the overall effect of patents, i.e., whether they provide incentives for innovation and have a net social benefit, is likely to be dependent on the specific industry or endeavor. We next discuss how characteristics of research in genetics may affect the pros and cons of patenting DNA.

4.2.2.2 Economic Considerations and DNA Patenting

Many of the incentives for and effects of patenting related to genetic research and material, including human genome sequences, are similar to those shared by all inventions. There are clearly incentives for patenting DNA which are created by the problems of uncertainty, inappropriability, and nonrivalry described previously. However, special characteristics of genetic research need to be taken into account in evaluating patenting of DNA sequences. Innovations in genetics sometimes build heavily on previous innovations that act as "research platforms." Thus, unlike other industries, for example many pharmaceutical innovations, genomic research is often highly sequential in nature (Rai, 2002).[32] This creates a tension between the

---

[32] In the pharmaceutical industry, unlike in other industries, large-scale surveys of research and development laboratory managers have shown that patents are indeed viewed as a very important means of capturing the benefits of production innovations (Levin et al., 1987; Cohen et al., 2000). First, many pharmaceutical discoveries are not dependent on upstream research platforms. Second, the pharmaceutical research and development differs in several ways from genetic research, and thus we use the development of new drugs as a counter-example to illustrate how industry differences may lead to different optimal patenting solutions. Third, patent claims in pharmaceuticals tend to define products precisely. Fourth, new innovations must be rigorously – and expensively – tested through clinical trials to prove safety and efficacy, adding large costs to the research and development effort. Finally, the cost of imitation in pharmaceuticals is particularly low relative to the initial innovation, which often entails very high costs. For example, DiMasi et al. (2003) survey the research and development costs of 68 randomly selected new drugs from ten pharmaceutical companies, and find that the total RandD costs per new approved drug (including capital costs) is US$802 million in US$2,000. Thus absent patent protection, invention costs will likely be difficult if not impossible to recoup. Thus, several authors argue that the patent system "seems ideally designed for the pharmaceutical industry" (Murashige, 2002, 1329).

Of course, there are disadvantages to patenting of pharmaceuticals (Schweitzer, 1997). For example, new drugs may not be available as widely as they could be, because the patent holder – in order to maximize profit – will set the price of the new drug to maximize profits. Thus, some people who could make productive use of the patented item will not be able to. Consequently, some of the potential social gain of use of the new drug will not be realized. However, if the pharmaceutical company sets a low (marginal cost) price of the drug for everyone, the company will not be able to recoup the research and development costs. The ability to use price discrimination – charging different prices to different people (based on ability to pay/who is buying/different countries, etc.) – is one way to get around this problem. Thus, the problem of patient access to drugs might best be addressed through modifications to the system for financing and delivering health care rather than through modifications of the patent system.

needs and incentives of developers of the "research platforms" and subsequent researchers who need to use these platforms for subsequent research.

To encourage upstream innovation, the overall benefits of the "upstream" discovery should include all the extra benefits, net of production and research and development costs, that are realized by the downstream inventions. However, in an open market without patent protection or other policies, this benefit will not be realized by the initial discoverer, and thus there may be insufficient incentives for individuals to innovate. This is an example of the inappropriability problem.

Patents do provide a solution for these problems, allowing upstream innovators to require subsequent inventors to pay royalties for using the patented product or process. However, patenting of upstream innovations can also lead to an inhibition of subsequent research. When numerous patented technologies are precursors to an innovation, the weight of the transaction costs in contracting for licenses (both direct royalty costs and also time costs) may inhibit downstream innovation. Each upstream patent holder has an incentive to extract royalties from a downstream researcher, yet if the accumulated costs of this are too high, innovation will be stifled. New innovations and products that could have come into existence do not, and there may be a net welfare loss due to the patent protection.

As a solution to this problem, Scherer (2002) recommends that genome patent claims be interpreted narrowly, and that research-only use of patented technologies be exempted from royalties. However, Rai argues that not all genomic research should carry this special status, and that only "broadly enabling research platforms" and not more downstream "research tools" should be the focus of a narrow patent scope.[33] Goldstein and Golod (2002) argue that the transaction costs of conducting licensing agreements cannot be that high, simply because there have been very few instances of patent litigation. Therefore, they argue against trying to make distinctions for genes or uses that are and are not patentable, and conclude that the current patent system works well enough.

4.2.3 Nature, Property and Patents: A More Extensive Exploration

Nature has figured in legal and economic understandings of property and patents in a number of ways. In the West, from the Romans forward, there was recognition of a realm of *res communes*, things in nature that by their nature are open to all, such as the oceans and the air. Carol Rose notes that the Romans had several other categories of nonexclusive property, including *res divini juris*, things that are unowned by any human being because they are sacred, holy or religious (e.g., temples, tombs); *res publicae*, things belonging to the public and open to the public by operation of law (e.g., roads,

---

[33] Rai acknowledges that making this distinction will be difficult in practice, and that – in addition – transaction costs can be a problem when the number of upstream patents is large, even if any individual patent is narrow.

4 Ethical Challenges of Patenting "Nature": Legal and Economic Accounts

harbors); and *res universitatis*, things belonging to a limited community or group in its corporate capacity (e.g., theaters for civic performances) (C. Rose, 2003).

After the fact, jurists and philosophers wrestled with the question of how individuals or groups legitimately come to appropriate any part of nature as their private property. (The question of how humans come to have property rights in nature was not of much practical or moral concern; God had put nature at the disposal of humankind.[34]) Many of the accounts of the origin of private property begin with a state of nature in which *all* things are held in common. For 17th century natural law theorist Samuel Pufendorf, God both ordained this original state and offered a way out of it, by allowing humans to form societies for their mutual self-preservation and then allocate portions of common things among individuals or groups by agreement, provided the terms were equitable (Travaglini, 2001, citing Pufendorf, 1994).

Pufendorf's ideas influenced John Locke, who in turn exercised a profound influence on the Western philosophical, political, and legal traditions. Contemporary legal scholars credit Locke with developing two "theories" of property that are mutually reinforcing (e.g., Schaffner, 1995). The first theory is instrumental. Property rights provide the incentive for labor, productive work that would not otherwise occur, thus benefiting society as a whole. For Locke, in the state of nature the fruits of nature might indeed have been common property, put there by God for use according to need, but God could not have intended that much of nature remain uncultivated, as would surely be the case were no private appropriation permitted. Yet the concurrence of all in each removal from the common could never be practicable. Hence an individual's investment of labor must be sufficient to establish rights. Locke imposed two conditions to ensure that private property rights serve the public interest: the common must be maintained so that "enough and as good" is available for use by others, and individuals must limit what they appropriate to what they can use lest resources be wasted. The second Lockean theory is deontological. God has given each man rights over his body, and by extension, rights in the fruits of his labor. Property rights in the products of one's industry or ingenuity flow out of these "natural" rights and cannot be withheld without injustice.[35]

---

[34] For more on this subject, see Chapter 1 of the present volume, "Spiritual and Religious Concepts of Nature."

[35] In Locke's exposition, the two theories are intertwined: "God, who has given the world to men in common, has also given them reason to make use of it to the best advantage of life and convenience. The earth and all that is therein is given to men for the support and comfort of their being. And though all the fruits it naturally produces and beasts it feeds belong to mankind in common, as they are produced by the spontaneous hand of nature; and nobody has originally a private dominion exclusive of the rest of mankind in any of them, as they are thus in their natural state; yet, being given for the use of men, there must of necessity be a means to appropriate them some way or other before they can be of any use or at all beneficial to any particular man.... Though the earth and all inferior creatures be common to all men, yet every man has a property in his own person; this nobody has any right to but himself. The labor of his body and the work of his hands, we may say, are properly his. Whatsover then he removes out of the state that nature has provided and left it in, he has mixed his labor with, and joined to it something that is his own, and thereby makes it his property..." (Locke, 1952, 17, para. 26, 27). See also Glitzenstein (1994).

Several contemporary legal scholars suggest that modern systems of intellectual property nicely accord with this theoretical framework.[36] In the case of patent law, Joan Schaffner argues that the exclusion that renders laws of nature, natural phenomena, and abstract ideas ineligible for patenting can be linked to Locke's emphasis on labor as the basis for claiming property rights in what would otherwise belong to the common, nature. The utility and nonobviousness requirements may be viewed as ways of assuring that the inventor has contributed something of value and merits reward. The disclosure requirement and the time limit help to guard against waste, although as noted above there is no requirement that an inventor exploit her invention during the term of the patent. Justin Hughes suggests that the laws of nature and natural phenomena exclusions should be construed as rough matches for two kinds of inventions that should not be subject to monopolies if the "enough and as good" requirement is to be satisfied: everyday ideas, encompassing ideas or concepts that are spread so effectively that they become woven into ordinary social intercourse/discourse, and fundamental ideas, including scientific ideas or insights that are of "foundational" importance (Hughes, 1988).

These scholars believe that Locke's ideas about nature and property and the public interest, already dominant in the 18th century, directly influenced the drafters of the early patent statutes. For example, Schaffner writes that "strains of Lockean labor theory principles are evident in the writings of Madison and, to a lesser extent, Jefferson, two principal architects of our patent laws" (Schaffner, 1995, 1100–1101). This is likely the case, but a review of the history of the patent laws in England and the US does more than prove the point. For example, many of the themes dominating current debates over biotechnology patents are anticipated in debates that took place in the 16th century, when the English patent system became a subject for serious reflection. The continuity with the present may be evidence of the futility of debate. On the other hand, the lesson could be that certain tensions are inherent in any patent system and must be managed in line with economic and political circumstances and the temper of the times. The disappearance of debate, as a consequence of a refusal to recognize that there are trade-offs, would then be a sign of failure rather than success.

According to historian Christine MacLeod, in England a normative theory of monopoly privilege as a "just reward for services rendered to the commonwealth" was in wide circulation by the late 16th century (MacLeod, 1988). Francis Bacon was

---

[36] See, e.g., Schaffner, at 1089–1090. Interestingly, when Locke himself had occasion to address issues of intellectual property (copyright), he objected strongly to the grant of monopolies to publish texts from ancient authors and supported limits of 50–70 years from death or publication for the works of living authors. See Rose, M. (2003, 78).

a leading proponent (MacLeod, 1988, 204).[37] Bacon presented the mechanical arts as a model for "natural philosophy," the branch of inquiry that evolved into modern science. The ancient hierarchy of knowledge, culminating with disinterested contemplation of truth, was toppled in pursuit of the practical improvement of the human condition. But Bacon also believed that collaboration and the unhindered dissemination of information, rather than the pursuit of individual glory and gain, would be the key to progress in natural philosophy. Likely due to his influence, few fellows of the Royal Society sought patents for their inventions (MacLeod, 1988, 186).

The normative "reward" theory was frequently joined to two theses suggesting reinforcement by a nascent instrumental theory: patent monopolies would act as an incentive to invent and would serve as the quid pro quo for public disclosure of the inventor's secret. The actual relationship between the operation of the patent system and economic development or welfare was not systematically studied. There were emerging concerns about the term of patents – extensions of patent terms were viewed as potentially inhibiting improvements to patented inventions. But according to MacLeod, "[m]ost opposition was of an ethical rather than an overtly economic nature," motivated by "concern that the benefits of a discovery should be widely disseminated" (MacLeod, 1988, 183). Opponents suggested that alternatives to the patent system could result in wider dissemination of inventions while preserving the incentive and reward features of the patent system. Concerning one proposal advanced in the early 18th century, MacLeod writes:

> In outlining a plan for the improvement of hospitals and medical care John Bellers, the Quaker philanthropist, proposed that new medicines should be tested in controlled conditions and, if they succeeded repeatedly, "a gratuity should be given by the government, to the owner of such medicines, to make a public discovery of them"; the inventor should be further rewarded with the offer of a post as physician in a hospital "where his medicine is proper." Bellers also contended that the Royal Society should be endowed by the state, to enable it to continue improving natural knowledge, and to offer annually "a prize to every mechanic that shall produce the best piece of work, or anything new" (MacLeod, 1988, 192).

---

[37] For example, consider Bacon's role in the patent debate of 1601. A bill had been introduced in Parliament in response to the alleged abuse of patents. At the time, "patent" had a very broad meaning, covering virtually any kind of state-granted monopoly. The greatest outrage was reserved for patents touching on necessities such as salt. (Patents affecting cards, dice, starch, and the like, stated one Dr. Bennett, were, simply as monopolies, "very Hateful," but "not so Hurtful.") In the debate in the House of Commons, Francis Bacon criticized the bill as unfocused, defended royal prerogatives generally, and suggested that distinctions among different types of patents were important. Among the worthier types: "If any Man, out of his own Wit, Industry, or Endeavour, find out any thing beneficial for the Common-Wealth, or bring any New Invention, which every subject of this Realm may use; yet in regard of his Pains, Travel and Charge therein, Her Majesty is pleased (perhaps) to grant him a Privilege, to use the same only by himself, or his Deputies, for a certain time." Bacon noted that the courts could void privileges later found not "good for the Common-Wealth"; Queen Elizabeth had also directed her attorney general to seek repeal of troublesome patents. Bacon concluded with a plea for petitions regarding particular patents rather than a general abolition of the prerogative of granting patents. See "The Debate on Monopolies in the House of Commons, 1601" (Tawney and Power, 1924, 272–273).

Bishop Berkeley was another proponent of direct payments to inventors. Between 1750 and 1825, at least eight laws were passed authorizing monetary awards to particular inventors, most notably £30,000 to Edward Jenner for the smallpox vaccine.

Patents were also problematic for those with strong beliefs in divine Providence. "If the inventor was no more than God's instrument in bringing His gifts to the community, then he could at most claim user's rights over them" (MacLeod, 1988, 198). For Protestants, it seemed clear that God planted the idea of the printing press when it was time for the Reformation. Newtonian physics changed the view of divine intervention. God the clockmaker went into retreat after winding up the mechanism. This suggested that human agency was the key element in invention, and also that there was no divine or natural limit to the progress possible with sustained human effort. In a 1774 court case concerning property rights in inventions under the English common law, Lord Camden could still fulminate that those asserting such rights "'forget their Creator, as well as their fellow creatures'" in trying to monopolize "'his noblest gifts and greatest benefits'" (MacLeod, 1988, 221). But this view was already on the wane. If invention was a product of individual industry or genius, rather than a gift from God, individual rights seemed to follow.

Still, utilitarian reasoning rather than natural law or natural rights dominated in the defense of patents and other legally-protected monopolies in England. The French National Assembly, not the English Parliament, declared in 1790 that "'it would be a violation of the Rights of Man…not to regard an industrial discovery as the property of its author'" (MacLeod, 1988, 199). The US patent system was based on the English system but reflected a hybridization of the two kinds of justification and discourse. Early state copyright statutes and, later, laws related to patents, incorporated the concept of "natural equity and justice." The preamble to the first copyright law adopted in Massachusetts included the following language:

> Whereas the Improvement of Knowledge, the Progress of Civilization, the public Weal of the Community, and the Advancement of Human Happiness, greatly depend on the Efforts of learned and ingenious Persons in the various Arts and Sciences: As the principal Encouragement such Persons can have to make great and beneficial Exertions of this Nature must exist in the legal Security of the Fruits of their Study and Industry to themselves; and as such Security is one of the natural rights of all Men, there being no Property more peculiarly a Man's own than that which is produced by the Labour of his Mind… (Bugbee, 1967, 114).

In 1788, a private act of the South Carolina Assembly awarding a 14-year patent to the inventors of an improved steam engine recited that

> the principles of natural equity and justice require that authors and inventors should be secured in receiving profits that may arise from the sale or disposal of their respective writings and discoveries, and such security may encourage men of learning and genius to publish and put in practice such writings and discoveries as may do honor to their country and service to mankind (Bugbee, 1967, 95).

Yet zeal for patenting, even among inventors, was moderated by ethical and practical considerations. In his autobiography, Benjamin Franklin wrote that he declined the offer of a patent on an open stove from the colonial Governor of Pennsylvania, on the principle "that, as we enjoy great advantages from the inventions of others, we should be glad of an opportunity to serve others by any invention of ours, and this we should do freely and generously." Franklin also mentioned his dislike of disputes in explaining his personal policy against patenting.[38]

It is significant that early American intellectual property laws sometimes incorporated public or social welfare-oriented protections against waste and price-gouging through a form of compulsory licensing. For example, the general copyright-patent law in South Carolina permitted the state to intervene if the author of a protected work either neglected to publish it or set an exorbitant price. Further, where the author persisted in the neglect or abuse, another individual could apply to the state court of common pleas for a license (Bugbee, 1967, 94). The US Senate sought to include a compulsory licensing provision in the first federal patent statute (Bugbee, 1967, 143).

Although the compulsory licensing provision was not, ultimately, incorporated in the version of the patent statute signed into law on April 10, 1790, leaders expressed reservations concerning the English system. Enthusiasm for intellectual property protections was tempered by antipathy to state-created and -enforced monopolies. A letter from James Madison to Thomas Jefferson presents protections as necessary evils rather than matters of natural right: "With regard to Monopolies, they are justly classed among the greatest nuisances in Government. But is it clear that as encouragements to literary works and ingenious discoveries, they are not too valuable to be wholly renounced? Would it not suffice to reserve in all cases a right to the public to abolish the privilege at a price to be specified in the grant of it?" (quoted in Bugbee, 1967, 130). Jefferson was skeptical of the use of nature to

---

[38] The surrounding text more fully conveys the ethics of the stove case, as presented by Franklin: "[H]aving, in 1742, invented an open stove for the better warming of rooms, and at the same time saving fuel, as the fresh air admitted was warmed in entering, I made a present of the model to Mr. Robert Grace, one of my early friends, who, having an iron-furnace, found the casting of the plates for these stoves a profitable thing, as they were growing in demand. To promote that demand, I wrote and published a pamphlet, entitled *An Account of the new-invented Pennsylvania Fireplaces; Wherein their Construction and Manner of Operation is Particularly Explained; their Advantages Above Every Other Method of Warming Rooms Demonstrated; and all Objections That Have Been Raised against the Use of Them Answered and Obviated, etc.* This pamphlet had a good effect. Gov'r. Thomas was so pleas'd with the construction of this stove, as described in it, that he offered to give me a patent for the sole vending of them for a term of years; but I declin'd it from a principle which has ever weighed with me on such occasions, viz., *that, as we enjoy great advantages from the inventions of others, we should be glad of an opportunity to serve others by any invention of ours; and this we should do freely and generously.*

An ironmonger in London however, assuming a good deal of my pamphlet, and working it up into his own, and making some small changes in the machine, which rather hurt its operation, got a patent for it there, and made, as I was told, a little fortune by it. And this is not the only instance of patents taken out for my inventions by others, tho' not always with the same success, which I never contested, as having no desire of profiting by patents myself, and hating disputes. The use of these fireplaces in very many houses, both of this and the neighbouring colonies, has been, and is, a great saving of wood to the inhabitants" (Eliot, online).

justify property rights generally, and he was especially concerned about claims on ideas. In a letter composed following his tenure as Secretary of State and administrator of the patent system, he wrote:

> That ideas should freely spread from one to another over the globe, for the moral and mutual instruction of man, and improvement of his condition, seems to have been peculiarly and benevolently designed by nature, when she made them, like fire, expansible over all space, without lessening their density in any point, and like the air in which we breathe, move, and have our physical being, incapable of confinement or exclusive appropriation. Inventions then cannot, in nature, be a subject of property (Walterscheid, 1998, 73, quoting a letter from Jefferson to Isaac McPherson dated August 13, 1813).

Jefferson, and his contemporaries, would likely have been astonished by the notion that patents could be issued on new knowledge in natural philosophy or science (Bugbee, 1967, 194, n. 13).

The evolution of US patent law shows that Jefferson did not carry the day; through the power of the state, ideas have become property. At the same time, the restrictions on patentable subject matter reflect sensitivity to Jefferson's concerns. Even as the US Supreme Court declared organisms patent-eligible subject matter, it affirmed that nature sets certain limits: "The laws of nature, physical phenomena, and abstract ideas have been held not patentable. Thus, a new mineral discovered in the earth or a new plant found in the wild is not patentable subject matter. Likewise, Einstein could not patent his celebrated law that $E = mc^2$; nor could Newton have patented the law of gravity. Such discoveries are 'manifestations of ... nature, free to all men and reserved exclusively to none" (*Diamond v. Chakrabarty*, 1980, 309). The properties of bacteria, solar energy, electricity, and metals were all "part of the storehouse of knowledge of all men" (*Funk Brothers Seed Co. v. Kalo Inoculant Co.*, 1948, 130).[39] While the language suggests that knowledge of nature's workings is easy to come by, considerable effort and ingenuity may be required for some discoveries. Thus, a labor or value-added theory would not necessarily support the Court's line drawing. Locke's public welfare conditions may provide the missing justification.

It is important to conclude this discussion by noting an absence. In the early debates over patenting, nature does not figure as a realm that should be shielded from human manipulation and use. Hence, there is little precedent here for some aspects of the current debates over biotechnology patents. Patents may serve as catalysts for controversy because they make activities or at least the premises for activities visible – public disclosure is required as an aspect of obtaining a patent – and so rally groups whose core concerns are affected by these activities. The patenting of substances or information derived from human DNA has become a flashpoint for controversy because it creates a focus for those concerned about the regard due the human body, the kinds of control that should (or should not) be

---

[39] In what is regarded as the classic statement of the doctrine, Ex parte Latimer, 1889 December Comm'r Pat. 123 (Comm'r Patents 1889), the patent commissioner asserted that "nature" intended plant fibers and the like "to be equally for the use of all men."

exercised over human bodies and, potentially, human nature, and the role of industry or commerce in all of this. In a similar manner, the patenting of a mouse genetically engineered to reproduce characteristics of human cancers has become a flashpoint for controversy because it creates a focus for those concerned about the integrity of species, the appropriate relations between humans and other sentient beings, and more particularly the status of higher organisms, and the role of industry or commerce in all of this.[40]

Borrowing from the nuanced taxonomy of non-exclusive property developed by the Romans, contemporary critics of DNA patenting seem to fall into at least four categories: those who believe DNA (whether isolated and purified or converted to a sequence) is or should be regarded as *res communes* ("it is part of nature and therefore unpatentable"), *res publicae* ("it is in the public interest to put such fundamental discoveries in the public domain"), *res universitatis* ("research tools should be open to use by members of the scientific community without worries about infringement"), and *res divini juris* ("the commodification of the human genome is at odds with proper piety toward God, nature, or human dignity").

4.2.4 Framing the Normative Discussion

We have already touched on some of the legal and economic arguments for and against liberal approaches to patenting in general and patents on DNA in particular. In the next three sections, we will explore in somewhat greater depth the two themes that largely define the normative discussion of patenting in biotechnology: concerns about commodification and concerns about access.

4.2.4.1 Concerns About Commodification

A commodity is commonly defined as an article of trade or commerce. Commodification refers to the reduction to a mere commercial good of something that should be treated or valued in other ways. The exclusion of "natural phenomena" from patent-eligibility has been of little help to those who believe that some things should not be patentable because (a) they should not be treated as goods or any kind of property, period, or (b) they should not be treated as purely private property. DNA patenting may arouse commodification concerns because people think that a patent holder is granted a kind of global ownership of the DNA in our bodies. The guidelines published by the US Patent and Trademark Office are technical in nature, but they do try to address fears that human bodies are being colonized and commodified through the patent system. Yet these kinds of fears are stoked by the growing trade in human tissue and

---

[40] The two strands of concern and protest come together in preliminary skirmishes over the possible patenting of human organisms. See, e.g., President's Council on Bioethics (2002).

stories of patents on cell lines or gene sequences awarded with indifference to the feelings of violation and betrayal expressed by the tissue "donors."

Patenting of higher organisms may arouse commodification concerns for somewhat different but related reasons. To some the assertion of control over the creation and use of any cancer-prone transgenic mouse seems much more problematic than the assertion of control over, say, the creation and use of any device meeting a particular description, or this or that existing mouse. Terms like "hubris" and "arrogance" recur in critical accounts, suggesting that the problem is a lack of proper humility combined with crass commercialism. The following is a typical statement of the anti-commodification viewpoint: "Biotechnological innovations must be disciplined by a prior respect for all of creation, by a reverent concern to understand the contribution each part makes to the whole, and by an awareness that we did not invent and must not monopolize the living matrix within which life continues" (Canadian Council of Churches, online, 15).[41]

It is important to underline here that the grant of a patent is not equivalent to legal permission to engage in the underlying activity. Further, even a product patent on a living organism does not give the inventor full and unrestricted ownership and control over that organism wherever it comes into being. However, the issuance of a patent does seem to have some expressive significance; it may be perceived by those in favor of or opposed to the underlying activity as a kind of social endorsement. Further, it may make the activity more likely or encourage its spread, especially in commercial contexts where property rights have the greatest significance. This is a problem for those who believe the activity is ethically questionable but unlikely to be prohibited directly, or that it is best conducted in the non-commercial sphere or with non-market-related motivations.

Some critics of DNA patenting and patenting of higher organisms as commodification trace their concerns to their religious convictions, and we will examine statements from representatives of a number of religious traditions in later sections of this chapter. Concerns about commodification also appear in texts written from the standpoint of developing world bioethics.[42] Further, they are taken very seriously in Europe, where they have influenced official pronouncements. For example, Article 21 of the Council of Europe's Convention on Human Rights and Biomedicine states that "the human body and its parts shall not, as such, give rise to financial gain." Of course, application of this principle is hardly straightforward. It is fairly clear that this article of the Convention prohibits payments to sources, apart from reimbursement for expenses and perhaps compensation for time and inconvenience.

David Resnik has examined the various arguments put forward by those who worry about commodification in relation to DNA patenting. He concludes that "patents on human DNA do not violate human dignity because they do not treat human beings as complete commodities." At the same time, "since human DNA patenting uses market rhetoric to describe human body parts, it does treat human beings as

---

[41] For an extended argument on the same points in wholly secular terms, see Andrews and Nelkin (2001).

[42] E.g., Robin Bunton's chapter on "Global genes" (Peterson and Bunton, 2002) includes a section entitled "Patenting and the commodification of health."

4 Ethical Challenges of Patenting "Nature": Legal and Economic Accounts

incomplete commodities and, therefore, may threaten human dignity by taking us further down the path of human commodification." However, "since market rhetoric already pervades many aspects of science, medicine, and society, the threat posed by DNA patenting is neither new nor unique" (Resnik, 2001, 152). Hence the controversy over biotechnology patents in general and DNA patents in particular is linked to the question of gains and losses from the growing commercialization of science. Although some critics of biotechnology are straightforwardly anti-science, many are at pains to express their support for properly bounded and motivated science.

The increasing involvement of science with commerce has contributed to the erosion of trust in academic institutions participating in biomedical research. David Korn attributes the commercialization of science to three events: the invention of recombinant DNA technology, the decision in *Chakrabarty*, and the passage of Bayh-Dole Act. Concerning Bayh-Dole, he remarks: "While dramatically increasing the flow of revenues into research institutions, and driving the interests of both institutions and faculty toward ever-more-vigorous commercialization of their intellectual property, it may have created a troubling intoxication with and dependency upon 'cashing in' on academic biomedical research" (Korn, 2002, 1091–1092). Korn highlights a widely-shared ambivalence. US culture is characterized by a general preference for market transactions combined with a desire for speedy translation of scientific discoveries into biomedical products, and so it seems "natural" to rely on private enterprise in the health sector. This entails acceptance of close ties between academic researchers and industry as a matter of course. At the same time, money, or the love of it, is the "root of all evil," and the flow of money is believed to threaten the norms of science, corrupt institutions and individuals entrusted with public or pooled resources, and place the lives of research subjects at risk.

A recent meta-analysis of studies of the prevalence and power of industry affiliations concluded that financial relationships among industry, investigators, and academic institutions are "pervasive," but data concerning their impact on biomedical research is limited, apart from strong, consistent evidence that industry-sponsored research tends toward pro-industry conclusions (Bekelman et al., 2003, 463). The evidence also shows that industry ties are associated with publication delays and data withholding; problems mostly arise when an investigator or institution has a stake in the process of bringing research to market (e.g., through holding stock in the sponsor).[43] The commercialization of research also complicates efforts to ensure accountability through regulation. Government agencies are hard pressed to find scientists without industry ties to serve

---

[43] David Blumenthal's work is also relevant here, pointing as it does to costs such as barriers to pursuit of potentially fruitful lines of investigation, redundant work, and wasted work on unfruitful lines of investigation as a consequence of secrecy and data-withholding. See Blumenthal et al. (1997). More recently, see Campbell et al. (2002). (Eighty percent of geneticists cited the effort required to actually produce materials or information as a very important or important reason for withholding postpublication, versus 27% and 21% for the need to honor the requirements of an industrial sponsor and the need to protect the commercial value of the results, respectively.) The same practices may hinder effective oversight, e.g., industry may also pressure reviewers and regulators (e.g., the RAC, the FDA) to conduct proceedings in private and withhold information from the public. See Wright and Wallace (2000).

in an advisory capacity. Questions are also being raised about the effects of commercialization on the motivational structure of scientists engaged in biomedical research. Surely the desire for material gain is a motivating force for individuals, but so are compassion, curiosity, and the desire for recognition. It is possible that these qualities are additive – the more the better – but it is also possible that a stress on narrow, material self-interest will erode the other springs of scientific activity. One commentator expresses the concern that "[c]ontinued pressure to extend patent law's money economy at the expense of science's traditional economy of reputational credit could create impediments to future progress while providing less effective, or at least insufficiently effective, countervailing rewards" (Golden, 2001, 110).[44]

4.2.4.2 Concerns About Access

The access concerns that resonate most with the public appear at the end of the research and development trajectory, where the question is whether patients will be able to afford tests and treatments. However, as noted above, concerns about access first arise in connection with the ability of scientists to avail themselves of fundamental scientific knowledge and research tools and the ability of competitors to conduct follow-on research, areas where impediments may ultimately manifest in diminished access by patients because products are never developed due to higher costs. There are many ways in which property arrangements may prove dysfunctional. While the Lockean justification for private property rests on the assumption that common property (or nature absent a means of private appropriation) will be underdeveloped, the phrase the "tragedy of the commons" has been used to describe the overuse and degradation of common property. Some now argue that a multiplication of intellectual property rights is likely to lead to a "tragedy of the anticommons" as the many private owners of pieces of the genome and other research tools hinder each others' efforts (Heller and Eisenberg, 1998). An OECD report on gene patenting summarizes the results of two studies looking for an anticommons effect, one in the US and the other in Germany. The consensus seemed to be that concerns exist, but to date industry actors have for the most part been able to find "working solutions." On the other hand, problems have been identified with the argument that patents for pre-commercial inventions will lead to a beneficial coordination of follow-on development.[45]

---

[44] Golden suggests that there is a symbiotic relationship between industry and academia/government that would be weakened were either sector to collapse into the other: "the publicly funded sector gains from having privately funded outlets for vitality, imagination, and rapid growth; the private sector builds upon the base of knowledge, talent, and competitive-but-cooperative spirit that the publicly funded sector supplies" (2001, 131). Of course, many private firms adhere to traditional norms of science, if only to attract good people, and evidence suggests that firms that behave like academic research institutions and encourage open communication enjoy greater commercial success than their more generically business-like counterparts (Golden, 2001, 153–162).

[45] "The unpredictability of biotechnological development makes the coordination of subsequent invention implausible; allowing multiple firms to pursue a variety of commercial 'spin-offs' seems a better strategy" (Golden, 2001, 166).

A telephone survey of US institutions holding US patents covering diagnosis of human genetic disorders found that in all cases where an invention had been licensed, the license was exclusive. On the other hand, a majority of respondents (78%) either did not require licenses for research activities, made an exception for academic researchers, or did not charge any royalty for use of their inventions under license (Schissel et al., 1999, 118). Of course, this was before *Madey*. Institutions may have refrained from requiring licenses for research activities in the belief that these uses were exempt. A related survey found that for-profit firms tended to adopt broad patenting strategies, building patent portfolios to attract investors, block competitors, and protect the firm's ability to work in a particular area. Non-profits were more likely to assess inventions individually for potential market value before filing and to grant exclusive licenses. In keeping with our discussion of economic considerations, many respondents distinguished between research tools and research targets, stating that "tools useful to performing research should be made broadly available [like the Cohen-Boyer patent on recombinant DNA], while exclusive licensing may be necessary to promote investment in downstream development" (Henry et al., 2002, 1279, 2003).

In Europe, it is much clearer than in the US that the health sector has a special status. When patents affect the health sector, both kinds of access issues are high in the hierarchy of concerns.[46] The priority given to access concerns by Europeans is reflected in more stringent standards for patenting, more generous allowances for non-infringing uses in the realms of medicine and research, and in greater comfort with compulsory licensing. In its report on DNA patenting, the UK's Nuffield Council on Bioethics, citing access concerns, proposed that governments go further in tightening up the requirements for product patents asserting rights over DNA sequences, and where patents have been granted for diagnostic tests based on genes, consider compulsory licensing to enable development of alternative tests (Nuffield Council on Bioethics, 2002).[47]

This is not to say that access concerns are absent in the US. Indeed, at least one study has attempted to provide empirical backing for claims that DNA patenting has diminished patients' access to appropriate genetic testing services. Mutations in the HFE gene are responsible for hereditary hemochromatosis, an autosomal recessive disorder of iron metabolism that, untreated, can lead to cirrhosis, liver cancer, diabetes, and heart disease. Patents were issued covering clinical HFE testing in early 1998, and the patent holder entered into an exclusive license with a national chain of laboratories. In September 1999, researchers set out to assess the effects of the HFE patent and licensing scheme on access by conducting a telephone survey of laboratories with the technical capacity to offer HFE testing. Thirty percent of respondents reported that they were not performing the test, or had stopped

---

[46] See, e.g., Opinion of the European Group on Ethics, Ethical Aspects of Patenting Inventions Involving Human Stem Cells, May 7, 2002.

[47] The Council reviewed commodification concerns, but these were for the most part dismissed as either premised on a misunderstanding of what a patent is, or too vague to provide much practical guidance.

performing it. A majority of laboratories in this group reported that patent concerns were "the reason" for their lack of activity in this area. The investigators note that hospital-based laboratories have an economic incentive to develop in-house testing, if justified by demand, since sending samples off-site generates expenses that are not reimbursed by payers (Merz et al., 2002).[48]

At the policy level, in 2004 the National Academy of Sciences published a report on the patent system that concluded with seven recommendations, including reinvigorating the non-obviousness standard, instituting an open review procedure that would permit third parties to challenge patents in an administrative proceeding, and shielding some research uses of patented inventions from liability for infringement (Merrill et al., 2004). Princeton President Shirley Tilghman chaired a panel that looked specifically at the impact of patents on genetics, genomics, and biotechnology. The panel concluded that intellectual property rights rarely impose significant burdens on biomedical research, but it also expressed concern about their effects on scientific advances in the future.[49]

Further, recall that Federal agencies themselves hold a significant number of patents. For a time, NIH included a reasonable pricing clause in its licenses. This policy was changed in 1995.[50] A report generated in July 2001 suggests that access to health

---

[48] The story is, however, complicated by a series of mergers and waxing and waning patent enforcement efforts. At the time of the survey, the licensee was selling a test kit. It was also offering to sublicense the invention to laboratories interested in developing and performing their own tests, although with up-front payments and per test fees that could be expected to encourage purchase of the test kit.

[49] The final report, titled *Reaping the Benefits of Genomic and Proteomic Research: Intellectual Property Rights, Innovation, and Public Health*, was published in 2006.

[50] A not-entirely-disinterested history from a report to Congress defends the change in policy:

"In the years following passage of Bayh-Dole, members of Congress continued to express concerns about an appropriate monetary return for taxpayers' investment in biomedical research. In response to those concerns, in 1989 the NIH adopted a policy stating that there should be 'a reasonable relationship between the pricing of a licensed product, the public investment in that product, and the health and safety needs of the public.' It was applied in Cooperative Research and Development Agreement (CRADA) negotiations between NIH intramural laboratories and potential private collaborative partners interested in engaging in collaborative research. The 'reasonable pricing' clause was required in exclusive licenses to inventions made under NIH CRADAs. Shortly after the policy of 'reasonable pricing' was introduced, industry objected to it, considering it a form of price control. Many companies withdrew from any further interaction with NIH because of this stipulation. Both NIH and its industry counterparts came to the realization that this policy had the effect of posing a barrier to expanded research relationships and, therefore, was contrary to the Bayh-Dole Act."

NIH convened a series of panels to consider the issues. "The panels concluded that the policy did not serve the best interests of technology development and recommended to the Director, NIH, that the language be rescinded. The Director, NIH, accepted the recommendation, and the policy was revoked in 1995. The consequences of NIH's 'reasonable pricing clause' policy can be seen in the relatively flat growth rate of CRADAs that occurred between 1990 and 1994, and the subsequent rebound in CRADAs following revocation of the policy" (NIH, 2001).

As early as February of 1993, Bernadine Healy was testifying to a congressional committee about the difficulties NIH was experiencing in enforcing the reasonable pricing clause. In summary, the NIH had no way of forcing contractors and grantees to divulge the information required for analysis,

services is currently a low priority for NIH. The agency convened a series of panels to assess "what kind of return on the public investment is appropriate." According to the report, "the panels agreed on the following hierarchy, from most-to-least important: fostering scientific discoveries; rapid transfer of discoveries to the bedside; accessibility of resulting products to patients; and royalties" (NIH, 2001).[51]

In February of 2003, Floyd Bloom, the outgoing president of the American Association for the Advancement of Science, "shocked" delegates to the association's annual conference by suggesting that some of the dollars flowing into cutting-edge biomedical research should be shifted to salvaging the system that delivers the benefits of research to people in need. In an interview with a reporter, Bloom noted that scientists' efforts to increase public funding for research were premised on the potential for an eventual conversion of advances into "practicable treatments and diagnostics for people" (Tuma, 2003; see also Bloom, 2003).

4.2.4.3 Case Studies

Much of the normative discussion of patents and biotechnology is bound up with cases that land in the courts and in the news. It is possible for individuals with quite different positions to find support for their views in the same case. We do our best to present these cases in a balanced manner, although we make no pretense of offering facts free of interpretation.

OncoMouse and Humouse

The controversy over patenting of higher life forms has largely been concentrated on two cases. In the first, two Harvard University researchers and their commercial

---

the NIH in any event lacked the expertise to perform the analysis, and there was no consensus on the concept of reasonable price. Harold Varmus made the decision to eliminate the clause. See Brody (1996, 7). In June 2000, the House passed a bill containing a reasonable pricing requirement for products developed with federal funding, but the bill died in the Senate.

Baruch Brody has argued that attaching a reasonable pricing provision to public largesse as a strategy for addressing access concerns is both overly ambitious and not ambitious enough. It is overly ambitious because it assumes consensus on the concept of reasonableness and the expertise and technical capacity to collect and analyze vast quantities of information related to cost. It is not ambitious enough because many drugs are developed without public funding, i.e., the reasonable pricing clause approach does not remedy "the lack of any social mechanism for generally controlling unreasonable prices for new drugs" in the US (Brody, 1996, 9).

[51] The NIH position is perhaps defensible in light of its restricted mission, but NIH could call on its parent agency, the Department of Health and Human Services, to attend to problems of affordability. That is not the approach taken in a recent follow-up to the 2001 report, which asserts that better NIH service to industry "should…permit industry to price the drugs lower than they would otherwise" and then finds reasons to absolve the NIH director of virtually any responsibility for promoting the affordability of products (NIH, 2004). NIH *has* taken several steps to address access to research tools. See note 30 supra and accompanying text.

partners have sought patent rights in the methods to create a transgenic mouse *and* the transgenic mice themselves, as well as methods to create other transgenic non-human mammals and these other transgenic non-human mammals. Proceedings in Europe and Canada have been especially protracted, with Greenpeace, animal rights groups, and church groups as participants opposing patent rights. The opposition has focused on the inappropriateness of product patents on "life" or living creatures, bracketing or tacitly accepting the grant of process patents for methods of creating transgenic animals.

After a nine year legal battle, the European Patent Office upheld the patent in July 2004, albeit with a restriction to mice. As noted above, in the European context moral issues are clearly legitimate factors in decision making about patents, but little guidance exists on implementation. Also, the Biotechnology Directive by negative implication *permits* patents on genetically modified non-human animals, as well as processes for genetic modification of animals, where there is substantial medical benefit.

The Canadian Supreme Court came to the opposite conclusion when it considered the patentability of the transgenic mouse. Interestingly, the Court decided that a statutory provision describing patentable subject matter with language virtually identical to that in the US statute did not encompass higher life forms by a 5-4 majority, the same slim margin that the US Supreme Court used to validate such patents in *Chakrabarty*. Citing "the unique concerns and issues raised by patentability of plants and animals," the majority concluded that a higher life form is not a "manufacture" or "composition of matter" (*Harvard College v. Canada*, 2002, para. 199). They said it belonged to Parliament rather than the patent office or the courts to extend (and attach appropriate conditions to) intellectual property protection for higher life forms.

In the second case, one that abounds with ironies, two opponents of biotechnology are trying to patent the creation of chimeras through processes such as merging human and non-human embryos or introducing human embryonic stem cells into an early stage non-human embryo; creatures such as a "Humouse" and "Humanzee" could be useful in the study of embryonic development, the generation of organs for transplant, and the testing of drugs and other products. Activist Jeremy Rifkin and developmental biologist Stuart Newman, a founding member of The Council for Responsible Genetics, filed their patent application in December of 1997. As one sympathetic commentator notes, the strategy was clever:

> Granting a patent for a half-human chimera would throw religious, bioethical, animal-rights, and constitutional activists into high dudgeon. And the biotech industry would boil over the approval of what is clearly a preventive patent. But decline it and the agency is in court, eventually the highest court in the land. The last time that happened [in *Chakrabarty*] the Patent Office lost its case (Dowie, 2004).[52]

---

[52] The patent would be preventive in two senses. It would both preempt a patent on the basic concept by someone else, and it would give Rifkin and Newman the power to quash efforts by others to engage in the kinds of efforts covered by the patent. Newman has already warned a group of scientists contemplating chimeric research involving human cells that if they proceed they will face litigation. In filing an application, Rifkin and Newman were not required to engage in the underlying activity, which they find abhorrent and hope to hinder. Patent law does not require an actual experiment, e.g., a Humouse prototype, as a condition to issuance of a patent, so long as the steps of the experiment are described and credible.

Further, were a refusal of a patent to Rifkin and Newman to be vindicated by the US Supreme Court, the biotechnology industry would still be the loser to the extent that others within the industry hoping to patent human-non-human chimeras could expect to meet the same fate.

The application has been pending for six years, with four formal rejections in that period. Although some back-and-forth between patent examiners and applicants is not unusual, the length and nature of the process in this case has been extraordinary. In 1987, the Patent Commissioner at the time, Donald Quigg, had announced a policy that all "multicellular living organisms, including animals," were patentable, with the exception of a " 'claim directed to or including within its scope a human being.' " Deborah Crouch, the first of four examiners to work on the Rifkin-Newman application, rejected it on grounds that, among other things, the invention was not patentable subject matter because it " 'embraces a human being' " (Slater, 2002).[53] There have been many tussles since over what this means, since "human being" is nowhere defined, and it is even less clear when one is being embraced. What, for example, separates the OncoMouse from the Humouse?[54]

Canavan Disease

Many of the issues surrounding DNA patents are raised in a particularly compelling way by the case involving families afflicted by Canavan disease. Canavan disease is a rare and devastating inherited neurodegenerative disorder that predominantly affects individuals of Ashkenazi Jewish ancestry. Daniel and Debby Greenberg lost two children to the disease. In 1987, Daniel Greenberg contacted Dr. Reuben

---

[53] This restriction was attributed to the US Constitution, likely the 13th Amendment prohibition on slavery. See also President's Council on Bioethics (2004, 157–163). Some are interested in assuring that any exclusion reaches back before birth; the President's Council supports legislation instructing the US Patent and Trademark Office "not to issue patents on claims directed to or encompassing human embryos or fetuses at any stage of development" (President's Council on Bioethics, 223–224). Attempts to amend the patent statute to date have not succeeded, but Congress has used its appropriations power to prohibit use of federal funds to issue patents on claims directed to or encompassing a human organism (President's Council on Bioethics, 162–163).

[54] Within the European Union, the line seems to be drawn between the insertion of a few genes from another species and the use of "germ cells or totipotent cells of humans and animals." See Directive 98/44/EC (para. 38).

The journey of Bruce Lehman, Patent Commissioner at the time of the filing of the Rifkin-Newman application, adds another layer to the story. Violating a long-standing agency policy of refusing public comment on pending applications, Lehman issued a written statement on the Rifkin-Newman application and then held a press conference in which he announced that no "monsters" or other "immoral inventions" would be patented on his watch. Yet in 2004, having returned to private life, he reportedly told a writer for *Mother Jones* "that had Advanced Cell Technology, Geron Corp., or almost any biotech firm applied for the same patent Newman did, he would not have stood in their way" and that he rejects " 'a prohibition against a human patent' " (Dowie, 2004). Lehman now says that his objection was to the idea of using the U.S. Patent Office to make a point or stop the advancement of science.

Matalon to request his assistance in locating the gene responsible. Greenberg and the National Tay-Sachs and Allied Diseases Association located affected families and persuaded them to provide tissue samples, financial support, and aid in locating other families. They also created a disease registry with epidemiological and medical information. Another non-profit organization, Dor Yeshorim, provided about 6,000 stored blood samples.[55]

In 1993, Matalon and his research team isolated the gene. At the time, Matalon was working at the Miami Children's Hospital Research Institute, and in September 1994, Miami Children's filed a patent application covering the Canavan disease gene and any related activities, including diagnostic testing. The patent issued in October 1997. The Greenbergs, and the Greenbergs' partners in the effort to promote Canavan disease research and testing, said they first learned of the existence of the patent in November 1998 when centers offering Canavan testing began receiving "threatening" letters. They said they were never told of the intention to commercialize the fruits of the research, but rather expected testing to be widely accessible along the lines of the Tay-Sachs model. Miami Children's strategy of patent licensing and enforcement has restricted access to testing by increasing the cost of testing – owing to the addition of a royalty fee for each test performed – and by limiting the number of laboratories performing testing. Miami Children's has argued that its financial investment dwarfs the contribution of the families and it needs to recoup that investment.

The Greenbergs were not silent about their outrage over the turn of events or resigned to what they saw as a clear injustice. They and others affected took their grievances to the media. For example, in September of 2002, the CBS Evening News featured an interview with the parents of a child who died from Canavan disease, who remarked: "For somebody to say in the name of the almighty dollar that they're gonna limit the number of people that can have access to this test, to – to me, is just outrageous.... Obviously, people are entitled to make money, but there's got to be a limit on what they do." Reporter Sharyl Attkisson concluded the piece portentously: "With the increasing value of our parts and virtually no protection from the government or courts, there's no way to know whether you're viewed as patient or commodity" (CBS Evening News, 2002).[56]

---

[55] See the case study prepared by Jon Merz, available at http://www.med.upenn.edu/bioethic/programs/benefit/canavan.shtml (visited June 13, 2003).

[56] It is clear, then, that the Canavan case aroused both access and commodification concerns. In the latter regard, the case recalls an earlier lawsuit resting on the creation and patenting of a cell line derived from the tissues of John Moore, a patient undergoing treatment for hairy-cell leukemia, without his knowledge or consent. Moore reported feeling "violated for dollars" upon learning of the patent, and initiated a lawsuit that eventually reached the California Supreme Court. See Andrews and Nelkin (2001, 28). The majority of the members of that court rejected Moore's claim that his excised tissue and associated information were his property, and therefore their appropriation without his agreement amounted to a kind of theft, while allowing a claim that Moore's physician breached his fiduciary duty in failing to disclose his complete motivation for obtaining numerous samples from Moore.

Apparently in response to public criticism, Miami Children's agreed to halve its per-test fee (OECD, 2002, 17). The Greenbergs and their allies also went to court seeking an injunction restraining enforcement of the Canavan gene patent and damages equal to all royalties collected for testing plus all of their own financial contributions. Their lawsuit advanced six causes of action: breach of informed consent, breach of fiduciary duty, unjust enrichment, fraudulent concealment, conversion (i.e., theft), and misappropriation of trade secrets. On May 29, 2003, Judge Frederico Moreno granted a motion by Miami Children's and the other defendants to dismiss all causes of action except unjust enrichment (*Greenberg v. Miami Children's Hospital*, 2003; see Anderlik and Rothstein, 2003). The outcome of a trial on the unjust enrichment claim will never be known, as the case was settled in September of 2003.

Families affected by devastating diseases with a likely genetic component, prodded by the Canavan experience, are beginning to create a new model of research collaboration that includes contracts and explicit attention to issues such as patenting and benefit-sharing (see, e.g., Solovitch, 2001). In seizing greater control, they will likely find themselves facing questions such as: Is it possible that the injection of too much "humanitarian concern" into a "profit-motive world" will hinder research and development, resulting in delays in bringing new tests and treatments to market?[57] If the gene of interest turns out to have commercial value, perhaps even a connection to a common condition such as heart disease or diabetes, to what extent should they seek to generate profits that can then be plowed back into their research or otherwise used to benefit their group? How might they determine a reasonable profit on their and their collaborators' investment, or a fair price for any product that eventuates from the research?

The majority opinion scarcely embraced the anti-commodification argument, as it was extremely solicitous of the biotechnology industry and suggested that excised cells might well be the property of someone. However, a concurring opinion upbraided Moore for entreating the court "to regard the human vessel—the single most venerated and protected subject in any civilized society—as equal with the basest commercial commodity" (*Moore v. Regents of the University of California*, 1990, 148). One dissenter, Justice Broussard, took up the same theme with a different emphasis, arguing that while it might be best to create a buffer between the body and commercial activity, "[f]ar from elevating...biological material above the marketplace, the majority's holding simply bars plaintiff, the source of the cells, from obtaining the benefit of the cells' value, but permits defendants, who allegedly obtained the cells from plaintiff by improper means, to retain and exploit the full economic value of their ill-gotten gains free of their ordinary common law liability" (*Moore v. Regents of the University of California*, 1990, 150). The other dissenter, Justice Mosk, noted that at the time of excision Moore "at least had the right to do with his own tissue whatever the defendants did with it," such as contracting with others to exploit its commercial potential, and that right should not simply evaporate upon excision (166). For Justice Mosk, the strongest public policy considerations were not facilitating the use of human tissue by researchers and the biotechnology industry, but rather "a profound ethical imperative to respect the human body as the physical and temporal expression of the unique human persona" and the inequity in refusing to recognize the contribution to research and to biotechnology made by patients through their very flesh (173–135).

[57] A prospect suggested by M. Fleischer (2001, 87).

Breast Cancer

An estimated 5% of all breast cancers are associated with certain inherited genetic defects. The screening test for mutations in the BRCA1 and BRCA2 genes known to be associated with heightened risk costs about $2,400. Myriad Genetics holds the relevant patents (Krimsky, 1999, 37, citing Merz et al., 1998, 4).[58] The screening test for BRCA1 and BRCA2 is often compared to the screening test for Tay-Sachs disease. The Tay-Sachs test, covered by a patent held by the US Department of Health and Human Services, costs about $100. It is not surprising, then, that criticism has been directed at the pricing of the BRCA tests, as well as the breadth of Myriad's patent claims and its approach to licensing, which is basically to perform all testing in the North American region itself and to give laboratories in other parts of the world exclusive licenses for their regions (Andrews, 2002, 804). Also, the Myriad testing method has been criticized for missing newly-discovered disease-linked mutations in the BRCA genes, perhaps totaling 10–20% of BRCA1 and BRCA2 mutations (OECD, 2002, 8).

The Institut Curie and others challenged Myriad's European patents, alleging lack of novelty, inventiveness, and adequate description. In May 2004 the European Patent Office revoked a key Myriad patent on BRCA1 (L'Institut Curie, online; Abbot, 2004, 329).[59] The province of Ontario has announced plans to offer a test using a process different from Myriad's, which it says produces results faster and will cost two-thirds less. A Canadian bioethicist suggests an analogy between Myriad and the inventor of a mousetrap who claims that he has rights over any device that traps mice (Ernhofer, 2003).[60] On the other side, Myriad and its defenders point to the substantial investment the company has made in testing, and also the benefits of centralizing testing, which include tighter quality control and a more coordinated approach to further research.

HIV/AIDS

HIV/AIDS has occasioned some of the most impassioned disputes about the normative significance of patents. Controversies about patenting have shadowed

---

[58] The basic research to identify BRCA1 was supported by private money but also more than $5 million in federal funds. See Williams-Jones (2002, 131).

[59] The ruling was based on the discovery of a number of small errors in the gene sequence as disclosed in Myriad's series of filings up until the fifth U.S. application, filed in March 1995, well after information on the sequence had been published. Hence, at the time the correct information was included in a filing, the content was no longer novel. Such a discovery, and aftermath, are not unprecedented in the biotechnology industry. For example, Genentech lost its priority dates for patents claimed on tissue plasminogen activator and the HIV virus due to sequence errors, resulting in a narrowing of the scope of the patents. Another BRCA1 patent was upheld by the European Patent Office in 2005, but with a significant amendment. It now relates to a gene probe for the detection of a specific mutation and no longer includes claims for diagnostic methods. (L'Institut Curie, online.)

[60] For a lengthy journalistic account of the attacks on Myriad, see Blanton (2002).

the pandemic at every point from basic research to the supply of drugs to those in need. In the initial stages of the pandemic, Human Genome Sciences patented a chemokine receptor known as CCR5 on the basis of a speculative use. Thereafter, researchers at another institution demonstrated that this receptor was the main entry point for the HIV/AIDS virus. Given the way US patent law works, anyone pursuing applications based on this discovery, such as the development of tests or drugs for HIV/AIDS, would have to obtain a license from Human Genome Sciences (and, should the researchers who found the HIV/AIDS connection patent their work, a license from them as well) (Barton, 2002; Rai, 2002). Hence, there was the potential for major roadblocks to research. However, Human Genome Sciences issued several licenses for drug development projects and has indicated that it will not sanction academic researchers who proceed with their investigations without a license (OECD, 2002, 13).

If research in the West was complicated or compromised by licenses, in Africa attempts at developing therapies for the virus was not recognized by many in the scientific and pharmaceutical industry. In Cameroon, Professor Victor Anoma Ngu developed a vaccine which he tried on some of his patients but his research did not get the recognition he anticipated in the scientific community and he did not receive any funding for his work (Ngu, 1997). Ngu, and Cameroonian philosopher, Godfrey Tangwa, presented his findings at a Bioethics congress in London, September 2000. They claimed that HIV antigens whose envelope has been destroyed cannot infect a person and hence could be an effective vaccine. Ngu tested it on some patients and it "provoked immune responses that kill[ed] the virus, confirming its effectiveness as a vaccine" (Ngu, 1997, 224). Tangwa and Ngu estimated that mass production of such a vaccine could eventually bring down the cost to US$0.10 per vaccine. Ngu thus presented a candidate for the VANVAHIV vaccine. As Tangwa argues, the West did not expect this development. They did not expect such an African initiative. They also questioned why western agencies did not fund this project. Back in Yaounde, Cameroon Ngu continued with his research even though some continued to doubt the possibility that he could have made significant headway on a vaccine, even if had discovered a cure for AIDS (Ngu, 1997, 225). Ngu's case should be of interest to scholars in bioethics because it is a local initiative that is being stifled by funding and patents. In addition to savings, active research in Africa could offer an opportunity for pharmaceutical companies to pursue both profit and human values. Tangwa argues:

> As commercial enterprises, multinational corporations cannot be blamed for pursuing profit as such. The question, however, is whether the nature of the catastrophe in Africa does not require imaginative emergency measures, where such profit pursuit can be combined with more morally sensitive global co-operation aimed at helping those in dire need and towards discovering and manufacturing a cheap and affordable vaccine (Tangwa, 2002, 226).

The collaborative imperative is crucial because in Nigeria, for example: "quality antiretroviral (ARV) medication for HIV/AIDS is accessible to less than 1% of those in need."[61] Kris Peterson and Olatubosun Obileye have argued that researchers and policy makers need to consider several issues, including government communication and accountability, quality of drugs, regulatory issues, drug distribution, cost of drugs and factors that influence drug prices, drug donation, manufacturing of drugs in Africa (Peterson and Obileye, 2002, 2–4). Access to drugs by people living with HIV/AIDS in Africa remains a major problem. On April 28, 2003 GlaxoSmithKline (GSK) announced that it would again reduce the not-for-profit prices of its HIV/AIDS for the poorest countries by up to 47%. This was a significant step forward although critics argued it was not enough. Several customers ranging from the governments of the Least Developed Countries in sub-Saharan Africa, Non Governmental Organizations, and United Nations agencies engaged in the fight to combat HIV/AIDS, were eligible to buy at these reduced prices. Legal action had previously been taken against GSK and Boehringer Ingelheim, which also manufacture AIDS drugs, to stop them from charging exorbitant amounts for their drugs and gaining excessive profits from misfortune. The various parties to the complaint before the Competition Commission, such as AIDS activists, trade unions, health care professionals and people living with HIVAIDS, accused the companies of charging prices in excess of the economic value of their products. The complainants charged that the price of AZT drugs were very expensive and did not reflect the cost of production. The South African group, Treatment Action Campaign (TAC), indicated that even at generic prices, the drug companies still made profit. "According to the complainants, a 300 mg pill of AZT costs 2.58 times the economic value, a 150 mg pill of Lamivudine is 4.01 times the economic value, combined pill with 300 mg AZT and 150 mg Lamivudine is 2.24 times the economic value, and a 200 mg nevirapine pill is 1.7 times the economic value. The syrup form of AZT (used for treating children) is 2.27 times the economic value, and the syrup form of Lamivudine is 1.97 times the economic value."[62]

The complainants compared prices charged in the South African private sector with these branded drugs, and the prices of their generic equivalents in other parts of the world. Generic antiretroviral medicines were not sold in South Africa because patent law protected pharmaceutical companies, although Glaxo had issued a conditional license to Aspen Pharmacare for the manufacture of generic AIDS drugs under the terms that they can only be sold to the public sector. The complainants argue that patent protection did not entitle a firm to charge an unreasonable price that has no bearing on the economic value of the product. Since antiretrovirals at that stage could not be substituted, patent protection strengthened the manufacturers' domination of the market for its patented drug, regardless of market share. Therefore, the complainants alleged that the prices charged by Glaxo and Boehringer are "excessive to the detriment of consumers" within the meaning of section 8(a) of the Competition Act.

---

[61] The following discussion is taken from Peterson and Obileye (2002, 2).
[62] See allAfrica.com, 2002. allafrica.com/stories/200209200225.html.

Before taking this action in South Africa, Médecins San Frontiéres (MSF) had announced their intention to import generic drugs into South Africa to challenge patent laws, which protected the pharmaceutical giants (Virus Weekly, 2002, 13). MSF took this action because the generic versions of the drugs made in Brazil cost $1.55 a day, compared to $3.20 a day. Furthermore, activists pointed out that there were provisions that allowed the Brazilian government to ignore patent laws in crises or an emergency. The Congress of South African Trade Unions (COSATU) recommended this challenge in South Africa because their own findings established that generic drugs manufactured in Brazil would be more affordable at R$450 a month than the going price of R$1,000 per month for the non-generic drugs.[63]

The coalition demanded that companies issue voluntary licenses to other companies to produce generic versions of the drugs in South Africa. If the pharmaceutical companies did not do that, the activists would call on the South African governments to grant non-exclusive compulsory licenses. They rejected President Mbeki's position that the generic version of the drugs was toxic. The generics manufactured by Farmanguinhos, for the Brazilian government, were safe products. The huge difference was the cost of medications and the generic versions were clearly cheaper and would serve a larger portion of the population.[64] Although Boehringer had donated about R$1.5 million worth of Nevirapine free, activists pointed out that this drug, used to prevent mother to child transmission, could only serve about 355 people per year in South Africa. The activists claimed that as a signatory of the World Trade Organization Agreement on Trade Related Aspects of Intellectual Property Rights or TRIPS Agreement, South Africa was under obligation to uphold those treaties. However, they noted that the treaties make provision for compulsory licenses, and the TRIPS Agreement could not inhibit a country's health concern.[65] Peterson and Obileye argue that in Nigeria the price of drugs is a crucial issue in the fight against HIV/AIDS.[66]

---

[63] COSATU and the Treatment Action Campaign or TAC called this move a defiance campaign they were compelled to take because the South African Constitution gives protection and rights to life and human dignity, and the importation of such drugs was a fulfillment of their constitutional obligation to South Africans and people living with HIV/AIDS. See COSATU (2002).

[64] Cost remains a great concern because one cannot understand the difficulties of dealing with the HIV/AIDS pandemic in Africa without considering the issue of poverty which provides a social context for the disease. See van Niekerk (2001). Van Niekerk also points out that we cannot understand the problems only from the perspective of poverty because African countries must make a concerted effort to address sexual mores, make condoms available, co-operate with pharmaceutical companies and western governments to make drugs available to people, explore ways of importing generic drugs, introduce sex education, and use role models in society to educate people, p. 147.

[65] China has also given licenses to companies to market low-cost generic drugs as ways of fighting the costly marketing practices of the pharmaceutical giants. In a story filed September 24, 2002, by United Press International, a Chinese official from the office of Drug Administration claimed that the Chinese government had given permission to Desano to produce generic drugs, a Chinese cocktail version of Nevirapine, that would cost about $600 a year per patient, a small fraction of the cost of drugs they import from overseas, which cost between $3,600 and $12,000 per year.

[66] The most important drugs for people living with HIV/AIDS in Nigeria at the time of the study were:

Several factors influence drug pricing in Nigeria. These include the global pharmaceutical industry, which justifies high prices on research, production, length of clinical trials, distribution and advertising cost. However, they point out that sometimes in the United States, public funds support some research that leads to the production of drugs (Peterson and Obileye, 2002, 20). Furthermore, despite the arguments made by the pharmaceutical companies, their drugs have brought in a huge amount of money – for Glaxo Wellcome (now GlaxoSmithKline), total sales of AZT, 3TC and Combivir between 1997 and 1999 exceeded US$3.8 billion, while Bristol-Myers Squibb's sales of d4T and ddI added up to more than US$2 billion during the same period (Médecins Sans Frontières, 2000). Further factors include monopoly rights, few generic options, tariffs and taxation, and differential pricing. African states should study factors that drive up costs closely and make sure that tariffs and VAT promote the availability of drugs to those who need them most. International organizations should continue to work with governments to ensure that patents do not become a barrier to drug accessibility.

Lawyers know the importance of the burden of proof. Surely each side in the legal and public relations battle over DNA and other biotechnology patents can claim some share of the truth, and each side can offer legal, economic, and moral arguments in support of its position. Conflicting empirical claims cannot be resolved as the relevant evidence is ambiguous or nonexistent, and this is likely to be the case for the foreseeable future. Who, then, should have the burden of proof? Those urging "departure from established law" – the charge leveled at the critics of current practices of patenting of DNA and of living creatures? Or those urging the expansion of monopoly rights – one way of characterizing the promoters of patents on DNA and living creatures? And what of the barriers to access to life-saving drugs in developing countries? Should patent holders be absolved of moral

---

1. Anti-infective agents to treat or prevent opportunistic infections
2. Anti-cancer drugs for Kaposi sarcoma and lymphoma
3. Palliative medications to relieve physical and mental pain
4. Antiretroviral drugs to limit the destruction of the immune system and reduce the viral load in the blood system

Peterson and Obileye point out that Médecines Sans Frontiéres (Doctors Without Borders) has proposed the following priorities for poor countries:

Drugs that will prevent opportunistic infections and recommended by the World Health Organization Joint United Nations Program on AIDS such as isoniazide and contrimoxazole
Palliatives such as analgesics, antidiarrheals
Antiretrovirals which prevent opportunistic infection, reduce viral load and improve the quality of life
Antiretrovirals such as AZT, NVP, which could prevent mother to child transmission and could serve as post-exposure prophylaxis (Peterson and Obileye, 2002, 9)
See also Pérez and Boulet (2000).

responsibility for the consequences when they exercise their legal rights? It may be that a wider vision and a deeper kind of "satyagraha" (the term is derived from Sanskrit and was used by Mahatma Gandhi to refer to the effort to pursue, or act on the basis of, truth) is necessary for a comprehensive understanding and response to conflicts linked to contemporary Western patent regimes.

4.3 Religious Perspectives on Nature, Property, and Biotechnology

Since space will not allow for a comprehensive review of religious perspectives on the appropriation of nature as property, and patenting in the realm of biotechnology, we offer two case studies. We have selected Protestantism as the initial case study for a number of reasons, including the asserted connections among Protestantism, capitalism, and contemporary Western property regimes and the unmatched volume of texts on DNA patents emanating from the Protestant churches. The views of indigenous peoples are singled out for attention because they reflect a fundamentally different understanding of nature and the relationship between human beings and nature and because they been subject to marginalization and misunderstanding.

4.3.1 The Protestant Perspective

4.3.1.1 Protestantism and Property

The Hebrew Bible contains many rules that regulate commercial activity, and it suggests at least the aspiration to an economic system that promotes social equality. There is a strong emphasis on the needs of the poor and those on the margins of society, an emphasis found also in the sayings of Jesus. Christian texts expressing caution about commerce and wealth include the Sermon on the Mount (Matt. 6:19–34); the counsel that it is difficult for the rich to be saved (Mark 19:20–5); the especially strong condemnation of the rich who defraud the poor (James 5:1–6); Paul's statement that "The love of money is the root of all evil" (I Tim. 6:10); the parable of the rich man and Lazarus (Luke 16:19–31); and the story of Jesus driving the buyers and sellers from the temple (Matt. 21:12). On the other hand, the parable of the talents expresses approval for hard work and profitable ventures (Luke 19:12–26), Paul also wrote "If any would not work, neither should he eat" (II Thess. 3:10), and other stories speak approval of conspicuous consumption, at least on certain occasions (Gilchrist, 1969, 51).

The early Christian fathers adopted the view of nature as a common, a foundation for the later natural law theories of property. Some spoke clearly about the implications of this view for the distribution of goods. For example, Ambrose of Milan stated:

> Our Lord God wished this earth to be the common possession of all men, and its produce to supply the needs of all. But avarice has assigned private rights of possession. It is just, therefore, if you have anything of your own, which belongs in common to the human race, or rather to all living things, at least to give something from it to the poor; in order that you may not deny nourishment to those to whom you owe partnership in your right (Charlton et al., 1986, 224).

Gregory the Great expressed similar sentiments in harsher terms:

> The land is common to all men, and therefore brings forth nourishment for all men. Those people, then, are wrong to think themselves guiltless when they claim what God has given to men in common as their private property; who because they do not distribute what they have received, proceed to the slaughter of their neighbours, since they kill each day as many of the dying poor as could have been saved by what they keep to themselves. For when we give necessities to people in need, we are returning to them what is theirs, not being generous with what is ours; we are rather paying a debt of justice than performing a work of charity (Charlton et al., 1986, 225).

In his history of Christian social teaching, German theologian Ernst Troeltsch asserted that the ascriptions of private property to sin were never meant to be taken seriously, i.e., to change existing social institutions. Rather, they were intended to overcome resistance to charitable giving. Attitudes toward property were linked to attitudes toward work. Work was prized as a means of inculcating virtue and keeping men and women occupied and therefore out of trouble, but it was also regarded as a consequence of the Fall and a punishment for sin. The early Church crafted a "twofold solution" to the problems of property and poverty. On the one hand, there was "an ethic of compromise, with relative standards"; private property and even riches were permitted as a means to social prosperity, with the provisos that superfluous possessions be shared and property holders cultivate detachment. On the other, there was the complete abolition of private property in monastic communities (Troeltsch, 1931, 115–117).[67]

Medieval commentators generally evaluated profit in light of labor and expense and intention, and they counseled restitution in cases of overreaching. They adapted the concept of just price from Roman law; it was invoked primarily to condemn price discrimination against the weak or desperate, or artificially high prices made possible by monopoly (Gilchrist, 1969, 59–61).[68] For example, Aquinas argued that a seller should not take advantage of the special need of a buyer (*Summa Theologiae*, IaIIae 77.1–77.4). In reality, "most merchants pursued profit for its own sake and made restitution for the good of their souls in their wills" (Gilchrist, 1969, 56). The system worked quite well, though, as direct responsibility for meeting the needs of the poor belonged to the Church rather than the merchant. The giving of alms to fund these efforts was, for individuals, a matter of conscience, but even rulers who

---

[67] Of course, in the course of time many monasteries acquired vast holdings. As noted by Gilchrist, the medieval Church exercised economic influence not only through its teaching and preaching, but also through the management of its own substantial economic resources.

[68] The Romans allowed a seller to sue for recovery where his bargain brought him less than half than the prevailing market price, but bargainers were generally free to get the better of one another.

were debt-ridden and at odds with the Church over other matters honored these obligations. Social support was not regarded as charity by recipients, but neither was it an unconditional entitlement; in keeping with Paul's strictures, no grants were made to the able-bodied who refused to work (Gilchrist, 1969, 78–81).

Aquinas and other medieval moralists recognized that the just price might fluctuate. Some asserted that a just price was likely to be reached under a regime of freedom of contract, since the striking of a bargain was an indicator of satisfaction on both sides. In the later middle ages, Spanish theologians argued that free competition was best, based on their observation that price regulation in times of economic hardship often led to a total disappearance of scarce goods. Unfortunately, the compensating system of relief for the poor faltered due to a growing gap between rich and poor, which diminished sympathy across social groups (Gilchrist, 1969, 116–119).

The Reformation did not immediately introduce any dramatic change in moral teaching on economic matters. The British historian R.H. Tawney emphasized the friction between Protestantism and the changing economic order. On usury in the narrow sense of lending at interest, for example, Tawney noted that Calvin hemmed it in with conditions, including a requirement that loans be made free of interest to the poor (Tawney, 1926, 106). Tawney also considered the case of Robert Keane of Boston, accused of making excessive profits, at some length. Since " '[t]here was no positive law in force limiting profits; it was not easy to determine what profits were fair; the sin of charging what the market could stand was not peculiar to Mr. Keane; and, after all, the law of God required no more than double restitution,' " the elders let him off with a large fine. Keane then had the temerity to offer a defense, asking " 'how could he live, if he might not make up for a loss on one article by additional profit on another?' " In the next public sermon, the minister of Boston thundered: " 'When a man loseth in his commodity for want of skill, etc., he must look at it as his own fault or cross, and therefore must not lay it upon another' " (Tawney, 1926, 129–130). Keane narrowly escaped excommunication.

Under Queen Elizabeth, conventional religious teaching sought to "moralize economic relations, by treating every transaction as a case of personal conduct, involving personal responsibility," even as new classes of merchants and new business structures were promoting impersonality (Tawney, 1926, 184). Also, the word "nature" had begun to connote human appetites rather than divine ordinances (Tawney, 1926, 180). Still, on through the reign of the Puritans there was little change in the actual content of moral guidance for the masses. For example, Bunyan's *The Life and Death of Mr. Badman* illustrated the evils of extortion and covetousness, especially in the form of exploitation of the poor. But the capitalist spirit, which certainly predated the Puritans, was nourished by the individualism that emerged triumphant when the collectivist aspect of Puritanism fell away. The economic policy of Charles I promoted capitalism, but of a kind resting on the award of privileges such as patents, and subject to government meddling guided by a "spasmodic" paternalism. This regime led to a backlash favoring complete autonomy in the economic sphere (Tawney, 1926, 236–237).

In the late 1600s, new manuals of morality tailored to the trades and professions appeared. These judged the traditional standards for calculating a reasonable profit, such as cost of production, standard of living, and customary price, of little practical use. "'An upright conscience must be the clerk of the market'" (Tawney, 1926, 244–245, quoting Steele, 1684). According to Tawney, religious individualism "led insensibly, if not quite logically, to an individualist morality, and an individualist morality to a disparagement of the significance of the social fabric as compared with personal character." One outcome was a diminished sense of social responsibility for those in need; "admonitions which had formerly been turned upon uncharitable covetousness were now directed against improvidence and idleness" (Tawney, 1926, 255, 265). Of course, Christian and Protestant social teaching always reflected concern about the effects of aid on the character of the poor. But in the 17th century and beyond, "in sharp contrast with the long debate on pauperism carried on in the sixteenth century," there was a "resolute refusal to admit that society had any responsibility for the causes of distress" (Tawney, 1926, 271).

Troeltsch and Tawney, pioneering scholars of religion and society, suggested that the stance of Protestantism vis-à-vis contemporary capitalism should be critical. Troeltsch offered a qualified but strong endorsement of Max Weber's famous thesis concerning the contribution of ascetic Protestantism of the Calvinist variety to the rise of capitalism, but he believed there was nothing direct or intentional about this contribution, and capitalism as he understood it was no credit to its Protestant roots.[69] Tawney concluded that "unless industry is to be paralysed by recurrent revolt on the part of outraged human nature, it must satisfy criteria which are not purely economic" (Tawney, 1926, 284).

There are no assessors of the American religious and economic scenes of quite the same stature. In the 19th century, the ideological spectrum in the US ranged from the Social Gospel to the Gospel of Wealth, and it is easy to identify the heirs of these two Protestant traditions. The basic creed of the Social Gospel movement was that churches "must stand for equal rights and complete justice for all men in all stations of life" and "for the most equitable division on the product of industry that can ultimately be devised" (Federal Council of the Churches of Christ in America, 1908, quoted in De Vries, 1998, 214). The Gospel of Wealth promoted commercial activity, in some cases of a rather brutal kind, so long as the fruits supported worthy causes. Andrew Carnegie was a great proponent and exemplar.

At mid-century, some Protestant tracts offered condemnations of laissez-faire economics, while others were more concerned about the threat of Communism. In *Christian Responsibility in Economic Life*, Albert Terrill Rasmussen urged a balance between goals of increased production and fairer distribution. He wrote that the businessman should be guided by "the basic Christian recognition of the supremacy of persons over things and, therefore, they are never to be reduced to

---

[69] For example, Troeltsch wrote: "Above all, the imposing but terrible expansion of modern capitalism, with its calculating coldness and soullessness, its unscrupulous greed and pitilessness, its turning to gain for gain's sake, to fierce and ruthless competition, its agonizing lust of victory, its blatant satisfaction in the tyrannical power of the merchant class, has entirely loosed it from its former ethical foundation; it has become a power directly opposed to genuine Calvinism and Protestantism" (Troeltsch, 1986, 74).

exploitable resources" and a recognition of the importance of community, "the relational network of intersupporting life in which no one can be excluded or rejected from its benefits and opportunities without damage to and constriction of all of the 'selves' involved." Rasmussen argued that a central insight of the Calvinist traditions was the "doctrine of limitation," meaning "no human being escapes being influenced by the value systems and the prevailing expectancies of the organizations in which he works." This doctrine dictated that Christians work for voluntary codes of ethics for industry to establish patterns of fairness and responsibility and support government intervention to create a "balanced pressure system" as a context for responsibility (Rasmussen, 1965).

John R. Richardson's *Christian Economics* had a quite different emphasis and tone. According to Richardson, liberty is the value central to Christianity as a religious system and capitalism as an economic system. Christianity is not concerned with groups or collectivities; it speaks to the individual. Increased attention to the individual would in fact serve as a corrective for many contemporary social problems. For example: "In dealing with poverty, many overlook the fact that these people should be taught thrift, temperance, godliness, and hygiene." In America, limited government, economic freedom, private property and the right to profit from it, and a competitive market "have combined to give men the incentive to forge ahead, to build up their businesses, to invent and to develop better tools, to produce more, thereby providing more and better jobs for more people." All property is held in trust, belonging ultimately to God. Even so, "it is entirely proper for a Christian to desire money for good purposes…. In fact, if he can earn great sums of money honestly, it is his Christian duty to do it," so long as he remembers that "material transactions should be guided by spiritual principles" (Richardson, 1966, 19, 35, 125).[70]

The same themes continue through more recent work. Perhaps the greatest change is the addition of a new global and ecological consciousness. A 1992 text on Christian faith and economic life linked to one of the more liberal Protestant denominations includes an extensive treatment of issues in development and the responsibilities of multinational corporations (Blank, 1992).[71] Proponents of a liberal ecology-theology find standard liberal approaches too anthropocentric, i.e., too focused on human welfare and purposes. They tend to have a favorable, and perhaps slightly romanticized, view of life uncorrupted by modern economic and social ideas: "American Indian tribes, much of tribal Africa, and indigenous populations in the Amazon rain forest all have long histories of living in balance with nature, of forming effective human and natural communities that establish long-term ecological and economic sustainability. For them, the good life cannot be conceived of independently of the common good" (King and Woodyard, 1999, 129).

---

[70] "The only ones who are ultimately benefited by government grants and aid advocated by misguided humanitarians are those on relief who live off the earnings of others, bureaucrats, thieves, corrupt politicians, and vote-buyers" (Richardson, 1966, 37).

[71] Blank's text builds on the United Church of Christ Pronouncement on Christian Faith (1989).

There is a basis here for dual and somewhat opposing views of DNA patents from the perspective of contemporary Protestantism. On one view, the problem is with the distribution of the goods produced by Western commercial civilization, and the stress is on concerns about access. On another view, it is the whole thrust of Western commercial civilization, of which patenting is a part, that is the problem (for example, as a threat to nature or natural piety). In the DNA patenting debate the stress shifts to concerns about commodification.

4.3.1.2 Protestantism and Patenting

In 1983, on roughly the third anniversary of the *Chakrabarty* decision, Jeremy Rifkin, a Reform Jew, persuaded 50 religious leaders to sign a resolution opposed to "efforts to engineer specific genetic traits into the germline of the human species." In 1988, the US Patent and Trademark Office issued the US patent for the Harvard transgenic oncomouse. Four years later, a task force of the United Methodist Church came out against patents on animals and individual genes, at the same time expressing its support for process patents on recombinant DNA techniques. An end to a five-year lull in the issuance of patents on genetically engineered animals, in 1993, triggered a round of phone calls from staff of the United Methodist Church to other denominations and to Rifkin. On May 18, 1995, nearly 200 religious leaders joined Rifkin in a "Joint Appeal against Human and Animal Patenting." A statement released in connection with a press conference stated: "We believe that humans and animals are creations of God, not humans, and as such should not be patented as human inventions." The statement did not address genetic engineering or patents on techniques of genetic manipulation (Stone, 1995, 1126). The leaders of many Protestant denominations continue their opposition to patenting of humans and animals as a matter of ethics and public policy.[72]

---

[72] For example, in 2002, the Canadian Council of Churches and the Evangelical Fellowship of Canada filed a factum (brief) opposing the OncoMouse patent. The factum argued against the creation of rights in higher life forms without corresponding responsibilities, asserted that the patenting of mammals creates questions about the patenting of human beings, and concluded that if Patent Act was to be expanded to higher life forms the issues were best addressed by Parliament, arguments embraced by the Canadian Supreme Court. A study guide prepared after the decision in the case quotes a section of the factum:

Granting patents on animals would be indicative of a morally problematic shift in humans' perception of the natural world. The notion that a part of a species of complex animal life should be viewed as an invention…a mere industrial product…is based on the metaphysical position which holds implicitly that nature and/or the environment is simply composed of manipulable data—a 'standing reserve' of calculable forces, completely subject to human manipulation…[Such a notion] fundamentally objectifies the natural world and would inevitably objectify humans, as they are part of the natural world (Canadian Council of Churches, online, at 4).

On the cover of the study guide is a reproduction of a painting by Lynn Randolph entitled the "The Laboratory, or, The Passion of OncoMouse," which portrays the OncoMouse as a Christ-like figure. (For more on this subject, see Chapter Five of the present volume, "Aesthetic and Representational Concepts of Nature." Interestingly, according to the study guide, intervener

4 Ethical Challenges of Patenting "Nature": Legal and Economic Accounts 249

In his discussion of the response to DNA patenting by religious groups, chiefly the Protestant denominations that have been most vocal on the issue, Mark Hanson argues that religious leaders have, by and large, accepted the industry case for biotechnology and for patents, a case resting on values of self-interest, utility, and desert. They have focused on how patenting of DNA and animals impinges on other values, and on the possible existence of less offensive paths to industry goals. Some themes are prominent across the ideological spectrum. Hanson quotes Richard Land and C. Ben Mitchell of the Southern Baptist Convention:

> Human beings are pre-owned. We belong to the sovereign Creator. We are, therefore, not to be killed without adequate justification (e.g., in self-defense) nor are we, or our body parts, to be bought and sold in the marketplace.... Admittedly, a single human gene or a cell line is not a human being; but a human gene or cell line is undeniably human and warrants different treatment than all non-human genes or cell lines. The image of God pervades human life in all of its parts. Furthermore, the right to own one part of a human being is ceteris paribus the right to own all the parts of a human being. This right must not be transferred from the Creator to the creature. (Hanson, 1997, 8)

Here and in the work emanating from more liberal groups, there are three sets of interrelated claims: about the unique status of DNA or biological material, about patenting as a challenge to God's sovereignty, and about patenting as commodification.

Hanson writes: "If one understands DNA to be a building block of life, a gift from God – bearing, in a sense, the grace of God – it would not be the sort of thing that proprietary rights of any sort – even those short of full ownership – could or should apply to" (Hanson, 1997, 9). (Hanson himself believes that sacredness should be extended with caution.) A distinct but related understanding of genes as the common heritage of humanity puts DNA and genetic information in the category of things that ought not to be considered *private* property. True, a patent in itself does not create a right to make or to use, or ownership of any material thing or being, but it can be the source of enormous power. Land and Mitchell invoke Leon Kass's argument that "'It is one thing to own a mule; it is another to own *mule*'" (Hanson, 1997, 10).

---

status for the church groups was opposed by Harvard's lawyers, who argued that the theological and ethical issues likely to be raised by such bodies would not be relevant to statutory interpretation.

Another example is worthy of mention here. A group including Carl Anderson (the Supreme Knight of the Knights of Columbus), Gary Bauer, Robert Bork, Ben Carson (a new appointee to the President's Council on Bioethics), Charles Colson, James Dobson, and Bill Kristol, as well as Richard Land and C. Ben Mitchell of the Southern Baptist Convention, two men who have long been active in this area, has signed a "Manifesto on Biotechnology and Human Dignity" titled "The Sanctity of Life in a Brave New World." Citing the work of C.S. Lewis, the Manifesto warns that power over ourselves and our own nature "will always tend to turn us into commodities that have been manufactured." This translates into a call for a "wide-ranging review of the patent law to protect human dignity from the commercial use of human genes, cells and other tissue." The Manifesto is posted to the site of the Council for Biotechnology Policy, chaired by Colson, http://www.pfm.org/. This may seem a surprising stance for many who are normally pro-market, and it is worth re-iterating that this is an area where one can find alliances across the usual left-right divide.

Following Margaret Jane Radin, Hanson suggests that market rhetoric creates discomfort for some because it seems to assault human dignity and because it tends "'to crystallize a social worry'...that is often linked to other kinds of social wrongs, such as injustice" (Hanson, 1997, 11). Employing Baruch Brody's typology, the human dignity objection can be recognized as having four forms: "'ownership of human genes infringes on human dignity [1] because it is equivalent to ownership of humans, [2] because it commercializes body parts which should not be commodified, [3] because it cheapens that which defines human identity, and [4] because it would lead to inappropriate modifications in our genetic integrity.'" (Hanson, 1997, 12) Hanson notes that Brody dismissed the first because patented genes lack a particularized connection to a person or body, the second because gene patenting does not raise concerns analogous to commercialization of organs (e.g., exploitation of poor people, reduction in possibilities for altruism), the third as premature absent a patent application on the full, unique genome of a human individual, and the fourth because, while hinting at eugenics, it is too vague. Hanson himself concludes that it is important to avoid the reductionism implied in extending market and materialist values and rhetoric to all areas of human life *and* the reductionism implied in equating human or individual identity with DNA. This is in line with Resnick's conclusions regarding commodification concerns.

Audrey Chapman asserts that reflection on the wider social context is a distinctive contribution of religious ethics:

> [I]n contrast with those who claim that decisions about genetic developments should be left solely to the individual, moral theologians consistently argue that societal well-being goes beyond questions of private ethics to wider issues of public decision making and morality. Religious thinkers are also more likely to have sensitivities for the justice implications of genetics, gene patenting, and cloning, particularly for those persons and groups who are unable or unlikely to speak for themselves (Chapman, online).

Sections of a study document from the National Council of Churches serve as illustrations. Inspired by the symbol of the cross of Christ, understood as expressing divine suffering with and for all humanity and hope in the face of suffering, but also gratitude to God for human inventiveness and its contribution to individual and social welfare, the report's conclusions include:

> Economic benefits from salable products derived from publicly funded research should be shared with the public.... Even if we hesitate to advocate the payment of monies by businesses to the government in this context [e.g., due to practical difficulties], we reaffirm the need for public access to the beneficial products and techniques resulting from genetic research.... Private companies that have developed or have a license to produce biogenetically engineered products should expect a reasonable return on their investment. We must also respond to the request that biogenetic innovations be available to Third World persons at costs relative to their income.... The question of what is a fair and just distribution of the world's resources and wealth enters this discussion as well. How do we justify spending vast amounts on genetic research for procedures and products that by cost or need will benefit only a few? (Panel on Bioethical Concerns, 1984, 35, 39, 43–44).

A 2001 report from the Evangelical Lutheran Church in America is unsympathetic to the "unique status" cluster of concerns. No resolution of the DNA patent debate

4 Ethical Challenges of Patenting "Nature": Legal and Economic Accounts

is attempted. Rather, the authors commend a set of four principles, drawn from the work of Ted Peters. The first principle is:

> Disregard the attempt to draw a connection between divine creation and a ban on patents as proposed by the Joint Appeal Against Human and Animal Patenting. There is no warrant for treating nature prior to human technological intervention as the sole domain of divine creation. Nature is not less sacred after being influenced by human ingenuity. Rather, pursue a theological vision that affirms continuity between humanity and the rest of nature, and affirms our divine call to be stewards. The goal of stewardship is to relieve suffering and make the world a better place.

Other principles include "[a]void simplistic and misleading rhetoric," e.g., the mystification of genes, and "[t]ake seriously the concern...that the dignity and integrity of life be safeguarded" (Evangelical Lutheran Church in America, 2001, 44).[73] The ELCA document does not include Peters' principles five through seven, which concern recognition of a "distinction between discovery and invention, between learning what already exists in nature and what human ingenuity creates," adoption of a federal policy that aims "primarily to enrich the public domain of knowledge," and promotion of an egalitarian philosophy for government-funded medical research that entails attention to orphan diseases and other research areas ignored by industry (Peters, 1997, 140–141). A chapter in a 2003 book generated by a task force of the Episcopal Church focuses on themes of stewardship and justice, emphasizing the Bible's injunction to attend to the problems experienced by persons on the margins of society. The authors also make the case for laws or public processes that increase transparency and enhance the role of the public in oversight of the biotechnology industry (Anderlik and Heller, 2003).

As even this brief discussion makes clear, there are some significant differences across Protestant denominations on substantive issues. Yet in most cases the analysis follows a similar format, moving from interpretation of Scripture or the Cross to an interpretation of contemporary developments in light of the same, with perhaps a word or two from a founding figure such as Luther or a nod to the traditions of the Church. One commonality across the major religious groups is worth underlining. As Courtney Campbell has noted, members of these groups are likely to be troubled by the "reductionist account of the body embedded in genomic research," and also in the contemporary legal and economic culture facilitating the commercial exploitation of genomic research. Yet for most, acts that would be morally wrong considered in the abstract may be permissible or even required if necessary to relieve serious human suffering or

---

[73] Interestingly, an official working group of the European Ecumenical Commission for Church and Society rejected the notion of a "sacred" nature while also rejecting a purely instrumental view of nature: "As Christians, we regard nature as God's creation, but this does not make nature 'sacred' or untouchable. On the contrary, God has given use the dual responsibility to develop and also take care of what he has created. But we are stewards and companions to nature, not its owners" (Working Group on Bioethics and Biotechnology, 1996).

avert death (Campbell, 2000). This sensitivity to justification is reflected not only in official documents and scholarly works emerging from within specific traditions, but also in the opinions of individuals who identify themselves as "very religious."[74]

4.3.2 Attitudes and Approaches of "Indigenous" Religions

4.3.2.1 "Indigenous" and "Religions": Issues of Definition

According to Grim: "The term 'indigenous' is a generalized reference to the thousands of small scale societies who have distinct languages, kinship systems, mythologies, ancestral memories, and homelands. These different societies comprise more than 200 million people throughout the planet today" (Grim, 1998). Even if one wants to use the word, "religion," John Grim argues that "to analyze religion as a separate system of beliefs and ritual practices apart from subsistence, kinship, language, governance, and landscape is to misunderstand indigenous religion" (Grim, 1998).

Following these themes and in elaboration of them, the Independent Commission on International Humanitarian Issues issued a report in 1987 which tried to give more depth to that understanding:

> There are four major elements in the definition of indigenous peoples: pre-existence (i.e. the population is descendant of those inhabiting an area prior to the arrival of another population); non-dominance; cultural difference; and self-identification as indigenous. Other terms are often used to refer to indigenous peoples: autochthonous, ethnic minorities, tribal people, first nations, fourth world (Independent Commission, 1987, 6).

Weighing in somewhat on the other side of the discussion, Stan Stevens has added that other terms used are "native peoples," "primary peoples," "ecosystem peoples," and "aboriginal peoples." "The term, indigenous peoples," he says, "is not in universal use, however, and there are disagreements over how the term should be defined and applied" (Stevens, 1997, 299, n. 1).

There have been several attempts at a more legally recognized definition. The Special Rapporteur on the Problem of Discrimination against Indigenous Populations for the United Nations Sub-Commission on Prevention of Discrimination and Protection of Minorities produced this definition in 1986:

> Indigenous communities, peoples and nations are those which, having a historical continuity with pre-invasion and pre-colonial societies that developed on their territories,

---

[74] In a national survey conducted in 1987, 52% of those self-identified as "very religious" felt that on balance changing the genetic makeup of human cells is morally wrong, versus 23% of those self-identified as "not very religious," but religiosity did not affect willingness to use gene therapy to save the life of a child (US Office of Technology Assessment, 1987).

consider themselves distinct from other sectors of the societies now prevailing in those territories, or parts of them. They form at present non-dominant sectors of society and are determined to preserve, develop and transmit to future generations their ancestral territories, and their ethnic identity, as the basis of their continued existence as peoples, in accordance with their own cultural patterns, social institutions and legal systems (United Nations/Document No. E/CN.4/Sub.2/1986/7/Add. 4, para. 379, quoted in Independent Commission, 1987, p. 8, n. 3).

The International Labor Organization Convention 169 defines indigenous and tribal peoples as follows:

> a) [T]ribal peoples in dependent countries whose social, cultural, and economic conditions distinguish them from other sections of the national community, and whose status is regulated wholly or partially by their own customs or traditions or by special laws and regulations; and b) peoples in independent countries who are regarded as indigenous on account of their descent from the populations which inhabited the country, or a geographical region to which the country belongs, at the time of conquest or colonisation or the establishment of present state boundaries and who, irrespective of their legal status, retain some or all of their own social, economic, cultural and political institutions (Beltran, 2000, Annex 3, p. 17).

The Native American/American Indian scholar, Vine Deloria, Jr., has written about the difficulty of drawing comparisons between the Western concepts of "religion" and the belief systems of what other scholars have labeled "primitive" peoples. "Primitive peoples," he says, "do not differentiate their world of experience into two realms that oppose or complement each other. They seem to maintain a consistent understanding of the unity of all experiences" (Deloria et al. (eds.), 1999, 354). They "maintain a sense of mystery through their bond with nature; the world religions sever the relationship and attempt to establish a new, more comprehensible one" (Deloria et al., 1999, 360).

According to Deloria, two perceptions are the key to conceptualization of religion and religious experience:

> [P]rimitive peoples' perceptions of reality, particularly their religious experiences and awareness of divinity, occupy a far different place in their lives than do the conceptions of the world religions, their experiences, and their theologies, philosophies, doctrines, dogmas, and creeds. Primitive peoples preserve their experiences fairly intact, understand them as a manifestation of the unity of the natural world, and are content to recognize these experiences as the baseline of reality. World religions take the raw data of religious experience and systemize elements of it, using the temporal or the spatial dimension as a framework, and attempt to project meaning into the unexamined remainder of human experience. Ethics becomes an abstract set of propositions attempting to relate individuals to one another in the world religions while kinship duties, customs, and responsibilities, often patterned after relationships in the natural world, parallel the ethical considerations of religion in the primitive peoples. Primitive peoples always have a concrete reference—the natural world—and the adherents of the world religions continually deal with abstract and ideal situations on an intellectual plane. 'Who is my neighbor?' becomes a question of great debate in the tradition of the world religions, and the face of the neighbor changes continually as new data about people becomes available. Such a question is not even within the worldview of primitives. They know precisely who their relatives are and what their responsibilities toward them entail (Deloria et al., 1999, 364–365).

The very use of the word, "religion," may be inappropriate with regard to indigenous peoples. Deloria argues that

> [O]nce we step outside the Western framework, we realize that religion is not a universal concept and is probably not even a proper category when examining human cultures and their perspectives on the world. Murray Wax, sociologist of religion, among other scholars, asserts that "only the languages of the modern West contain a term responding to religion." Tracing the concept backward in time, Wax found that "even the ancient Hebrew Scriptures contain no word corresponding to religion." Except in the modern West, then, there are virtually no "religious" traditions that separate religion from the rest of human knowledge and experience (Deloria, 2002, p. 120).

Even if one wants to use the word, "religion," John Grim suggests that "to analyze religion as a separate system of beliefs and ritual practices apart from subsistence, kinship, language, governance and landscape is to misunderstand indigenous religion" (Grim, 1998).

For example, when comparing Native American/American Indian tribal religions and Christianity, Deloria found that while they:

> agree on the role and activity of a creator....there would appear to be little that the two views share (Deloria, 1994, 78).... The overwhelming majority of American Indian tribal religions refused to represent [the] deity anthropomorphically (Ibid., 79).... Indian tribal religions also held a fundamental relationship between human beings and the rest of nature, but the conception was radically different (Ibid., 81).... In the Indian tribal religions, man and the rest of creation are cooperative and respectful of the task set for them by the Great Spirit. In the Christian religion both are doomed from shortly after the creation event until the end of the world (Ibid., 82).
>
> ...[T]raditional religions do not have the point-counterpoint recitation of beliefs that we find in the Near Eastern traditions. Singers and individual medicine people sing specific songs that compose the ceremony. While there is the expression of humility as humans stand before the higher spiritual powers, the Indian tradition lacks the admission of individual and corporate guilt which Near Eastern religions make the central part of their doctrines (Ibid., 85).... The task of the tribal religion, if such a religion can be said to have a task, is to determine the proper relationship that the people of the tribe must have with other living things and to develop the self-discipline within the tribal community so that man acts harmoniously with other creatures. The world that he experiences is dominated by the presence of power, the manifestation of life energies, the whole life-flow of a creation (Ibid., 88).... Tribal religions find a great affinity among species of living creatures, and it is at this point that the fellowship of life is a strong part of the Indian way (Ibid., 89).... Very important in some of the tribal religions is the idea that humans can change into animals and birds and that other species can change into human beings. In this way species can communicate and learn from each other.... But many tribal religions go even farther. The manifestation of power is simply not limited to mobile life forms. For some tribes the idea extends to plants, rocks, and natural features that Westerners consider inanimate (Ibid., 90).... [Finally] [t]he Indian tribal religions would probably suggest that the unity of life is manifested in the existence of the tribal community, for it is only in the tribal community that any Indian religions have relevance Ibid. (Ibid., 93).

Scholars of African indigenous or traditional religion have made similar observations. Jacob Olupona has noted that religious life in Africa is expressed through cultural idioms such as music and arts, involves ecology, and cannot be studied as if religion exists in isolation from its socio-cultural context (Olupona, 1991, 30). Laurenti Magesa has argued that African traditional religion is communal and *"embraces the whole life.* People do not distinguish between a religious and secular leader. Religious leaders maintain a close relationship with departed ancestors so that things will go well in the visible and invisible world" (Magesa, 1997, 71). Writing about the Akan of Ghana, Elizabeth Amoah writes that there is

> also a demonstrable cultural disposition that emphasizes the spiritual dimension of total human life. This finds expression in a certain specific conception of nature, of humanity and Nana Onyame's (God's) relations with creatures of various types. It also finds expression in the relationships between persons and their physical environment and in the ways in which explanations are sought for the major problems of life, the problem of the meaning of life, the problem of suffering and the problem of evil (Amoah, 1998).

The natural world, human community, and health are important issues in African religious traditions. The universe is inhabited by God, spirits, living things, and non-living things, which interact with each other. Yoruba cosmology demonstrates that the created order consists of *aye*, the visible concrete world that is the habitat for humans, wildlife and all things in nature. The invisible world is called *orun*, and is the home of God, the spirits, and the departed ancestors (Drewal et al., 1989, 14). The Sotho-Tswana believe that God maintains a close relationship with the universe and uses natural forces, like thunderbolts, to penetrate into the universe (Setiloane, 1976, 82). Thus, humans are part of the rhythms of nature. Life is a journey to the other world and a successful journey requires a right relationship with nature, divinities, and other inhabitants of the universe. In the social world humans are expected to use *ase* (power or authority) to ensure the well being of nature and other human beings. In order to do this they must employ wisdom, knowledge, and understanding. "Theoretically, every individual possesses a unique blend of performative power and knowledge – the potential for certain achievements. Yet because no one can know with certainty the potential of others, *eso* (caution), *ifarabale* (composure), *owo* (respect), and *suuru* (patience) are highly valued in Yoruba society and shape all social interactions and organization" (Appiah-Kubi, 1981, 16). Living on earth and exploring the natural world calls for a balance that recognizes the realm and power of nature as well as the unique gifts that each individual has received from the creator. One maintains right relationships through mediation and employs certain graces and act with loving care to manage and alter nature.[75] As part of a community, humans are expected to use the things God has created for the benefit of other people. This calls for a different attitude towards the acquisition and use of natural resources, especially in sub-Saharan Africa where deforestation and depletion of resources continues at a very fast pace.[76]

---

[75] Appiah-Kubi argues that among the Akan: "Proper land use is evidence of faithfulness to God and Mother earth and it is reflected in the health of the soil and nature and economic health or prosperity of the society. Unjust land use on the other hand spells social, economic, spiritual disaster, crop failure, and epidemics" (Appiah-Kubi, 1981, 9). In addition to the natural environment, religious thought in Africa emphasizes the social community. One lives, exercises power, controls, and manages nature in solidarity with other people. Community is the center for fecundity, friendships, hospitality, and healing (Nyamiti, 1973, 9–11).

[76] In a country like Cameroon which is richly endowed with natural resources and an extensive biodiversity, poor management is contributing to the decline of these resources (Blaike, 1996; Atyi, 1998).

African religious ideas include health and medicine, an area where humans are constantly altering nature. Researchers point out that spiritual healers claim that God has given the gift of healing and shown them the natural products they use in healing so that they can help the entire community (see Janzen, 1978; Ngubane, 1977; Warren, 1974; Young, 1975). Early practice of medicine and healing was not commercialized, as is the case with contemporary biomedical approaches to healing. Ideally, "healers" accepted gifts from their clients and sometimes these gifts could exceed what they could have charged.[77] Godfrey Tangwa argues that in Cameroon, since medicine was considered a gift for the entire community, no one was to be denied access to medicines because he or she could not afford them (Tangwa, 2002). In Cameroon, traditional healers have formed an association to coordinate their activities. Murray Last states that such professionalization has not resolved all issues, which include discussions on whether their practice can be linked to health departments in different countries (Last, 1986). Were such a relationship to state health departments to occur, it would raise new questions about patient/doctor relationships, the status of traditional medicine in such a system, the status of traditional medicine compared to biological medicine, and ways of determining the effectiveness of traditional therapies. Further questions include the nature of the working relationship between traditional associations and professional guilds. What really is "traditional," especially when these systems include the social world in their definition of disease, and the World Health Organization definitions include in its understanding of healing the cure of the whole person? In spite of these questions, traditional healing associations have opened a space for healers to share information, network, and compete in the development of therapies.

Healing in Africa is also part of a global system because healers are competing in the search for new remedies in an age of HIV/AIDS. Perhaps the difference here is that the level of competition is not as complicated as it is in the west, which Tangwa argues: "is fraught with danger [because] Western industries and businesses[es] … have an understandable tendency to be driven by morally blind pure economic logic and the lure of profit. This tendency is often on the lookout for and sees economic opportunities and advantages in even the most distressing human calamities" (Tangwa, 2002, 220). Medicines which for most Africans came directly from nature, have become complicated because patents, which change the ownership of resources, have introduced complex questions about the resources of nature.

4.3.2.2 Attitudes of Indigenous People Toward and Changes That Threaten That World View

Much of the literature regarding indigenous people and ecology incorporates understandings of attitudes of those people toward nature and their relationship to it. Grim has written that "[t]hemes which provide orientation for understanding the

---

[77] "Healers" is used here to refer to a variety of medical practitioners who are often called "traditional" doctors.

4 Ethical Challenges of Patenting "Nature": Legal and Economic Accounts 257

relations between indigenous religion and ecology are kinship, spatial and biographical relations with place, traditional environmental knowledge, and cosmology." He offers as examples the Lakota [Native American] "memory of rocks and stones as persons;" the belief of the Temiar people of Malaysia that a cool healing liquid called *khayek* (the "form taken by the upper soul of a spiritual being from the local Malayan rainforest") "can be imparted to human beings through dreams," specifically the songs in dreams; and the cosmologies of the Dogon peoples of Mali "which elaborate the close relationships which living Dogon share with their ancestors, their land, and their animals among which the soon to be living reside." Other examples are those of the Salish speaking Colville peoples on the Columbia River of eastern Washington who "weave ritual forms of knowing and proper approach into the taxonomic discussions of plants," and among the Yekuana people of Venezuela the "ethics of limits with regard to nature consumption" based on the pragmatic use of plants and roots (Grim, 1998).

As Deloria sees the problem:

> [W]e have reduced our knowledge of the world and the possibility of understanding and relating environment to a wholly mechanical process. We have become dependent, ultimately, on this one quarter of human experience, which is to reduce all human experience to a cause-and-effect situation. When we look at nature and environment through Western European eyes, that is really what we are looking at. That is not the "nature" Indians understand. Indians never made any of those divisions.
>
> In the Indian tradition we find continuous generations of people living in specific lands, or migrating to new lands, and having an extremely intimate relationship with lands, animals, vegetables, and all of life. As Indians look out at the environment and as Indians experience a living universe, relationships become the dominating theme of life and the dominating motif for whatever technological or quasi-scientific approach Indian people have to the land (Deloria, 1999, 225–226).

When this environment is threatened, the indigenous people potentially are threatened not only physically but culturally and "religiously." An example can be seen in the description by Julio César Centano and Christopher Elliott of the Yanomami peoples, described as "the largest group of indigenous people still living their traditional life-style in the Americas" and who live on either side of the border between Brazil and Venezuela. They are said "to exist in an interdependent relationship, living in harmony with the environment, practicing low-intensity shifting cultivation, fishing, hunting, and gathering. *Urihi* and *uli* are Yanomami words for forest. Translated from the Amerindian language, they mean 'home' or 'the place where one belongs.'" As that forest area becomes threatened by outsiders who want to take their land, by scientific expeditions who ignore the values of the Yanomami to follow their own agendas, by the Brazilian and Venezuelan armies that are arguably trying to "protect" the Yanomami in what is now a legally protected area, and by ecotourists seeking a "unique jungle experience," the survival of the Yanomami peoples is threatened (Kemf, 1993; "Forest Home: The Place Where One Belongs").

Another example can be seen in the beliefs of the Navajo for whom "the complex network of effect makes the human individual part of and dependent on the kinship group and the community of life, including plants, animals, and aspects of the cosmos that share substance and structure" (Schwartz, 1997, 12). For example, the Navajo "have a sentimental attitude toward plants, which they treat with incredible respect" (Reichard, 1950, 22).

In another part of the world

> [T]he Maori, native peoples of New Zealand, speak of themselves as *tangata whenua*, "people of the land" .... Land is the connection both to larger mythologized cosmic forces as well as to the source of personal life...Interactions with the environment bring humans in contact with the mana, or inherent understanding of all reality.... All creatures possess *mana* suggesting that all reality has intrinsic worth, as well as a personal life force, *maori* (Grim, 2001, 41–42).

These relationships have been recognized by the World Conservation Union which has adopted the principle that "[i]ndigenous and other traditional peoples have long associations with nature and a deep understanding of it. Often they have made significant contributions to the maintenance of many of the earth's most fragile ecosystems, through their traditional sustainable resource practices and culture-based respect for nature" (Kemf, 1993, 7).

4.3.2.3 Manifestations of Beliefs in Legal and Economic Realms

Legal Recognition of Indigenous Rights and Values

The most significant evidence of international legal recognition for the rights and values of indigenous peoples occurred on September 13, 2007, when the United Nations General Assembly adopted the United Nations Declaration on the Rights of Indigenous Peoples. After more than 20 years of negotiations between governments and indigenous peoples, the Declaration passed by a vote of 143-4 with 11 abstentions. Four countries with large indigenous populations voted against the adoption of the Declaration: Australia, Canada, New Zealand, and the United States. Those countries abstaining were Azerbaijan, Bangladesh, Bhutan, Columbia, Georgia, Kenya, Nigeria, the Russian Federation, Samoa, and the Ukraine.

A lengthy document with a preamble and 46 articles, the Declaration recognizes a broad range of basic human rights of indigenous peoples. Among these are an acknowledgment that indigenous peoples have the right to full enjoyment, collectively and individually, of all human rights and fundamental freedoms recognized in the U.N. Charter, the Universal Declaration of Human Rights, and international human rights law. Other topics include a right to ownership, use and control of lands, territories and their natural resources; rights to maintain and develop their own political, religious, cultural and educational institutions; access to financial and technical assistance from nation-states; and the right to the recognition, observance and enforcement of treaties and agreements with nation-states.

Article 31 of the Declaration gives indigenous peoples "the right to maintain, control, protect and develop their cultural heritage, traditional knowledge and traditional cultural expressions, as well as the manifestations of their sciences, technologies, and cultures, including human and genetic resources, seeds, medicines, knowledge of the properties of fauna and flora, oral traditions, literatures,

4 Ethical Challenges of Patenting "Nature": Legal and Economic Accounts 259

designs, sports and traditional games and visual and performing arts. They also have the right to maintain, control, protect and develop their intellectual property over such cultural heritage, traditional knowledge, and traditional cultural expressions." (See the discussion in subsection "Legal and Economic Assertions of Indigenous Rights and Values" below for earlier efforts to assert such intellectual property rights).[78]

The International Labor Organization (ILO) has also taken note of issues involving indigenous peoples in its Convention 107 and related Recommendation 104 (1957) (Convention and Recommendation "concerning the protection and integration of indigenous and other tribal and semi-tribal populations in independent countries").[79] In 1989, the General Conference of the ILO adopted the Indigenous and Tribal Peoples Convention, better known as "Convention 169", which entered into force in 1991.

A number of developments have also taken place with regard to the Agreement on Trade-Related Aspects of Intellectual Property Rights (TRIPS), Annex 1C of the Marrakesh Agreement Establishing the World Trade Organization, signed in Marrakesh, Morocco, on April 15, 1994. In October 1999, four Latin American countries (Cuba, Honduras, Paraguay, and Venezuela) tabled a proposal at the United Nations General Council for the protection of intellectual property rights regarding the traditional knowledge of local and indigenous communities. The proposal, made in connection with the Third Ministerial Conference of the World Trade Organization (WTO) held at Seattle from November 30 to December 3, 1999, asked specifically for "the carrying out of a detailed study of how to protect the moral and economic property rights relating to the traditional knowledge, medicinal practices and expressions of folklore of local and indigenous communities" (Raghavan, 1999).

Another TRIPS-related example involves the Association of Southeast Asian Nations (ASEAN) and the recommendation to its ten member governments that emerged from a report of an ASEAN regional working group meeting in Jakarta, Indonesia, in May 2002 entitled "Biodiversity, Biotechnology, Traditional/ Indigenous Knowledge, and Traditional Medicine." The report concerned the TRIPS Agreement and its impact on pharmaceuticals. Its recommendations noted that the TRIPS Agreement does not "include specific provisions related to the protection of traditional/indigenous knowledge (systems, practices, naturally occurring plants, products." The report suggested that new instruments needed to be developed, both internationally and nationally, and suggested the development of "an ASEAN position on the protection of traditional/indigenous knowledge" (Raghavan, 2000).

---

[78] A full copy of the Declaration may be accessed online at http://www.iwgia.org/graphics/ Synkron-Library/Documents/InternationalProcesses/DraftDeclaration/07-09-13ResolutiontextDe claration.pdf

[79] Extracts can be found in Independent Commission on International Humanitarian Issues, 1987, Annex I, 142–163.

Africa on the TRIPs Review Process

African states have responded to the debate by the World Trade Organization's review of TRIP by establishing a model law (ML) "to ensure the conservation, evaluation and sustainable use of biological resources, including agricultural genetic resources, and knowledge and technologies in order to maintain and improve their diversity as a means of sustaining all life support systems" (See http://www.grain.org/br/oau-model-law-en.cfm, p. 2). Furthermore, the Model Law was to:

> [R]ecognize, protect and support the inalienable rights of local communities including farming communities over their biological resources, knowledge and technologies;
> Recognize and protect the rights of breeders;
> Provide an appropriate system of access to biological resources, community knowledge and technologies subject to the prior informed consent of the state and the concerned local communities.

Overall, the ML seeks to reconfigure bioprospecting in a manner that would ensure equity, involve local communities, and increase the participation of women in the process. The ML also requires governments and organizations promoting the use of biological resources to use the products to increase capacity building, establish institutions and structures that would promote scientific and technological conservation, put in place intuitions that would protect the rights of local people, promote sustainability and protect food security.

J.A. Ekpere, former Executive Secretary of the OAU Scientific, Technical and Research Commission, grounds African perspectives on TRIPS on the Rio Summit, and WTO protocols (Ekpere, 2001, 3). The Scientific and technical commissions of the African Union (AU) have recommended the drafting of a new law to protect indigenous knowledge and medicinal plants, and coordinate national policies and interregional interests with regards to ownership, access, utilization and conservation of medicinal plants. The AU called on its member states to understand the implications of TRIP on pharmaceutical products, the bio-resources of Africa and the anticipated harmonization of TRIP by 2000 and 2005 (Ekpere, 2002, 5). Earlier, the 1998 OAU Summit in Burkina Faso considered draft legislations dealing with community rights, access to biological resources, protection, conservation and sustainable use of biological diversity and genetic resources, and recommended that member states regulate access to resources, study the TRIP document, adopt a model law, negotiate an African convention on biological Diversity, and formulate a common position to safeguard member states' rights.

Ekpere claims that the TRIPS protocol as it now stands has inequities because it legitimizes and sanctions global creativity through the patent system, while local creativity has not received any such protection. "This inequity does not only derogate and threaten the validity of the biological resources and knowledge system of local communities and indigenous people, but the value of their technologies, innovation and practices" (Ekpere, 2002, 5). By giving monopoly rights through Intellectual Property rights, (Article 27.3b) the international community ignores local creativity, an action that contradicts the recognition that is given to states by the Convention on Biological Diversity.

4 Ethical Challenges of Patenting "Nature": Legal and Economic Accounts

Ekpere argues that what is at stake is "appropriation of knowledge, innovation, technologies and practices of local communities associated with their biodiversity as well as equitable sharing of benefits associated with the sustainable use of these resources" (Ekpere, 2002, 6). Conventions and protocols extend rights of use and ownership, therefore locals should exert control over these resources. However, he cautions that the pursuit of economic goals must not sacrifice sustainable development of natural resources. Further, such economic pursuits cannot disrupt life and endanger "seed and other planting materials; traditional medicinal plants, the basis of health care delivery service for majority of African people; natural fiber and dyes, the basis of African arts and crafts" (Ekpere, 2002, 6). It is in this light that most African states contest the patent system, which allocates exclusive rights to multi-national corporations.

The OAU model law therefore focuses on these core issues. We highlight here the summaries articulated by Professor Ekpere. First, food security must be safeguarded. Food production for many African states depends generally on their biological resources and diversity. Any changes that could disrupt such diversity must be studied carefully and provisions should be made for adequate conservation. Furthermore, the rights of the local farmers ought to be protected against the encroachment of multi-nationals. Second, conventions must guarantee the sovereign and inalienable rights of states provided for by the Rio declaration. In addition, rights of specific local communities must be protected. "Local communities are the custodians of their biological resources, innovations, practices, knowledge and technologies which are governed completely or partially by their own customary laws, written or orally transmitted" (Ekpere, 2002, 11). The law calls for the promotion of local knowledge and technology that has protected their resources. International conventions such as TRIP tend to ignore this knowledge. Third, conventions must guarantee that indigenous people will participate in decision-making. Fourth, states must also control access to biological resources. "The current trend towards privatization, commercialization, bioprospecting and biotrade promoted through TRIPs could erode local livelihood systems based on biological resources.... The model law provides for a system to regulate access subject to prior informed consent of the state and the concerned local community" (Ekpere, 2002, 12). Protocols must guarantee prior consent from the local community. Finally, the ML calls for fair and equitable sharing of the profits and benefits from bioprospecting and all commercial activity.

The core concerns of the ML address legitimate concerns about the cultural, moral, religious, and social values of local communities that face a challenge not primarily from science, research, and technology, but from commercial values anchored in modalities like privatization, the transfer of ownership to new owners through the legal stamp of patents and conventions like TRIPS. Therefore the issue is not whether African states are going to embrace new research and technologies, but for whose interest that research and technology is being used. Furthermore, and this is crucial, can this new research ensure local control and sustainability? It is imperative that African states develop mechanisms of protecting biodiversity. The concern that current protocols do not guarantee this has led African states to reject the notion of granting patents on life forms.

Legal and Economic Assertions of Indigenous Rights and Values

One example cited of successful challenges to intellectual property decisions involved the decision in May, 1995 by the US Patent Office to grant the University of Mississippi Medical Center a patent [#5,401,504] for "Use of Turmeric in Wound Healing."

> The Indian Council of Scientific and Industrial Research (CSIR) requested the US Patent and Trademark Office (USPTO) to re-examine the patent. CSIR argued that *turmeric* has been used for thousands of years for healing wounds and rashes and therefore its medicinal use was not novel. Their claim was supported by documentary evidence of traditional knowledge, including an ancient Sanskrit text and a paper published in 1953 in the Journal of the Indian Medical Association. Despite arguments by the patentees, the USPTO upheld the CSIR objections and revoked the patent. A second example involved extracts from an Indian tree called *neem* used to treat malaria, skin diseases and even meningitis. In 1994 the European Patent Office granted European Patent No. 0436257 to the US Corporation W.R. Grace and USDA for a "method for controlling fungi on plants by the aid of a hydrophobic extracted neem oil". In 1995 a group of international NGOs and representatives of Indian farmers filed a legal opposition against the patent. They submitted evidence that the fungicidal effect of extracts of neem seeds had been known and used for centuries in Indian agriculture to protect crops, and thus the invention claimed in EP257 was not novel. In 1999 the European Patent Office determined that according to the evidence "all features of the present claim have been disclosed to the public prior to the patent application... and [the patent] was considered not to involve an inventive step". The patent was revoked by the European Patent Office in 2000 (Commission on Intellectual Property Rights, 2002, ch. 4).

An example of a challenge to a specifically biotechnology patent – part of the Human Genome Development Program – concerned a US patent obtained on March 14, 1995, by the National Institutes of Health (#5,397,696) for a cell line containing unmodified Hagahai DNA and several methods for its use in detecting HTLV-1-related retroviruses. The Hagahai people are from Papua New Guinea and only came into consistent contact with the outside world in 1984. After international pressure, this patent was withdrawn. An earlier patent application in 1993 for another cell line, this involving human T-cell lymphotrophic viruses (HTLV) type 1 from a 26-year Guaymi woman from Panama, was also withdrawn after pressure from the Rural Advancement Foundation International (RAFI) and other groups of indigenous peoples and non-governmental organizations (Tauli-Corpuz, online).

On a slightly more positive note, an example of a successful negotiation between an indigenous community and a pharmaceutical company involves the hoodia cactus and the San community of South Africa (sometimes referred to by outsiders as the "bushmen" of the Kalahari).

> The San, who live around the Kalahari Desert in southern Africa, have traditionally eaten the Hoodia cactus to stave off hunger and thirst on long hunting trips. In 1937, a Dutch anthropologist studying the San noted this use of Hoodia. Scientists at the South African Council for Scientific and Industrial Research (CSIR) only recently found his report and began studying the plant.
>
> In 1995 CSIR patented Hoodia's appetite-suppressing element (P57). In 1997 they licensed P57 to the UK biotech company, Phytopharm. In 1998, the pharmaceutical company Pfizer acquired the rights to develop and market P57 as a potential slimming drug and cure for obesity (a market worth more than £6 billion), from Phytopharm for up to $32 million in royalty and milestone payments.

4 Ethical Challenges of Patenting "Nature": Legal and Economic Accounts   263

> On hearing of possible exploitation of their traditional knowledge, the San People threatened legal action against the CSIR on grounds of "biopiracy." They claimed that their traditional knowledge had been stolen, and CSIR had failed to comply with the rules of the Convention on Biodiversity, which requires the prior informed consent of all stakeholders, including the original discoverers and users.
>
> Phytopharm had conducted extensive enquiries but were unable to find any of the "knowledge holders". The remaining San were apparently at the time living in a tented camp 1500 miles from their tribal lands. The CSIR claimed they had planned to inform the San of the research and share the benefits, but first wanted to make sure the drug proved successful.
>
> In March 2002, an understanding was reached between the CSIR and the San whereby the San, recognized as the custodians of traditional knowledge associated with the Hoodia plant, will receive a share of any future royalties. Although the San are likely to receive only a very small percentage of eventual sales, the potential size of the market means that the sum involved could still be substantial. The drug is unlikely to reach the market before 2006, and may yet fail as it progresses through clinical trials (Commission on Intellectual Property Rights, 2002, ch. 4).

There are also examples of biotechnology companies attempting to recognize in advance of any claims the indigenous rights to medicinal plants and to provide some benefits to the indigenous peoples involved for the use of the plants, such as Shaman Pharmaceuticals, Inc. and Shaman Botanicals of South San Francisco, California (Bierer et al., 1996).

The Challenges of Biotechnology for Indigenous and World Religious Communities

For some observers such as Claire Smith, Heather Burke, and Graeme K. Ward, these issues are part of the challenge of globalization to indigenous people. They suggest that:

> The key issue here is control—control over land, control over knowledge, control over the past, present and future. The object of the struggle is not only Indigenous cultural and intellectual property but the continued future of Indigenous societies themselves. This conflict establishes an arena for radical change in the social and political environments of Indigenous peoples (Smith et al., 2000, 3).
>
> In countries with Indigenous peoples, legislative changes are being considered in order to accommodate Indigenous ways of knowing and curating knowledge. Such changes include revising patent legislation to recognize the contribution made by indigenous knowledge to the development of new medicines; changing copyright legislation to recognize the communal and multi-level ownership of designs and cultural knowledge; and changing legislation relating to performer's rights so as to recognize the secret and restricted nature of certain Indigenous performances.... Significant within the context of globalization is that these changes not only incorporate the recognition of Indigenous rights of ownership over, and control of, Indigenous intellectual and cultural property in accordance with traditional customary law, but that they also recognize the right of Indigenous peoples to benefit commercially from the authorized use of this property..." (Smith et al., 2000, 12).

These power issues, of course, while important – perhaps crucial – to the survival of indigenous peoples who are often typified by a lack of power, ignore the previously discussed value issues (whether one wants to call them "religious" or not) of needing to live generally in what many see as a harmonious relationship with nature and the environment, and the extent to which these scientific developments are seen as endangering that relationship.

4.4 Conclusion

The concerns of John Locke are for the moment honored more in the breach than in the observance. It would not have been inconceivable in Locke's time for the over-commercialization of the commons to be a concern. However, when the seas are over-fished, oil reserves are in the process of depletion, patients are dying for lack of access to drugs whose economic basis is modest but whose profit margin is substantial, and global warming has reached the stage of undeniability, even the profit-oriented government of the United States must take notice. The rest of the world seems to have already begun to factor these concerns into public policy.

Even more importantly, the voters in the United States seem to have taken interest in these issues. In the words of Benjamin Franklin (often attributed to P.T. Barnum): *You may fool all of the people some of the time; you can fool some of the people all of the time; but you can't fool all of the people all of the time.* We can be hopeful that the evidence of the depletion of the commons is at last becoming a catalyst for political action. As we noted earlier in this chapter: the Lockean justification for private property rests on the assumption that common property will be underused. It is by now abundantly clear that the "tragedy of the anticommons" is also a real danger and that its reach is international. Societal interests and private profits often march together, but in relationship with the commons they need not. One of the dawning perceptions of the 21st century is that interconnectedness has arrived. Solutions, as we illustrate below, must be global to be effective. To resort once more to a beloved quote, this time from Pogo: *we have met the enemy and he is us.*

The goal of this chapter has been to illustrate the various positions from which the issue of patenting or restricting access to nature can be viewed. We have highlighted:

1. Endorsement of many of the basic tenets of contemporary Western patenting regimes which could be grounded in the individualistic strand in Protestantism or the "liberal project" (see chapter 1 in this volume)
2. Support for reform to address access concerns – an agenda that could be grounded in religious worldviews that link private rights to social responsibilities
3. The effort to make the common good or the concern for the poor a priority – the welfare state version of the liberal project
4. The view that what is wrong with the treatment of indigenous peoples is the failure to share profits versus something more fundamental, which might be contrasted with
5. Support for radical change, an agenda that could be grounded in religious worldviews that reject the understanding of modern society in regard to the human/religious vision that undergirds attempts at control over nature

There are some hopeful signs for those who worry about access and equity. On June 29, 2004, the International Treaty on Plant Genetic Resources for Food and Agriculture entered into force. Sections of the treaty address capacity-building in developing countries and benefit-sharing. The Treaty also formally recognizes the "enormous contribution that the local and indigenous communities and farmers of all regions of the world...have made and will continue to make for the conservation and development of plant genetic resources" and calls on governments to protect and promote farmers' rights by safeguarding traditional knowledge and giving farmers opportunities to participate in decision making. A number of universities have joined in the creation of Public-Sector Intellectual Property Resource for Agriculture to facilitate technology transfer to developing countries, and the Rockefeller Foundation has joined with a number of corporations to create the African Agricultural Technology Foundation for similar purposes.

Another hopeful sign related to health care proper is the growth of product development partnerships, access partnerships, and coordination and financing mechanisms to redress the lack of interest among for-profit pharmaceutical companies in investing in the development of drugs for diseases that primarily affect the poor in developing countries. There are also efforts to ensure that those in developing countries have access to drugs already available for treatment of diseases such as HIV/AIDS. For example, OneWorld Health, a non-profit drug company based in San Francisco, recently received a $43 million grant from the Bill and Melinda Gates Foundation to support the development of a synthetic form of artemisinin, used in the treatment of malaria (Sharp, 2004; Anonymous, 2004, 792).

For religious communities, legal scholars and policymakers, our work in this chapter suggests that contemporary understandings of property are impoverished. Can these groups contribute to the development of a more complex understanding? To the creation of a set of expectations and incentives for property holders that will foster the responsible exercise of property rights? It also shows a range of meanings for nature in legal and economic discourse. Does the exclusion of phenomena or products of nature from patenting adequately capture the concerns that should shape deliberations about social institutions such as the patent system and the resolution of specific disputes such as the one over DNA patents?

For the religious and scientific communities, economists, and policymakers, our work conveys a sense of urgency related to the current debate about whether commercial values are invading spheres previously governed by other values and altering the motivational structure of scientists and others in unfortunate ways. How can each group make a contribution to preserving the core values of the academic research enterprise, if these values are indeed worth preserving, as we believe they are? Is it possible to open a discussion in which traditional religious concerns about the effects of money on individual and corporate character are brought to bear on current developments without sanctimony or naiveté? Is it possible to open a dialogue between those who voice traditional religious concerns about the effects of money on individual and corporate character and those who voice traditional economic concerns about incentives and innovation?

For the religious community and policymakers our work has highlighted suffering in many parts of the world, suffering that might be alleviated by access to the products of biotechnology. Our work, and that of our colleagues in this volume, has also touched on how religious and secular traditions of ethics, in affirming an awareness of our shared human nature and the importance of nurturing the springs of sympathy and compassion in that nature, converge in ruling out indifference to the plight of the other who suffers. How can religious communities educate themselves about global inequities and frame their concerns so as to awaken the public conscience? Will the policy makers elected or appointed to represent the public develop concrete proposals that translate conscience into action?

For policymakers our work covers historical examples and current proposals for alternatives to the patent system, at least in some sectors or areas. We also review an argument for a change in the law to empower federal agencies to take a more discriminating approach to researcher patents on work supported with public money. Are these proposals and arguments being seriously explored?

As the above indicates, this chapter is more in the nature of musings, case examples and questions than in the form of definitive comment on an area of intellectual and religious meaning. As the universe of commerce and of ideas shrink, these themes will need to be revisited to assist nations, peoples and religions to forge common ground.

References

Amoah, Elizabeth (1998). "African Indigenous Religions and Inter-religious Relationship: Autumn Lecture." Oxford International Interfaith Centre. Retrieved December 24, 2002 from the World Wide Web: http://www.interfaith-center.org/lectures/amoah98.htm.

Anderlik, Mary, and Jan Heller (2003). "The Economics and Politics of the New Genetics," in D. Smith (ed.), *A Christian Response to Our New Genetic Powers*. Rowman & Littlefield, Lanham, Md.

Anderlik, Mary A., and Mark A. Rothstein (2003). "Canavan Decision Favors Researchers Over Families," *Journal of Law, Medicine and Ethics* 31, 450–453.

Andrews, Lori B. (2002). "Genes and Patent Policy: Rethinking Intellectual Property Rights," *Nature Reviews Genetics* 3, 803–808.

Andrews, Lori, and Dorothy Nelkin (2001). *Body Bazaar: The Market for Human Tissue in the Biotechnology Age*. New York: Crown.

Anonymous (2004). "Gates Cash Lifts Hope of Cheaper Malaria Drug," *Nature* 432, 792.

Appiah-Kubi, Kofi (1981). *Man Cures, God Heals: Religion and Medical Practice Among the Akans of Ghana*. Totowa, NJ: Allanheld & Osumn.

Arrow, Kenneth J. (1962). "Economic Welfare and the Allocation of Resources for Inventions," in Richard Nelson (ed.), *The Rate and Direction of Inventive Activity*. Princeton, NJ: Princeton University Press.

Atlantic Works v. Brady, 107 US 192, 200 (1882).

Atyi, Eba'a (1998). *Cameroon's Logging Industry: Structure, Economic Importance and Effects of Devaluation*. Jakarta: Centre for International Forestry Research.

Baker, Dean (2004). *Financing Drug Research: What Are the Issues?* Center for Economic and Policy Research. http://www.cepr.net/publications/patents_what_are_the_issues.

Ball, Nan T. (2000). "The Reemergence of Enlightenment Ideas in the 1994 French Bioethics Debates," 50*Duke Law Journal* 545, 585.

4 Ethical Challenges of Patenting "Nature": Legal and Economic Accounts 267

Barton, John H. (2002). "Patent, Genomics, Research, and Diagnostics," *Academic Medicine* 77(Part 2), 1339–1347.

Bekelman, Justin E., Yan Li, and Cary P. Gross (2003). "Scope and Impact of Financial Conflicts of Interest in Biomedical Research: A Systematic Review," *Journal of the American Medical Association* 289, 454–465.

Beltran, Javier (ed.) (2000). *Indigenous and Traditional Peoples and Protected Areas: Principles, Guidelines and Case Studies.* Gland, Switzerland: IUCN – The World Conservation Union.

Bendekgey, Lee, and Diana Hamlet-Cox (2002). "Gene Patents and Innovation," *Academic Medicine* 77(December, Part 2), 1373–1380.

Bessen, James, and Eric Maskin (2000). "Sequential Innovation, Patents and Imitation." Working Paper 00–01. Department of Economics, Massachusetts Institute of Technology.

Bewley, Truman (2003). "Fair's Fair," Nature 422, 124–126.

Bierer, Donald E., Thomas J. Carolson, and Steven R. King (1996). "Shaman Pharmaceuticals: Integrating Indigenous Knowledge, Tropical Medicine Plants, Medicine, Modern Science and Reciprocity into a Novel Drug Discovery Approach." Network Science (NetSci). Retrieved January 9, 2003 from the World Wide Web: http://www.netsci.org/science/special/feature11.htm.

Biogen Inc. v. Medeva Plc, R.P.C. 1 (1997).

Blaike, P. (1996). *Environmental Conservation in Cameroon: On Paper and in Practice.* Occasional Papers Series. Copenhagen: Centre for African Studies, University of Copenhagen.

Blank, Rebecca M. (1992). *Do Justice: Linking Christian Faith and Modern Economic Life.* Cleveland, OH: United Church Press.

Blanton, Kimberly (2002). "Corporate Takeover," *Boston Globe Magazine,* April 18.

Bloom, Floyd E. (2003). "Science as a Way of Life: Perplexities of a Physician-Scientist," *Science* 300, 1680–1685.

Blumenthal David, Eric G. Campbell, Melissa Anderson, Nancy Causino, and Karen Seashore Louis (1997). "Withholding Research Results in Academic Life Science," *Journal of the American Medical Association* 277, 1224–1228.

Brody, Baruch (1996). "Public Goods and Fair Prices: Balancing Technological Innovation with Social Well-Being," Hastings Center Report, March–April, 5–11.

Bugbee, Bruce W. (1967). *Genesis of American Patent and Copyright Law.* Washington, DC: Public Affairs Press.

Campbell, Courtney S. (2000). "Research on Human Tissue: Religious Perspectives," in *Research Involving Human Biological Materials: Ethical Issues and Policy Guidance,* Vol. II. Rockville, MD: National Bioethics Advisory Commission.

Campbell, Eric G., Brian R. Clarridge, Manjusha Gokhale, Lauren Birenbaum, Stephen Hilgartner, Neil A. Holtzman, and David Blumenthal (2002). "Data Withholding in Academic Genetics: Evidence from a National Survey," *Journal of the American Medical Association* 287, 473–480.

Canadian Council of Churches (online). "Life: Patent Pending," available at www.ccc-cce.ca.

CBS Evening News (2002). Eye on America, "Battle over the ownership of human genes," September 18.

Chapman, Audrey (online). "The Contributions and Limitations of Christian Ethics on the Genetics Revolution," www.kalam.org/papers/audrey.htm, 21–22.

Charlton, William, Tatiana Mallinson, and Robert Oakeshott (1986). *The Christian Response to Industrial Capitalism.* London: Sheed & Ward.

Chiron v. Murex, F.S.R. 153 (1996).

Chisum, Donald S. (2002). *Chisum on Patents.* New York: Matthew Bender.

Clement D. (2003). "Creation Myths: Does Innovation Require Intellectual Property Rights?" *Reason Online,* March. www.reason.com/0303/fe.dc.creation.shtml.

Commission on Intellectual Property Rights (2002). "Integrating Intellectual Property Rights and Development Policy." Retrieved April 10, 2003, from the World Wide Web: http://www.iprcommission.org/papers/text/final_report/chapter4htmfinal.htm.

Cohen, Wesley J., Richard R. Nelson, and John P. Walsh (2000). "Protecting Their Intellectual Assets: Appropriability Conditions and Why US Manufacturing Firms Patent (or Not)." Working Paper, January. Carnegie Mellon University.

Congress of South African Trade Unions (2002). "COSATU Statement on the Importation of Generic Antiretrovirals from Brazil," January 29.
Cook-Deegan, Robert (1995). *The Gene Wars: Science, Politics, and the Human Genome.* New York: W. W. Norton.
Deloria, Barbara, Kristen Foehner, and Sam Scinta (eds.) (1999). *Spirit and Reason: The Vine Deloria, Jr., Reader.* Golden, Colorado: Fulcrum.
Deloria, Vine Jr. (1994). *God Is Red: A Native View of Religion.* Golden, Colorado: Fulcrum.
Deloria, Vine Jr. (2002). *Evolution, Creationism, and Other Modern Myths: A Critical Inquiry.* Golden, Colorado: Fulcrum.
Demaine, Linda J., and Aaron X. Fellmeth (2003). "Natural Substances and Patentable Inventions," *Science* 300, 1375–1376.
De Vries, Barend A. (1998). *Champions of the Poor: The Economic Consequences of Judeo-Christian Values.* Washington, DC: Georgetown University Press.
Diamond v. Chakrabarty, 447 US 303, 309 (1980).
DiMasi, Joseph A., Ronald W. Hansen, and Henry G. Gradowski (2003). "The Price of Innovation: New Estimates of Drug Development Costs," *Journal of Health Economics* 22, 151–185.
Directive 98/44/EC of the European Parliament and of the Council on the Legal Protection of Biotechnological Inventions, 6 July 1998, available at http://europa.eu.int/eur-lex/pri/en/oj/dat/1998/l_213/l_21319980730en00130021.pdf.
Dowie, Mark (2004). "Gods and Monsters," Mother Jones, January/February 2004, available at http://www.motherjones.com/news/feature/2004/01/12_401.html.
Drewal, Henry John, John Pemberton III, and Roland Abiodun, Yoruba (1989). *Nine Centuries of African Art and Thought.* New York: The Center for African Art.
Editorial (2003). "Patenting Pieces of People," *Nature Biotechnology* 22, 341.
Eisenberg, Rebecca S. (2002). "How Can You Patent Genes?," *American Journal of Bioethics* 2, 3–11.
Eisenberg, Rebecca S., and Richard R. Nelson (2002). "Public vs. Proprietary Science: A Fruitful Tension?," *Academic Medicine* 77(Part 2), 1392–1399.
Ekpere, Johnson (2001). "The OAU Model Law and Africa's Common Position on the TRIPS Review Process." International Centre for Trade and Sustainable Development Ad-Hoc Papers, available at www.ictsd.org/dlogue/2001–07–30/Ekpere.pdf.
Eliot, Charles W. (ed.) (online). *The Autobiography of Benjamin Franklin*, available at http://eserver.org/books/franklin/ (visited June 20, 2003). Original publication: New York: P.F. Collier, 1909.
Ernhofer, Ken (2003). "Ownership of Genes at Stake in Potential Lawsuit," *Christian Science Monitor*, February 27.
European Patent Convention (EPC), 5 October 1973, available at http://www.european-patent-office.org/legal/epc/index.html.
European Patent Office Guidelines for Examination, June 2005, available at http://www.european-patent-office.org/legal/guiex/.
Evangelical Lutheran Church in America (2001). "Genetics! Where Do We Stand as Christians?" Chicago, IL:ELCA, available at http://www.elca.org/dcs/genetics.study.html.
Federal Council of the Churches of Christ in America (1908). "Social Creed of the Church."
Fehr, Ernst, and Bettina Rockenbach (2003). "Detrimental Effects of Sanctions on Human Altruism," *Nature* 422, 137–140.
Fehr, Ernst, and Simon Gächter (2003). "Reply: The Puzzle of Human Cooperation," *Nature* 421, 911–912.
Fleischer, Matt (2001). "Pitfalls of Pro Se Patenting," *The American Lawyer*, June.
Foster, Charles (2003). "Current Issues in the Law of Genetics," *New Law Journal* 9, January 10.
Funk Brothers Seed Co. v. Kalo Inoculant Co., 333 US 127 (1948).
Gilbert R, Shapiro C. (1990). "Optimal Patent Length and Breadth," *Rand Journal of Economics* 21, 106–112.
Gilchrist, John (1969). *The Church and Economic Activity in the Middle Ages.* London: Macmillan.

4 Ethical Challenges of Patenting "Nature": Legal and Economic Accounts 269

Glennerster Rachel, and Kremer Michael (2000). "A Better Way to Spur Medical Research and Development," *Regulation* 200, 23(2), 34–39.
Glitzenstein, Kurt L. (1994). "A Normative and Positive Analysis of the Scope of the Doctrine of Equivalents," *Harvard Journal of Law and Technology* 281, 314.
Golden, John M. (2001). "Biotechnology, Technology Policy, and Patentability: Natural Products and Invention in the American System," *50 Emory Law Journal* 101, 110.
Goldstein, Jorge A., and Golod, Elina (2002). "Human Gene Patents," *Academic Medicine* 77, 1315–1328.
Greenberg v. Miami Children's Hospital Research Institute, Inc., 264 F. Supp. 2d 1064 (S.D. Fla 2003).
Grim, John A. (1998). "Indigenous Traditions and Ecology." Retrieved April 10, 2003, from the World Wide Web: http://environment.harvard.edu/religion/religion/indigenous.
Grim, John A. (2001). "Indigenous Traditions and Deep Ecology," in David L. Barnhill and Roger S. Gottlieb, (eds.), *Deep Ecology and World Religions: New Essays on Sacred Grounds*. Albany, NY: State University of New York Press, pp. 35–57.
Hall, Bronwyn H., and Rosemarie Ziedonis H. (2001). "The Patent Paradox Revisited: An Empirical Study of Patenting in the US Semiconductor Industry, 1979–1995," *Rand Journal of Economics* 32, 101–128.
Hanson, Mark J. (1997). "Religious Voices in Biotechnology: The Case of Gene Patenting," *Hastings Center Report* 27(6), 1–22.
Harvard College v. Canada (Commissioner of Patents), 21 C.P.R. (4th) 417 (S.C.C. 2002), at para. 199.
HARVARD/Onco-mouse (T19/90), E.P.O.R. 501 (1990).
Heller, Michael A., and Rebecca S. Eisenberg (1998). "Can Patents Deter Innovation? The Anticommons in Biomedical Research," *Science* 280, 698–701.
Henry, Michelle R., Mildred K. Cho, Meredith A. Weaver, and Jon F. Merz (2002). "DNA Patenting and Licensing," *Science* 297, 1279.
Henry, Michelle R., Mildred K. Cho, Meredith A. Weaver, and Jon F. Merz (2003). "A Pilot Survey on the Licensing of DNA Inventions," *Journal of Law, Medicine and Ethics* 3, 442–449.
Hirschman, Albert O. (1977). *The Passions and the Interests: Political Arguments for Capitalism Before Its Triumph*. Princeton, NJ: Princeton University Press.
Howard Florey/Relaxin, E.P.O.R. 541, 551 (1995).
Hughes, Justin (1988). "The Philosophy of Intellectual Property," 77 *Georgetown Law Journal* 287, 319–324.
Independent Commission on International Humanitarian Issues (1987). *Indigenous Peoples: A Global Quest for Justice: A Report for the Independent Commission on International Humanitarian Issues*. London: Zed Books.
Jacobs, Philippe, and Geertrui Van Overwalle (2001). "Gene Patents: A Different Approach," *European Intellectual Property Review* 23, 505.
Janzen, John (1978). *The Quest for Therapy in Lower Zaire*. Berkeley, CA: University of California Press.
Jenkins, Neil (2001). "The Impact of the EU Biotechnology Directive on the Patenting of Biotechnology," 128 *Patent World* 9, January, 9–10.
Kamien, Morton, and Schwartz, Nancy (1974). "Patent Length and RandD Rivalry," *American Economics Review* 64, 183–187.
Kemf, Elizabeth (ed.) (1993). *The Law of the Mother: Protecting Indigenous Peoples in Protected Areas*. San Francisco, CA: Sierra Club Books.
King, Paul G., and David O. Woodyard (1999). *Liberating Nature: Theology and Economics in a New Order*. New York: Pilgrim Press.
Kirin-Amgen Inc. v. Roche Diagnostics GmbH, R.P.C. 1 (2002).
Klemperer, Paul (1990). "How Broad Should the Scope of Patent Protection Be?," *Rand Journal of Economics* 21, 113–130.
Korn, David (2002). "Industry, Academia, Investigator: Managing the Relationships," *Academic Medicine* 77, 1089–1095.

Kortum, S., and Lerner J. (1998). "Stronger Protection or Technological Revolution: What Is Behind the Recent Surge in Patenting?," *Carnegie-Rochester Conference Series on Public Policy*, 48 247–304.

Krimsky, Sheldon (1999). "The Profit of Scientific Discovery and its Normative Implications," 75 *Chicago-Kent Law Review*, 15.

Lanjouw, Jean O. (1998). "The Introduction of Pharmaceutical Product Patents in India: 'Heartless Exploitation of the Poor and Suffering?'" Working Paper 6366. National Bureau of Economic Research.

Last, Murray (1986). "The Professionalisation of African Medicine: Ambiguities and Definitions," in Murray Last and G.L. Chavunduka (eds.), *The Professionalisation of African Medicine*. Manchester, England: Manchester University Press.

Lerner, Josh (2002). "Patent Protection and Innovation Over 150 Years." Working Paper 8977. National Bureau of Economic Research.

Levin, Richard, Alvin Klevorick, Richard R. Nelson, and Sidney Winter (1987). "Appropriating the Returns from Industrial Research and Development." Brookings Papers on Economic Activity: Microeconomics. Washington, 783–820.

L' Institut Curie, "Révocation par l'Office Européen des Brevets du 1er brevet de Myriad Genetics," http://www.curie.fr/upload/presse/190504_fr.pdf; "Another victory for opponents of patents held by Myriad Genetics: European Patent Office rejects the essential points of BRCA1 gene patents," http://www.curie.fr/upload/presse/myriadpatents310105.pdf.

Locke, John (1952). *The Second Treatise of Government*, in Thomas P. Peardon (ed.). New York: Bobbs-Merrill.

MacLeod, Christine (1988). *Inventing the Industrial Revolution: The English Patent System, 1660–1800*. New York: Cambridge University Press.

Magesa, Laurenti (1997). *African Religion: The Moral Traditions of the Abundant Life*. Maryknoll, NY: Orbis Books.

Marshall, Eliot (1997). "Secretiveness Found Widespread in Life Sciences," *Science* 276, 523–525.

Médecins Sans Frontières (2000). *HIV/AIDS Medicines Pricing Report*.

Merrill, Stephen A., Richard C. Levin, and Mark B. Myers (eds.), For the Committee on Intellectual Property Rights in the Knowledge-Based Economy, National Research Council (2004). *A Patent System for the 21st Century*. Washington, DC: National Academies Press.

Merz, Jon F. et al. (1998). "Patenting Genetic Tests: Putting Profits Before People," *Gene Watch* (October), Cambridge, MA: Council for Responsible Genetics.

Merz, Jon F., Antigone G. Kriss, Debra G.B. Leonard, and Mildred K. Cho (2002). "Diagnostic Testing Fails the Test: The Pitfalls of Patenting are Illustrated by the Case of Haemochromatosis," *Nature* 415, 577–579.

Moore v. Regents of the University of California, 51 Cal.3d 120 (Cal. 1990).

Murashige, Kate (2002). "Patents and research–an uneasy alliance," *Academic Medicine* 77(Part 2), 1329–1338.

Ngu, Victor (1997). "The Viral Envelope in the Evolution of HIV: A Hypothetical Approach to Inducing an Effective Immune Response to the Virus," *Medical Hypothesis* 48, 517–521.

Ngubane, Harriet (1977). *Body and Mind in Zulu Medicine*. New York: Academic.

NIH (National Institutes of Health) (2001). Report to the US Congress: A Plan to Ensure Taxpayers' Interests Are Protected, July.

NIH (National Institutes of Health) (2004). Report to Congress on Affordability of Inventions and Products, July 2004, available at http://ott.od.nih.gov/NewPages/211856ottrept.pdf.

Nuffield Council on Bioethics (2002). The Ethics of Patenting DNA (June), available at http://www.nuffieldbioethics.org/go/ourwork/patentingdna/introduction.

Nyamiti, Charles (n.d.). *African Tradition and the Christian God*. Eldoret, Kenya: Gaba, Spearhead, No. 49.

Nyamiti, Charles (1973). *The Scope of African Theology*. Kampala, Uganda: Gaba.

Olupuna, Jacob K. (ed.) (1991). *African Traditional Religions in Contemporary Society*. New York: Paragon House.

4 Ethical Challenges of Patenting "Nature": Legal and Economic Accounts 271

OECD (Organisation for Economic Co-Operation and Development, Genetic Inventions, Intellectual Property Rights and Licensing Practices) (2002). *Evidence and Policies.* Paris: OECD.
Panel on Bioethical Concerns (1984). National Council of the Churches of Christ/USA, *Genetic Engineering: Social and Ethical Consequences.* New York: Pilgrim Press.
Pérez, Carmen, and Pascale Boulet (2000). *HIV/AIDS Medicines Pricing Report. Setting Objectives: Is There a Political Will?* Geneva: Médecins Sans Frontiéres.
Peters, Ted (1997). *Playing God? Genetic Determinism and Human Freedom.* New York: Routledge.
Peterson, Alan, and Robert Bunton (2002). *The New Genetics and the Public's Health.* New York: Routledge.
Peterson, Kris, and Olatubosun Obileye (2002). "Access to Drugs for HIV/AIDS and related Opportunistic Infections in Nigeria," Policy Project, Nigeria.
Pollack, Andrew (2003). "University's' Drug Patent is Invalidated by a Judge," *New York Times*, March 6.
Pontifical Academy for Life (2003). "Concluding Communiqué on the 'Ethics of Biomedical Research for a Christian Vision," 24–26 February, http://www.vatican.va/roman_curia/pontifical_academies/acdlife/documents/rc_pont-acd_life_doc_20030226_ix-gen-assembly-final_en.html.
President's Council on Bioethics (2002). Transcript of Fifth Meeting, July 11, Session 2: Regulation 5: Patentability of Human Organisms, available at http://bioethicsprint.bioethics.gov/transcripts/jul02/session2.html.
President's Council on Bioethics (2004). "Reproduction and Responsibility: The Regulation of New Biotechnologies," March, available at www.bioethics.gov, 157–163.
Pufendorf, Samuel (1994). *Political Writings 177.* Michael J. Seidler (trans.), Craig L. Carr (ed.). New York: Oxford University Press.
Raghavan, Chakravarthi (1999). "Protecting IRPs of Local and Indigenous Communities," October 12. Third World Network. Retrieved April 10, 2003, from the World Wide Web: http://www.twnside.org.sg/title/local-cn.htm.
Raghavan, Chakravarthi (2000). "ASEAN for Protecting Indigenous/Traditional Knowledge," May 5. Third World Network. Retrieved April 10, 2003, from the World Wide Web: http://www.twnside.org.sg/title/asean.htm.
Rai, Arti K. (2002). "Genome Patents: A Case Study in Patenting Research Tools," *Academic Medicine* 77, 1368–1372.
Rai, Arti K., and Rebecca S. Eisenberg (2003). "Bayh-Dole Reform and the Progress of Biomedicine," 66 *Law and Contemporary Problems*, 289.
Rasmussen, A. Terrill (1965). *Christian Responsibility in Economic Life.* Philadelphia, PA: Westminster Press.
Rawls, John (1971). *A Theory of Justice.* Cambridge: Harvard University Press.
Reichard, Gladys A. (1950). *Navaho Religion: A Study of Symbolism.* Princeton, NJ: Princeton University Press.
Resnik, David B. (2001). "DNA Patents and Human Dignity," *Journal of Law, Medicine and Ethics* 29, 152–165.
Richardson, John R. (1966). *Christian Economics: Studies in the Christian Message to the Market Place.* Houston, TX: St. Thomas.
Rose, Carol M. (2003). "Romans, Roads, and Romantic Creators: Traditions of Public Property in the Information Age," 66 *Law and Contemporary Problems*, 89.
Rose, Mark (2003). "Nine-tenths of the Law: The English Copyright Debates and the Rhetoric of the Public Domain," 66 *Law and Contemporary Problems*, 75.
Sakakibara, M. Branstetter L. (2001). "Do Stronger Patents Induce More Innovation? Evidence from the 1998 Japanese Patent Law Reforms," *Rand Journal of Economics* 32, 77–100.
Schaffner, Joan E. (1995). "Patent Preemption Unlocked," 1995 *Wisconsin Law Review* 1081, 1088.
Scherer, F.M. (2002). "The Economics of Human Gene Patents," *Academic Medicine* 77, 1348–1367.

Scherer, F.M., and Sandy Weisburst (1995). "Economic Effects of Strengthening Pharmaceutical Patent Protection in Italy," *International Review of Industry Property and Copyright Law* 6, 1009–1024.

Schertenleib, Dennis (2003). "The Patentability and Protection of DNA Based Inventions in the EPO and the European Union," *European Intellectual Property Review* 124, 127.

Schissel, Anna, Jon F. Merz, and Mildred K. Cho (1999). "Survey Confirms Fears about Genetic Tests," *Nature* 402, 118.

Schwartz, M. Trudelle (1997). *Molded in the Image of Changing Woman: Navajo Views on the Human Body and Personhood*. Tucson, AZ: University of Arizona Press.

Schweitzer, Stuart O. (1997). "Patent Protection," *Pharmaceutical Economics and Policy*, 195–206.

Scotchmer, S. Green J. (1990). "Novelty and Disclosure in Patent Law," *Rand Journal of Economics* 21, 131–146.

Setiloane, Gabriel (1976). *The Image of God Among the Sotho-Tswana*. Rotterdam, The Netherlands: A. A. Balkema.

Sharp, David (2004). "Not-for-Profit Drugs—No Longer an Oxymoron?," *Lancet* 364, 1472–1474.

Slater, Dashka (2002). "Humouse™," *Legal Affairs*, November/December, available at http://www.legalaffairs.org/issues/November-December-2002/feature_slater_novdec2002.html.

Smith, Claire, Heather Burke, and Graeme Ward (2000). "Globalisation and Indigenous Peoples," in Claire Smith and Graeme K. Ward (eds.), *Indigenous Cultures in an Interconnected World*. St. Leonards, NSW, Australia: Allen & Unwin.

Solovitch, Sara (2001). "The Citizen Scientists," *WIRED Magazine* 9.09, September.

Steele, Richard (1684). *The Tradesman's Calling, being a Discourse concerning the Nature, Necessity, Choice, etc., of a Calling in general*. London.

Stevens, Stan (ed.) (1997). *Conservation Through Cultural Survival: Indigenous Peoples and Protected Areas*. Washington: Island Press.

Stone, Richard (1995). "Religious Leaders Oppose Patenting Genes and Animals," *Science* 268, 1126.

Tangwa, Godfrey (2002). "The HIV/AIDS Pandemic, African Traditional Values and the Search for a Vaccine in Africa," *Journal of Medicine and Philosophy* 27(2), 217–230.

Tauli-Corpuz (online). Biotechnology and indigenous peoples. Third World Network. Retrieved April 10, 2003, from the World Wide Web: http://www.twnside.org.sg/title/tokar.htm.

Tawney, Richard H. (1926). *Religion and the Rise of Capitalism*. London: John Murray.

Tawney, Richard H., and Eileen Power (eds.) (1924). "The Debate on Monopolies in the House of Commons, 1601," *Tudor Economic Documents*, vol. 2 of 3, 269–292. London: Longmans, Green.

Thomas, Sandy M., Michael M. Hopkins, and Brady, Max (2002). "Shares in the Human Genome—The Future of Patenting DNA," *Nature Biotechnology* 20, 1185–1188.

Travaglini, Arcangelo (2001). "Reconciling Natural Law and Legal Positivism in the Deep Seabed Mining Provisions of the Convention on the Law of the Sea," 15 *Temple International and Comparative Law Journal*, 313, 318.

Troeltsch, Ernst (1931). *The Social Teachings of the Christian Churches*, vol. 1, Olive Wyon (trans.). New York: MacMillan, pp. 115–117.

Troeltsch, Ernst (1986). *Protestantism and Progress: The Significance of Protestantism for the Rise of the Modern World*. Philadelphia, PA: Fortress.

Tuma, Rabiya S. (2003). "AAAS president's Surprise Plea for Low-tech Research," *BioMedNet News*, 13 February.

United Church of Christ Pronouncement on Christian Faith (1989). *Economic Life and Justice*, passed by General Synod 17 of the United Church of Christ in July.

University of Rochester Medical Center (2000). "University of Rochester Awarded Patent for New Class of Drugs Known as Cox-2 Inhibitors; Files Infringement Suit Against Searle, Pfizer," April 12, available at http://www.urmc.rochester.edu/Cox-2/pr.html.

USDHHS (US Department of Health and Human Services, National Institutes of Health) (1999). "Principles and Guidelines for Recipients of NIH Research Grants and Contracts on Obtaining

4 Ethical Challenges of Patenting "Nature": Legal and Economic Accounts

and Disseminating Biomedical Research Resources," 64 *Federal Register* 72090 (December 23, 1999).
USDHHS (US Department of Health and Human Services, National Institutes of Health) (2004). "Best Practices for the Licensing of Genomic Inventions," 69 *Federal Register* 67747 (November 19), at 67747–67748.
USGAO (U.S. General Accounting Office) (2002). "Intellectual Property: Federal Agency Efforts in Transferring and Reporting New Technology," GAO-03-47, October, at 5–7.
US Office of Technology Assessment (1987). "New Developments in Biotechnology: Public Perceptions of Biotechnology," May.
US Patent and Trademark Office (2001). "Utility Examination Guidelines," 66 *Federal Register* 1092, January 5.
Van Niekerk, Anton A. (2001). "Moral and Social complexities of AIDS in Africa," *Journal of Medicine and Philosophy* 27(2), 143–162.
Virus Weekly (2002). February 19–February 26, 2002.
Walterscheid, Edward C. (1998). *To Promote the Progress of Useful Arts: American Patent Law and Administration, 1787–1836*. Littleton, CO: Fred B. Rothman.
Warren, Dennis M. (1974). "Disease, Medicine and Religion Among the Techiman-Bono of Ghana: A Study in Cultural Change," Ph.D. dissertation, University of Indiana, Bloomington.
Waterson, Michael (1990). "The Economics of Product Patents," *American Economics Review* 80, 860–869.
Williams-Jones, Bryn (2002). "History of Gene Patent: Tracing the Development and Application of Commercial BRCA Testing," 10 *Health Law Journal*, 123.
Working Group on Bioethics and Biotechnology, European Ecumenical Commission for Church and Society (1996). "Clarification of the Submission on the EC Draft Patenting Directive," November 5, available at http://www.srtp.org.uk/eeccpt12.htm.
Wright, Susan, and David A. Wallace (2000). "Varieties of Secrets and Secret Varieties: The Case of Biotechnology," *Politics and the Life Sciences* 19(1), 45–57.
Young, Allan (1975). "Magic as a Quasi-Profession: The Organization of Magic and Magical Healing Among Amhara," *Ethnology* 14, 245–265.

Chapter 5
Technogenesis: Aesthetic Dimensions of Art and Biotechnology

Suzanne Anker, Susan Lindee, Edward A. Shanken, and Dorothy Nelkin

> ...*since the eye is the most perfect among the exterior senses, it moves the mind to hatred, love and fear.*
>
> (Ravenna, 1587, in Freedberg, 1989, 34)[1]

From material processes to elusive patterns, artists and scientists seek models of explanation (Kemp, 2000). Sometimes illusionally evocative, sometimes rigorously formulaic, and at other times sculpturally bounded, these conceptualizing tools have historically linked art and science. Bring to the fore new technologies, digitally driven, and a vast array of alternative schemes become possibilities. High resolution images of cells, scanned helical DNA structures and synaptic neural connections can presently be viewed in real time. Add to the mix embodied transgenic life forms and fabricated animal models, and our conceptualizing tools expand the possibilities for dimensional invention.

The accelerating dynamic between cultural and genetic evolution produces what can be termed a co-evolution between technical knowledge and living matter. And it is this co-evolution between technical expertise and animate matter we term *technogenesis*.[2] In other terms, *technogenesis* is the way in which the interactions between technology and biology impact our understanding of how nature exists, or would be, conceived and reconfigured in the future.

But how do art practices and the life sciences rely on the efficacy of images? And what part do these images play in the acquisition, comprehension, dissemination and even funding of visual or scientific study? In what ways do images reflect the socio/economic and cultural conditions of producing knowledge? Located somewhere between illusion, proof and cognitive projection, images, hence, become critical fictions operating within the cultural imaginary. They often traverse contested territories situated elsewhere on the axis between fact and fiction. These visualizing models,

---

[1] Quoted from G.B. Armenini, *De 'veri precetti della pictura* (Ravenna, 1587. In Freedberg, 1989, 34).

[2] For other references and meanings of the term *technogenesis* see Walby (2000); also see Mitchell (2002).

ubiquitously employed by artists, scientists, designers, corporate advertisers, journalists and politicians, clarify, mislead, aggrandize, stimulate and document. In brief, they are representations embedded in social structures, policy decisions and commercial ventures. As aesthetic devices such images perform their semiotic function activating thought and emotion by their salient powers of communication and circumscribed belief (Anker, 2004)

Since the mid-19th century the practice of art has repeatedly shifted its focus away from beauty as its primary and defining attribute. Whereas beauty was once an encompassing characteristic inherent in works of art, other aesthetic strategies have arrived to displace that domain. Most markedly, the incorporation and recontextualization of extant material, through the structural principle of "cut and paste," operates in unique ways to construct and alter meaning. Although originating as an early 20th-century modernist technique in picture-making, an enlarged concept of collage can readily include TV news reports, newspaper design and the Internet. In each case continuous time and contiguous space are diced and spliced to form a new representational space. Currently, this methodology is being applied to life forms in the lab, as gene sequences, transferred between discrete species, create novel combinatory possibilities of altered living matter.

With the dominant ascendancy of modern art, non-traditional materials have often been employed as well. Body fluids, such as urine, excrement, blood, semen and even harvested human ova have found their way into the illusive and expanding boundaries of artistic practice. Tissue engineering and scaffolding, fabricated laboratory animals and transgenic plants, have become, *ipso facto*, material resources for artists. Concrete instantiations of cells, bacteria and other microscopic entities appear in art galleries, museums, and international festivals as works of art. And the body itself is often engaged as a sculptural medium, as a site of aesthetic and protoplasmic investigation, in venues of performance art.

Whereas contemporary scientific iconography frequently aspires to and achieves aesthetic notions of beauty, for contemporary artists, the aesthetic dimension serves various manifold goals. For artists the ethical pillar of beauty as a marker of belief has been replaced by metacritical commentary and interdisciplinary means of thinking. These systemic modes for generating meaning also intersect with photography, video, telecommunications and a wide range of 2D and 3D design tools. In what follows, we explore more recent art practices involving the ways in which DNA and other bio-matter metaphorically and literally become part of the artist's palette. By suggesting some of the consistencies in the ways these images and life forms present, interpret or embody DNA, we elucidate the broader cultural meanings of innovative forms of bio-technological intervention.

As we consider the ethical implications of the new biotechnologies, it is essential to recognize social effects as values in practice, which are, furthermore, shaped by consensual beliefs and expectations. Developing a comprehensive grasp of the potential ethical dilemmas inherent in biotechnology demands that we take seriously all assumptions. Visual art and visual culture provide a domain – a nominal space – where ideologies and premises are likely to be more freely expressed than in concentrated scientific discourses. Visual practices have the proactive power to influence

5 Technogenesis: Aesthetic Dimensions of Art and Biotechnology

and signify in multiple directions; they profoundly affect the enterprise of science, its public reception and consumption. Building on a tradition of texts and exhibitions investigating intersections between art, technology, and the life sciences, especially since the early 1990s, unprecedented cultural investment has been made in this theme. Known under the rubric of "sci-art" or "bio-art", recent events have included exhibitions in art and science museums, art galleries, the media and advertising worlds, and have even found resonance in postage stamps (Figs. 5.1a and b).[3]

No single visualization, whether image or embodied living form, can provide ultimate insight into public, scientific and cultural assumptions. However, cumulative repeated representations across several scholarly and cultural fields will suggest

Fig. 5.1a Israeli stamp, 1964 (Unable to obtain permission. There are instances where we have been unable to trace or contact the copyright holder. If notified the publisher will be pleased to rectify any errors or omissions at the earliest opportunity.)

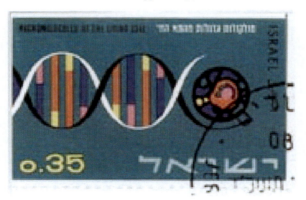

Fig. 5.1b Royal mail stamp (Peter Brookes, 1993) (Unable to obtain permission. There are instances where we have been unable to trace or contact the copyright holder. If notified the publisher will be pleased to rectify any errors or omissions at the earliest opportunity.)

---

[3] In the UK, the National Centre for Biotechnology Education maintains an extensive website containing images and other documentation of ephemera related to DNA. This is and extraordinary resource for further research on the influx of genetic science into popular culture through consumer products and visual culture.
See http://www.ncbe.reading.ac.uk/DNA50/ephemeral.html.

some broad tensions, concerns and ultimately beliefs. Some of these representations and performative practices are produced by artists critiquing, commenting upon and/or incorporating biotechnologies into the spectrum of their work. Other images, which we address in this paper, are advertisements of marketing campaigns and scientific product fairs which provide further insight into subliminal social and corporate messages.

Scientists as well join the aesthetic/genetic fray when they engage in manipulating their visual data as a means to enhance, communicate and persuade.[4] Is the best scientific image a reflection of the most profound science? How does aesthetic choice determine the importance of scientific data? Many scientists, enchanted by with new imaging technologies, refer to their microscopic visualizations as "art." And prestigious, peer-reviewed science journals compete with each other for the most attractively compelling cover image. Since the prevalent use of Photoshop and its allied software technologies, "seeing is believing" is no longer a truism. The role of visualization practices in scientific, cultural and artistic domains requires critical examination of the ways and means knowledge is produced and assessed in each area of study.

In addition to visualizations or bio-technological interventions into living forms, in the next section, we introduce molecular biology as a modernist science. Using literary tropes and biblical narratives, we look at some of the rhetorical contingencies associated with the current biological sciences. How do structural metaphors in language reflect cultural assumptions with regard to altering nature?

5.1 The Special Status of DNA

The ongoing dialogues supporting the genetic sciences conceptualize the body as a set of interacting signals. This abstract system of molecular data, comprising the mechanisms of inheritance, has in addition to its scientific data unraveled a multiplicity of molecular metaphors and rhetorical tropes. Although the cell contains approximately 500,000 molecules, DNA has been afforded unique status. Referred to by Martin Kemp as the "Mona Lisa" of molecules, DNA has achieved celebrity eminence as a recognizable and formidable popular icon. According to cultural critic Chris Rojek, the magnetic force of celebrity in a media-driven society, displaces religion as a supernatural power (Rojek, 2001, 51–100). Has the double helix, the celebrity molecule, become a secular visualization of divine presence? Is DNA, a supermolecule, a messenger that, like the archangel Gabriel, announced the word/spirit of the Christian God's immaculate conception? To what extent do the discoveries of genetics produce only scientific knowledge that aspires to unveil the mysteries of life but also visual symbols, metaphors and narratives that bring to bear human hopes and desires formerly associated within the provenance of religion?

---

[4] Felice Frankel works with scientists at MIT teaching them methods of designing more provocative and appealing images as visual data for their scientific articles. See Frankel (2002).

From genesis to gene, is the creation story embedded within this popular scientific icon?

In 1953, James Watson hailed DNA as "the most golden of molecules." In 2003, the discovery of the structure of DNA celebrated its 50th anniversary, its golden anniversary. Metaphorically speaking, gold has been the quarry of alchemists. Transmuting the profane into the sacred, turning lead into gold, one could catch a glimpse of the immortal. Ancient rulers, transfixed by their own immortality desires, looked towards the luminosity of this precious metal and its ability to resist decay, as a clear sign that something can last forever. Previously functioning as a standard of measure in the monetary system, gold was the agreed upon agent of exchange and everlasting value. DNA too, is a precious substance, a molecule of immortality in which genes are the archeological evidence of bio-historical ancestry, family identity and genealogical connections. Its magical powers are evidenced by its infinite and eternally driven self-replication and perpetuation.

As artist Larry Miller's *Genetic Code Copyright Certificates* (c. 1989) (Fig. 5.2) provocatively asks, in this golden age of biology, does my DNA belong to me? If not, to whom does it belong? Is my unique code equivalent to my being, spirit or elsewise? Or is it a natural resource, a family trust or an enzymatic recipe for surveillance? Is it a sacred secret of life or a deed to a body-part farm? Is it the book of life or the Holy Grail? Each description of DNA brings with it a range of ideological positions and emotional responses ranging from the awesome to the awful. So how can we measure the discordant meanings of this molecule and its impact on our personal lives, cherished values and as-yet-to-be clarified assumptions? (Anker, 1996, 90).

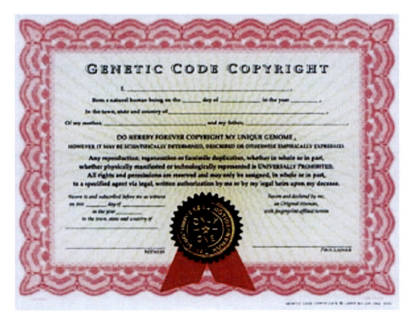

Fig. 5.2 *Genetic Code Copyright Certificate*, 1992, by Larry Miller; printed with permission from the artist

5.2 The Book of Life: Religious Metaphors and DNA

The history and ongoing developments in art reflect changing worldviews of religious and spiritual practice. Beginning as a magical-religious ritual, images created as icons or symbols were imbued with totemic value. They affectively operate as sensual models of thoughts and feelings, bringing to visualization the signification processes inherent in any symbolic order.

As identity becomes further linked with DNA as an intricate harbinger of coded, yet mutable meanings, symbolic inferences embedded within this "golden molecule" are being parsed out. In rhetoric as well, coded meanings are examined and find resonance with a broad range of religious texts. "In the beginning was the word, and the word was made flesh," a radical transcription of thought (and God), as postulated in Genesis (Shlain, 1999). In a post-genomic world, however, our mortal flesh has become grounded in a steam of nucleotide sequences represented as well by the alphabetical signs, namely A, G, T and C.

As early as 1992, Richard Lewontin, in "The Dream of the Human Genome," (a review of *Code of Codes: Scientific and Social Issues in the Human Genome Project*), identified an underlying religious narrative present in a collection of essays on molecular biology edited by Daniel Kevles and Leroy Hood (Lewontin, 1992). He states that "molecular biology is now a religion, and the molecular biologists are its prophets." Research scientists employ such metaphors when they describe molecular genetics as "the code of codes" or "the book of life", giving a quasi-religious overtone to DNA's molecular script (Kevles and Hood, 1992). For nucleic acid chemist Erwin Chargaff, "the double helix has replaced the cross in the biological alphabet" (Kevles and Hood, 1992, 83–97). Ian Wilmut's "The Second Genesis" invokes cloning as the Second Coming (Alexander, 2003). But how are these metaphors to be interpreted in light of the science and culture wars? Do these metaphors equate religion and science for a general audience? Are they helpful in bringing a clearer understanding of molecular biology to the public? Or, is this rhetoric only a partial and incomplete analogy of grandiose proportion?

There are myriad interpretations with regard to the influx of religious meanings into genetic iconography's place in narrative, popular culture, mass media and works of art. Molecular biologist Lynn Petrullo (2000) cites the conflicting interpretations of this science in the public domain. Dorothy Nelkin and Susan Lindee "primarily see the image of DNA as denoting genetic essentialism," a cultural message in which the genes are perceived as primarily deterministic (Nelkin and Lindee, 2004). For Bryan Appelyard, "holistic genetic thinking is a new metaphor of the sacred," underscoring the universality of life (Appelyard, 1998). The multiple meanings reflected in images of DNA are denominators of the varied and alternative ideological positions that genetic imagery can elicit. These variegated interpretations need not be mutually exclusive. Currently, conventional notions of reductionism are being expanded and nuanced to address questions concerning human identity, sacred beliefs, the practice of medicine and the politics of science.

In *The Religion of Technology*, David F. Noble discusses a dualism evident in millennium thinking: the rise of fundamentalist faith on the one hand, and an unprecedented embrace of technology on the other (Noble, 1997). Evidenced by this intense shift in geopolitical configurations between Western traditions and those theocratic societies currently operating under Islamic hegemony, American politics has in large measure usurped the "word" of God as a moral and political imperative. Simultaneously, Western cultures have, over time, integrated accelerating technologies (and their corresponding impact on society) into their probable ways of life and social organization. Alternatively, various other cultures in the world have not as yet experienced such technological "progress" as an aspect of modernity. Technological intervention, growing exponentially on all fronts in modern society, is, at least for the time being, an amalgamation of tools and techniques brought to the fore by Enlightenment thinking. But cultures throughout the world continually exist at various levels of development, industrialization and secularization.

Historically, religion's relationship to technological and iconographic power stimulated a wide range of theoretical issues culturally embedded in various theologies. St. Augustine in *City of God* warns against the dangers created by magic or illusion. In Book 21, Chapter 6 he states:

> But as to those permanent miracles of nature, whereby we wish to persuade the skeptical of the materials of the world to come, those are quite sufficient for our purposes which we ourselves can observe or of which it is not difficult to find trustworthy witnesses.[5]

In addition, in *De doctrina Christiana* Augustine takes heed of the falsely seductive qualities of beautiful rhetoric as a mechanism than confuses truth and the gospel.

By the 9th century, John Scotus Eriugena (John the Scot) calls upon the mechanical arts, including medicine and art, as a necessary part of the human restoration process. In their capacity to augment humans' inferior position to God, technology began to be seen as a way to recover Adam's fall from grace (Noble, 1997, 9–20). Contrasting the term "mechanical arts" from the "liberal arts" which included grammar, rhetoric, dialectic, arithmetic, astronomy and music, the medieval thinker states that the mechanical arts arise from "some imitation of human devising." Such a statement acknowledges technology as a restorative aid to the humble status of fallen beings. This sort of thinking becomes the groundwork for the Renaissance revival in which man became the measure of all things. Thus for Noble, "technology has come to be identified with transcendence, implicated as never before in the Christian idea of redemption" (Noble, 1997).

A further example of this attitude is demonstrated through the study of optics and optical analogy. During the Middle Ages, "the whole science of optics influences Christian thought," states Renaissance art historian Samuel Edgerton. Edgerton's history of optics begins in ancient Greece where optics was a "branch

---

[5] New Advent version of *City of God*. Many thanks to Christina L.H. Traina for this citation.

of Euclidean geometry." He goes on to explain that "in the 12th century the Greek word, *optika* was Latinized as *perspectiva*, meaning to 'see through.' Christian monks, fascinated by this science, could for the first time reveal how the mind of God works: how his divine grace transmits, just like light rays from an illuminating source."[6]

Whether through a prism or an alphabet, a gel or a painting, the tools of our invention make and remake our beliefs about what it means to be alive in all its diverse dimensions.

5.3 Typological Structure in Religion and Science

Further intersections and comparisons can be made between science, technology and religion within literary, biblical and scientific narratives. Mythical plots become working metaphors for bringing science and religion together in textual compositions (van Dijck, 1998). Literary critic Northrop Fry considers the Christian Bible a narrative, a form of literature, explicating through tropes and rhetorical devices the myths of the human condition. Fry's typologies are employed as analytic tools, systemic keys to the understanding of literary genres. Viewing the Bible as a work of literature in which plots and stories unfold, he perceives the Messiah as a hero and traces his development: "He enters the physical world at his Incarnation, achieves his conquest of death and hell in the lower world after his death on the cross. Then as noted he reappears in the physical world at his Resurrection and goes back into the sky with Ascension" (Fry, 1982, 174–176).

The story of Christ's death and resurrection can be thought of as an intrinsic cycle, because in Fry's words "however important for man, it involved no essential change in divine nature itself" (1982, 174–176). In Fig. 5.3a the viewer follows Christ's circular trajectory through incarnation and ascension into heaven. He travels through birth, death and finally rebirth as resurrection. What is relevant to our discussion is that Christ's transformation remains constant and complete, a transformation enveloped within everlasting and divine status. The structure and function of the Messiah become one. His actions collapse any distinction between form and content since only a Messiah can ascend through cycles of rebirth.

An analogous, yet secular process operates within the cell cycle (Fig. 5.3b). To construct or repair the host organism, a cell moves through a slow mitotic cycle. The so-called M phase leads to the G1 phase which leads to the S phase (where DNA synthesis takes place). From 6–10 hours later, the entire genome has been replicated and the cell goes into the G2 phase, that is the time it needs to prepare for mitotic division. In this constant process of renewal, the cycle of life is born and reborn through repetitive and structural action (de Duve, 1984, 319). The mechanism

---

[6] Conversation with Professor Samuel Edgerton, Professor of Art History at Williams College, Spring 2002. For further information see Edgerton (1975).

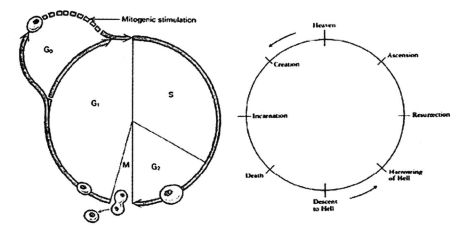

Fig. 5.3 a *Cell Cycle*, b *Ascension to Heaven* (Unable to obtain permission. There are instances where we have been unable to trace or contact the copyright holder. If notified the publisher will be pleased to rectify any errors or omissions at the earliest opportunity.)

of the cell cycle, like the Christian narrative, moves through the life processes of birth, death and rebirth inextricably linking form to function in a transfiguration tale (Anker, 1997).

5.4 The Birth of Molecular Biology

Whereas religious belief and artistic practice have ancient roots in civilizations, molecular biology is a modernist science. Coalescing as a scientific discipline during the late 1930s, molecular biology became a formal area of study during the 1940s and 1950s. The term itself, coined in 1938 by Warren Weaver of the Rockefeller Foundation, emphasized the minuteness of biological entities. It was an interdisciplinary fusion that would borrow methods from embryology, genetics, physiology, immunology and microbiology, heralding the biological sciences into "Big Science's" establishment (Kay, 2000). Physicists, now drawn to biology, began looking at the molecular machinery of the cell on the microscopic level. Could life itself operate under the "laws of the universe" as explicated by chemistry and physics? (Reichle, 2001).

In physicist Erwin Schrodinger's seminal text "What Is Life?" (1943), he identifies the chromosome as "the law code and executive power" of cellular structures. The implications of this metaphor reframe the cell as an entity that could be decoded, or deciphered, like any other information system (Anker and Nelkin, 2004). Perceiving the body as a decipherable text and information system, scientists increasingly employed linguistic tropes and communication models to describe the cell's molecular organization. Referring to the action of the gene, chemical reactions were considered to be transcriptions and translations, in a network of stored information. Converting the body from considerations of flesh and blood into communicating signals of microscopic parts recasts questions concerning the integrity of the body, its organs, tissues, cells, and molecules.

In this next section, we explore the characteristic manners in which contemporary visual artists picture the biological and genetic sciences. Drawing on visual metaphor and art's iconographic history, the following depictions speak to the cultural dimensions of DNA. Visual artists have always been interested in the tools, technologies and picturing devices of their day. The genetic sciences, for the contemporary artist, continue modern art's directive of exploring and making visible the invisible.

5.5 Artists Picture DNA: A Sampler

Picturing DNA, avatar of 20th century molecules, to a general audience is an act of blind faith. Rendered exclusively through instrumentation and intellectual grasp, this subdivisible entity is the matrix of all known life. But aside from scientific data, the optical gaze and its attendant pictorial signs bring into focus a secondary set of propositions: the social, cultural, and aesthetic dimensions bounded by this molecule.

During the late 1980s and early 1990s numerous art exhibitions related to the "body" were mounted, as the corporeal self was being scrutinized in terms of race, class and gender. Identity and its politics became key issues of discourse in contemporary art and a handful of artists began investigating genetics, its iconography, and mechanisms within a nature/culture dialogue. Additional focus came from an increasing awareness of digital technologies and the role they would play in our future, as in William Gibson and Bruce Sterling's fictional accounts of cyberspace. In addition, a growing Internet access, accelerated improvements in and reduced costs of computing speed and data storage along with powerful (user-friendly) software brought the virtual world into consumer's reach. At the same time, Photoshop, CAD 3D modeling programs, and intricate computer games delivered futuristic devices and digital tools to the artist's studio. Spurred by this convergence between concepts of a changing corporeality and the influx of personal computers and "new media" apparatus, a growing number of artists began working in biotechnology-inspired domains.

As employed by artists, sources for genetic, scientific and medical iconography are variegated. Extant material is often appropriated from visualizations in journals, textbooks or from the world-wide-web. Additionally, images are directly obtained from living (and dead) matter as visual artists increasingly work within the context of a scientific laboratory. Some images may, in fact, be rescued from personal charts of medical procedures and the like, such as X-rays, MRI printouts, sonograms, and endoscopic portraits. Many of these representations focus on diverse and differing levels of cellular and molecular organization. Add to this molecular image-bank, types of tools and techniques such as fluorescent markers, gels, scanning or atomic force microscopes, and what results is a full array of mediated pictorializations producing the body as more transparent than ever.

Ranging from the double helix to the chromosome to the gene, manifold images of genetic and cellular structures can be visually depicted or computationally mapped. Visualizations for the scientist are often employed as documents of sound scientific methods, an end point of empirical verification and authentic repeatable results. Images for the artist, genetic and otherwise, however, are starting points for the production and reception of inter-subjective communication. By compounding through multiple meanings and overlayed metaphorical tropes, the recontextualization of scientific imagery repurposes scientific icons to other ends. Polysemous in nature, visual art's interrogative status is articulated by its embodiment of human emotion and thought.

Representation carries with it a wide host of cultural assumptions, ideological constructions and subjective interpretations. What images mean, to whom and when, is part of dialogues within art history, literary theory and semiotics. Referring to images as pictorial signs, differences in modes of representation can be characterized using Charles Saunders Peirce's nomenclature of icon, index and symbol. The icon informs through mimetic resemblance. Like looking in a mirror, it is isomorphic in nature. The index, or trace, points to a cause and effect relationship, based on the manner in which one entity acts upon another. This type of representation is not a visual equivalent, but a logical connector causally calculated. For example, the residual tanning marks made from a bikini onto the skin of a young coed pictures the action of the sun. The third type of sign is the symbol which denotes culturally constructed meaning that is agreed upon by members of a social group. Examples include color coded gender preferences or fashion choices.

As the current century's iconic molecule, DNA has moved out of its confines in the scientific laboratory and into the social space of the world of art. As artists respond to their cultural milieu, images of the genetic sciences take their place along with other icons of popular culture, namely the double helix. In *Code Noah* (1988) (Fig. 5.4), for example, British sculptor Tony Cragg creates a helical spiral of industrially manufactured teddy bears, cast in golden bronze. As a molecular architecture playfully arranged, the artist transposes the microscopic molecule into a larger than life construction. Have the naturally and culturally constructed aspects of matter come to be interchangeable parts as the building blocks of life? Is DNA to be endlessly reconfigured at will as part of the utopian dream of biotechnology's status in a consumer society? Is DNA a new toy, an organic erector set, in which the possibilities of diverse industrial design lay dormant? Cragg's work underscores the multi-coded nature of visual art which brings to the fore the possibilities of its myriad interpretations. Unlike advertising or science, meaning is not targeted to a specific message, but instead resides in the poetic resonance of mutual possibilities. Within works of art, contradictory meanings can, and often do, share the same space.

The chromosome too, has moved out of the laboratory and into social space. Acutely recognizable by the public-at-large, the chromosome is heredity's ancient bio-marker.

Fig. 5.4 *Code Noah*, 1988, by Tony Cragg; courtesy: Marian Goodman Gallery, New York

To visualize this microscopic entity, a laboratory technician photographs cells in culture during their metaphase period, the phase in which chromosomes visually emerge. The lab worker's next task is to match identical pairs of chromosomes according to size and shape thus creating an artificial arrangement known as a karyotype. Congruent with an ideogram or a shorthand language, this synthetic assembly maps such biological characteristics as species or gender, among others. Additionally this technique is employed to examine differences in identity among divergent life forms, such as for example, a bluefish, a petunia or *E. Coli* (Anker, 1996). Like a form of primitive writing, these lexicons are the body's system of writing itself, witnessed by the way of a magnified instrumentalized vision.

Since 1990, Suzanne Anker has been incorporating images of chromosomes in her works of art. Intent on exploring the ways in which notations and experimental practices in science can be transferred to cultural spaces, Anker's work embeds the chromosome and its attendant optical apparatus into art historical parlance.

Art historian and curator Frances Terpek cites Robert Hooke in discussing the role of optics in Anker's sculptural installation *Zoosemiotics* (1993) (Fig. 5.5):

> In 1665, when Robert Hooke, director of experiments for the Royal Society in London, published his account of how to use a "round Globe of Water" to focus light, he was redeploying a device that had been used long before his time as a rudimentary instrument for magnifying objects. Anker's *Zoosemiotics* employs both these functions. It not only magnifies but also focuses the viewer's gaze on the four designs scattered across the walls of the installation. *Zoosemiotics* is emblematic of the intersection—past and present—of art and science. Literally and metaphorically the installation is "an ocular demonstration." Anker reconfigures this standard used by Hooke and his generation as empirical proof of physical phenomenon in an installation that, in Anker's words, offers an "abbreviated blueprint of cultural code summarizing the materialization of idea into visual form" (Stafford and Terpak, 2001, 220–222).

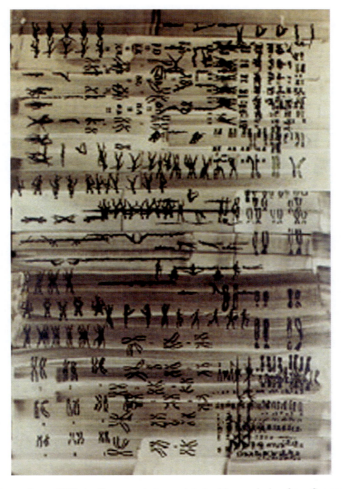

Fig. 5.5 *Zoosemiotcs*, 2001, by Suzanne Anker; printed with permission from Suzanne Anker

Anker's ongoing series of hand-painted silkscreen prints *Geneculture* (1991), *Scriptography* (1998), *Symbolic Species (1998)*, *Rebus* (1998), and *Mico-Glyph (Soma Font)* (2000) (Fig. 5.6) among others, ensconce the chromosome into a field of signs, ranging from silhouetted exercise figures to unusual animal karyotypes. Art critic Nancy Princenthal refers to them collectively as "glyphs" and describes them thus: "Precisely rendered in metallic gold and black against a brushy background, the human and chromosomal figures are aggregated and sequenced in ways that allude to the conventions of text (sentences, paragraphs) and also to scientific charts and diagrams" (Princenthal, 2000). Added to the mix, characters from diverse languages such as Arabic and Ethiopian meet on the picture plane, elucidating a further correlation between biological form and man-made language.

Moving from the iconic to the indexical, we encounter the scientific imaging device known as the autoradiograph. The autoradiograph, a technology made recognizable

Fig. 5.6 *Micro Glyph (Soma Font)*, 2000, by Suzanne Anker, printed with permission from Suzanne Anker

to popular audiences through forensics and police dramas, exemplifies Peirce's notion of the indexical sign, or trace. In this imaging technique, rows of light and dark bands in discrete lanes can be visualized as a "print-out" of an individual's unique genetic code. As a DNA fingerprint, this technique is widely used in criminal investigations, disputed paternity suits and by evolutionary biologists to trace family lineage, resemblance and evolutionary order. Dennis Ashbaugh was among the initial artists to employ this image in a work of art. Transforming laboratory iconography into color field painting, Ashbaugh's abstract pictures, such as *Designer Gene* (1992) (Fig. 5.7) become a discourse between two fields of inquiry, genetic evidence and Color Field painting. Despite their scientifically derived sources, his paintings possess an emotional quality that is reminiscent of abstract expressionist Mark Rothko's transcendental work of the 1950s. Both artists create a spiritualized world evoked by shades of color and vibrating light.[7] It is worthwhile to note that pictorializations speak to the context in which they operate, but they can also be critical devices for reframing dialogue. All images, be they scientific, commercial, or aesthetic, have historical resonance and point to modes of inquiry and concern at a given time and place.

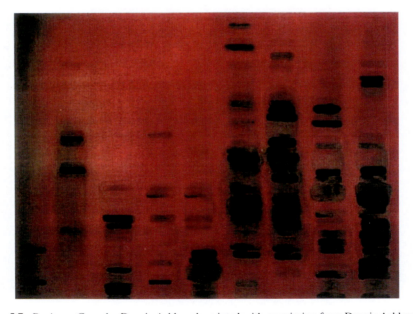

Fig. 5.7 *Designer Gene*, by Dennis Ashbaugh; printed with permission from Dennis Ashbaugh

---

[7] Mark Rothko (1903–1970), a prominent member of the New York School, studied biology, mathematics and physics at Yale University. See Weiss and Mancusi-Ungaro (2000). His early work with biomorphic forms employed many religious themes including baptism and resurrection.

The third type of sign categorized by Peirce is the culturally specific symbol. As symbols for the four-nucleotide bases comprising molecular DNA, the A, G, T, C sequence represents Adenine, Guanine, Thymine, and Cytosine, respectively. This unfolding sequence of nucleotides is particular for each individual, indicating differences in genomic variation. Kevin Clarke, working within the genre of the "genetic portrait," represents his sitters' "identities" by revealing their unique genetic codes. One of the caveats for his procedure is that the sitter must provide a blood sample, which is then sent to the lab to be genetically deciphered. For each sitter, Clarke chooses an identical area of the genome to be analyzed, so that the lab results, although individually varied, are also consistently represented. In addition to the coded scientific analysis, Clarke assigns a subjective image to the sitter's personality or sensibility. For example, for Jeff Koons, a stockbroker turned blue-chip artist, Clarke superimposes Koons' nucleotide sequence onto a slot machine (Fig. 5.8).

For the renowned scientist James Watson, Clarke chooses an image of parallel library shelves. For Clarke this architectonic emblem is a visual metaphor for the helical ladder of Watson's co-discovery of DNA's double helix. Set against a

Fig. 5.8 *Portrait of Jeff Koons* (Kevin Clarke, 1993) (Unable to obtain permission. There are instances where we have been unable to trace or contact the copyright holder. If notified the publisher will be pleased to rectify any errors or omissions at the earliest opportunity.)

complimentary field of green and orange looking very much like the colors of laboratory stains, Watson's bio-information becomes a read-out on public view. Although sometimes referred to as "genetic essentialism," Clarke's photographic portraits explore identity through macro-molecular difference.

5.6 Between Image and Substance

Marc Quinn's encompassing work treads on many intermediate zones, hovering between material reality and the symbolic order. Coming to prominence as a YBA (Young British Artist) during the 1990s with his signature piece, *Self* (1991), a portrait of himself cast in his own frozen blood, Quinn's work investigates the fragility of life and its processes.

Quinn's work employs once living materials, specimens of DNA, blood, pulverized placenta, and even feces while also embracing the traditional artistic materials of marble, plaster, lead, bronze, paint and photography. His unrelenting interest in life and life forms can be expressed in his own words:

> What is that emergent property called life, which occurs when a certain matrix of atoms are concentrated in a given space? What is it to be alive and know it, to exist only in this gravity, in this temperature, in this atmosphere, to be living in/trapped in a body, in time, in space. To be born and to die, to interact with others, with animals, with objects, with psychoactive substances, to feel emotions, obsessions, desires. To be a self-fueling organism, a living process, to be a potential object. To know that the atoms that make up your body will one day make up another. To transform energy and to be latent energy, to reproduce, have sex, stop, go, stop, what's it all about? That's what it's all about (Gruenenberg and Pomery, 2002).[8]

In 2001 Marc Quinn began using DNA to generate his images and sculptures, alluding to the power of DNA as a message in itself. Using a sperm sample garnered from Sir John Sulston, the former director of the Wellcome Trust Sanger Center in Cambridge, UK, Quinn's "portrait" of Sulston is composed of actual cultured cells. Looking like creamy blots, or the process paper goes through when it is "foxing," it is framed in reflective stainless steel. This work was initially exhibited at the National Portrait Gallery in London in 2001, along with many other traditional portraits. As an artist, Quinn is interested in the material and "invisible" world with all its attendant metaphysical questions.

For the artist, his portrait of Sir John Sulston cannot be any more realistic. He speaks of it as "the most realistic portrait in the Portrait Gallery," because it carries the actual genetic instructions that led to the embodiment of the subject (Sulston and Ferry, 2002; see also Anker and Nelkin, 2004). Quinn considers this kind of portraiture a bio-portrait of "every ancestor Sulston ever had back to the beginning

---

[8] Catalogue accompanying the exhibition *Marc Quinn*, organized by the Tate Liverpool. Curated and edited by Christoph Grunenberg and Victoria Pomery (2002).

of life in the universe. I like that it makes the invisible visible and brings the inside out." Sulston's response is more reserved: "The portrait contains a small fraction of my DNA, so it's only a detail of the whole, although there is ample information to identify me" (BBC News, 2001).

In *DNA Garden* (2002) (Fig. 5.9), Quinn comments on the continuity of all living matter. Within this stainless steel tryptych, 77 plates of cloned DNA, 75 plants and two human samples are contained. He discusses this work in terms of religious metaphor, invoking both the Garden of Eden and Hieronymus Bosch's morality painting *The Garden of Eartly Delight* (1504). In Quinn's own words he describes this sculpture as "a literalization of the Garden of Eden because if you follow back the DNA of all the plants and the two human beings there will be some single cell amoeba which is The Garden of Eden." He also comments on his use of bio-materials for this piece: "What's interesting to me is that reality should be real stuff and not illustrated. I always think that the literal is much more ambiguous than anything else." For Quinn, "science has become the religion of now."[9]

In an earlier piece, *Garden* (2000) (Fig. 5.10), commissioned by the Fondazione Prada, in Milan, Italy, Quinn brings together a variegated and copious assortment of flowers to be preserved in their splendor forever. Trying to do the impossible in this work, Quinn creates "an immortal" garden by technically freezing his flowers in silicon oil, a substance which stays liquid at −80°C. The multi-colored assortment of geographically diverse flora spring eternal in an installation with no seasonal

Fig. 5.9 Marc Quinn, *DNA Garden*, 2001. Stainless steel frame, polycarbonate agar jelly, bacteria colonies, 77 plates of cloned DNA - 75 plants, 2 humans.73 13/16 ×126 × 4 7/16 in. (187.5 × 320 × 11.2 cm). © the artist

---

[9] Ibid. Quoted from an Interview with Sarah Whitfield.

5 Technogenesis: Aesthetic Dimensions of Art and Biotechnology

Fig. 5.10 Marc Quinn, *Garden*, 2000. Refrigerating room, stainless steel, acrylic tank, heated glass, refrigeration equipment, mirrors, liquid silicone at –20 °C, turf, plants. 126 × 500 × 213 3/4 in. (320 × 1270 × 543 cm). © the artist

restraint or corporeal decay. For the artist, this process of freezing bio-matter can be compared to the way in which Andy Warhol's iconic portraits of movie stars are "ice-cold images, unrelated to the time of the flesh" (Celent, 2000). For Quinn "garden exists in the zone between biological mortality and symbolic immortality." The artist's cyrogenic/aesthetic technique, a technological intervention for life extension, not only conceptually suspends time and place but also underscores our nagging obsession with avoiding death and decay (Celent, 2000). Our dreams of perfection and immortality, underscored by a historically driven spiritual quest, encodes works of art as intermediate memory zones occupying the spaces between the present and the past. Like genetic heritage itself, our cultural lineage is but one node in the archive of life's continuity.

5.7 Feminizing the Relic

Artists are also addressing the role of gender in religion, scientific practice and visual art. This section will briefly look at two artists who investigate the politics of the body from feminist perspectives – Julia Reodica, a conceptual artist working with tissue culture technologies, and Orlan, a French performance artist who has employed surgical intervention in pursuit of bodily transformation.

Julia Reodica's *hymnNext*, composed of smooth muscle from rats, is tissue engineered to take the shape of designer hymens. Although not intended for human application at this time, these objects address feminine identity and worth with regard to social value. In many cultures, hymens are considered to be a badge of honor, a symbol of virginal purity and family pride. As a sexual site of religious ritual, the hymen, its absence or presence, is a membrane both biologically and culturally inscribed onto the bodies of women.

Reodica talks about this project as an example of the ways in which rules, protocols, and rituals in both science and theology address the issue of the clean/unclean divide. She describes this connection:

> In the laboratory, scientists go through a system of events to ensure the purity of their experiment or practice. Similar to attending church, attendants can observe the high priest prepare for the ritual. The flow-hood, a sterile air-flow area that serves as the operating platform, is the scientist' alter. Written protocols of preparing specimens and media are standardized in the field, just as sacred scrolls are copied and distributed to holy leaders (Reodica, 2004).

For Reodica, the fabricated hymen as soft sculpture raises questions about the redefinition of "new sexual beginnings for both men and women" since any orifice can become a site of physical attachment. As a possible component in regenerative medicine, her project opens up the possibilities of creating onself anew.[10] Julia Reodica is part of a growing trend of artists who consider themselves to be artist/researchers. Recently, Reodica's efforts have been substantially rewarded and recognized within the science community. She continues her projects on tissue engineering, which began as an art project, through a scientific research grant from the Rockefeller Foundation.

French performance artist Orlan employs her own body as living material to be manipulated as a sculptural form. Through repeated plastic surgeries which have been televised in real-time, and exhibited in galleries and museums worldwide, she discusses her work in the following terms: "My work and its ideas, incarnated in my flesh, interrogate the status of the body in our society and its evolution in future generations via new technologies and upcoming genetic manipulation. My body has become a site of public debate that poses crucial questions for our time" In her personal engagement with body transformation, biology for Orlan is no longer destiny (Anker and Nelkin, 2004).

In a series of small sculptures, produced in 1992 entitled *Les petits reliquaries (Small Reliquaries): Ceci est mon corps...Ceci est mon logicie (This Is My Body, This Is My Software)* (Fig. 5.11), Orlan employs 10 g of her own flesh suspended in resin. Obtained during her plastic surgery operations, this bio-matter is contained in a Petri-dish which is then overlaid on a tableaux of lexicons. In this series of work a variety of written languages are employed. Her flesh and accompanying words are foregrounded in a quotation from the contemporary French philosopher Michael Serres: "What can the common monster, tattooed, ambidextrous, hermaphrodite and

---

[10] For further discussion on this subject, listen to http//:www.wps1.org, "The Bio-Blurb Show", episode "The Two Cultures: Artists in the Lab" (2004–2005).

Fig. 5.11 © Orlan, 1992, *Small Reliquaries: This is My Body, This is My Software*, c/o ADAGP/Pictoright, Amsterdam 2008

cross-bred, show to us right now under his skin? Yes, blood and flesh" (Serres, 2004, 148–149).

These brief sets of examples serve to underscore the wide variety of molecular metaphors and performance practices employed by visual artists (Anker and Nelkin, 2004, 73–74). They reinforce the notion that DNA has stepped out of the laboratory and into the cultural domain. Bringing this analysis to where science and art meet at the intersection of genetics, its iconography and its material capabilities, Umberto Eco's invocation of code is helpful. In understanding current relationships between art practice and laboratory science, *code* as a tool of cultural encryption can be applied to a wide range of rule-governed systems, including aesthetics, advertising, and science. For Eco, the decipherability of cultural codes, such as the signifying and communicating power of visual representations, goes beyond traditional aesthetic theory.[11] Rejecting art as a personal and expressive activity uniquely created by an individual, Eco sees the underlying structure of art as a communication system. He states,

> The point is where there is rule and institution, there is a society and a deconstructable mechanism. Culture, art, language, manufactured objects are phenomena of collective interactions governed by the same laws. Cultural life is not a spontaneous spiritual creation, but rather is rule governed.... The code is not so much an isomorphic mechanism which allows communication, but it is a mechanism which allows transformations between two systems (Eco, 1984).

---

[11] For a description of the changing conceptions of art in contemporary practice see Staniszewski (1995).

As we move through the 21st century, novel technological interventions spark our imagination. In this accelerating dynamic between process and product, artists too are laying out their claims. Creativity, an unharnessible, yet resilient domain of mind, continues to add piercing dimensions to the ways we envision ourselves, in the natural world. Profound questions concerning the technological interventions into how we are born, how we die, and how nature is being transformed are regularly debated by religious leaders, politicians, ethicists and lay people. Artists too are addressing these issues. And it is in the free zone of aesthetic practice that hopes, fears and desires can be appraised, for better or worse, without doing harm.

We now turn to contemporary scientific images at the more creative margins, as they appear in advertising, scientific and medical journals. Pharmaceutical and biotechnology companies along with equipment manufacturers design such images carefully to conform to the values and expectations of the scientific community and to stimulate interest in their products. Such images provide insights into the subliminal messages generated by this community.

5.8 Imaging Art History as Science Commerce

In industrialized societies, the marketing of images has become the focus of tremendous creativity and expressiveness. Advertisements for laboratory technologies and biomedical interventions often feature striking reproductions of the body, its organs and cells. And even fantastic renditions of molecules have futuristic appeal. Decked out in a full regalia of color and design, such advertising furthermore, often refers to pictorializations appropriated from art historical parlance. The Renaissance masters are a favorite. Michelangelo's works appear repeatedly in contemporary scientific and medical journal advertising layouts, and Leonardo's Mona Lisa is an extremely popular icon in this venue as well. Often used ironically, these visualizations suggest that the scientific apparatus for sale is comparable to some of the most revered products of cultural evolution. The enduring and timeless qualities inherent in state-of-the-art masterpieces are employed to associate in the viewer's mind, a similar set of expectations. By intentionally transferring characteristics of one kind to another, a subliminal message analogically enjoins the two. In addition to the standard advertising strategies of humor, satire, hyperbole and fantasy to sell their products, those engaged in producing these campaigns seem to explicitly draw on the historical relationships between art and science.

Other artistic allusions to Michelangelo or Leonardo recurrently appear. In an ad for Operon synthesizers,[12] Michelangelo's David ponders a "revolutionary new

---

[12] Ad for Operon, *BioTechniques*, April 2000, 28(4), 647.

tool," scaleable synthesizers. And for the Pierce Science Company, which manufactures reagents and lab kits, the hand of God reaches out to Adam, as portrayed on the ceiling on the Sistine Chapel. In this illustration, however, the protagonists' hands and arms are entirely refigured as tightly wound coils of DNA. "Extraordinary interactions are within reach," the ad promises.[13] And an ad for titanium implants features Mona Lisa, modified so that she is frowning. "The difference is in the details."[14] And, again, Mona Lisa, this time with tape on her forehead, is an ad for surgical tape.[15] And finally, Leonardo's famous drawing of the Vitruvian man is superimposed over tissue cultures in an ad for Novagen, which reads "Human Tissues, normal and diseased." It is particularly intriguing, for our purposes, that so many of science's advertising images draw heavily on the history of art's traditional genres of portraiture and nudes. In some cases, the ads compare what is accomplished through biotechnology to what has been accomplished by great artists in the past. Thus in an ad for Bio-Rad, a company that makes gel electrophoresis tools, a young couple is pictured sitting in an art museum gazing intently at a large-scale, baroquely framed image of gel electrophoresis results. "Masterful results," the copy says, "advancing the art of precast gel electrophoresis."[16] In a similar ad, a youthful duo is absorbed by visual art in a gallery, an exhibition consisting only of scientific graphs and charts.[17]

And still again, in an advertisement for an Accuvix ultrasound imaging station, two observers discernly eye the ornately framed images produced by modern biomedicine. "Every picture is a masterpiece," the copy says.[18] Other images feature the undying art historical trope of the female nude, as for a scar cream advertisement announcing that "your patients are born works of art."[19] And a Criterion XT gels ad pictures what looks to be a 17th century portrait of an elegant, if not noble lady, holding up a gilt-edged frame which contains (what else?) a gel electrophoresis image.[20]

These allusions to the history of Western art may reference the sense that scientific achievement is now equivalent to the greatest achievements of art in the past. Science has, in this construction, replaced art as an arena for the exploration of beauty, symmetry, order and truth.

Indeed, the human body often symbolizes the technology being sold in these creative workings. But it must be noted that as in Leonardo's Vitruvian man, the human body is yet more than itself. The Vitruvian man represents not only the human body

---

[13] Ad for Pierce, *BioTechniques*, May 2003, 34(5), 1023.
[14] Ad for Kurz medical Inc., The Laryngoscope, April 2004, 114(4), 25.
[15] Ad for HyTape, AORN Journal, January 2004, 79(1), 41.
[16] Ad for Bio-RAD, *BioTechniques*, May 2002, 32(5), 993.
[17] Endnote ad, *BioTechniques*, September 2002, 33(3), 462.
[18] Ad for Medison Accuvix, *Journal of Ultrasound in Medicine*, May 2004, 23(5), inside back cover.
[19] Ad for Novagen, *BioTechniques*, February 2004, 36(2), 271.
[20] Mederma Ad, Journal of the American Academy of Dermatology, February 2004, 50(2), 59A.

but the cosmological relationship of man to the universe formulated with regard to humanistic perspectives. Can this imaging intersection be interpreted as a bringing together a closer, if only analogous, relationship between these distinctly different epistemological systems – science and art? Or conversely why use art to sell science? What dusting of grace do masterpieces of art give to innovative scientific practices?

5.9 Marketing Metaphors of Health and Fitness

What other metaphors are readily accessed by the advertising agencies to sell science? Many have to do with fitness and human action. For example, in an ad for Invitrogen, a massive and muscular male arm promises to "pump up your expression." Pictorializations of strong and healthy human muscle markets a "time tested Pichia Expression System."[21] Alternatively, in another ad from Promega, an infant climbing a stairway is presented as being the equivalent to a "cell proliferation assay" taking its first step.[22] And in a similar ad, a close up of a bawling baby's face is captioned, "Finally, an apoptosis assay you don't have to babysit," thus linking "mature" results with the assay.[23] A liquid handling system ad shows a silhouetted couple about to "take a spin with the latest innovation in automated liquid dispensing."[24] And an eight-armed man holds cloning kits in each hand, "for high thorough put cloning."[25]

In all these ads, human bodies, organic in nature, are presented as surrogates or stand-ins for the technologies being sold. The techniques, machines, kits and processors are "like" the actions of the human body when it dances, climbs, flexes or moves in other ways. The technology is "humanized," if you will, by images that equate its capacities with those of organic beings. Characterizing technology as an extension of the human will allows for an understanding of scientific iconography in literal terms, hence making self-evident to the viewer that the message intended is the message received. Imbuing technology with human characteristics further explicates technologies interface with nature, its repair, alteration and enhancement.

Further examples include the following. In the September, 2001 issue of *BioSciences*, a gymnast whose body is twisted around into a startling circle is pictured, targeted to represent, in physical form, the company's "genotyping services on flexible technology platforms." To be fit, increases one's chances of

---

[21] Bio-Rad for Criterion XT, *BioTechniques*, February 2004, 36(2), 271.
[22] Promega Ad, "One Step, One Solution", *BioTechniques*, October 2000, 29(4), 771.
[23] Promega ad, *BioTechniques*, March 2002, 32(3), 464.
[24] Robbins Scientific Corporation, *BioTechniques*, October 2000, 29(4), 795.
[25] Invitrogen, *BioTechniques*, November 2000, 29(5), 937.

survival. The limber young girl's body performs as a surrogate sign for "patented SNP-IT technology," and "provides SNP scoring and validation services tailored to meet your specific needs."[26] A similar ad features a middle-aged man in tie and dress pants, twisted like a pretzel and looking miserable, his ankles behind his neck. "Inflexible PCR technology got you tied up in knots?"[27] the copy asks. Here, the man's body represents the emotional state of the researcher, confronting an uncooperative natural phenomenon, rather than the technology. And in another ad promoting a gene transfer technology, a smiling deliveryman labeled "RNA" appears under the headline "Kill the messenger." The company's ad promises customers that its product can "knock out mRNA" and "silence target gene expression in vitro."

There are of course more comical cultural allusions in these marketing campaigns. For example, a man in a lab coat plays a flute to make a DNA strand rise like a snake from a wicker basket, thus appearing as a snake charmer.[28] Aliens are an exceedingly common subject in American popular culture, but they also appear in scientific advertising, announcing in an ad for a reagent, "we are here for gene Porter2."[29] And Dorothy's slippers from the Wizard of Oz promise that "wishes do come true" in an ad for scientific supplies.[30] Visually and verbally engaging advertisements clearly involve creative labor, and reflect informed calculations with regard to the market and culture of the scientific community.

These commercial images of the body are also often mediated by imaging devices that disclose the interiority of body to visual inspection. MRI, PET, endoscopic photography, and sonograms all make the invisible, hidden dimensions of the body accessible and visible (Kevles, 1998). As a result of this technological access, advertising images employ visualizations that are accessible through probes, arrays and scans. These mechanisms of standardized quality and control are officially licensed, with guarantees of reproducibility.

The images that appear in scientific and medical advertising suggest conceptions of the body as machine-like, technologically driven, and endowed with replaceable and renewable parts. Numerous images show figures that are porous, translucent, or open to visual inspection. The promise of complete control (of cell lines, gels, cloning techniques or experimental organisms) that underlies many of the ad campaigns and forms the basis of the fundamental appeal, replicates science's more general expectation of controlling the body, nature and technology. People who are out of control in these advertisements are symbolic of what is unscientific, problematic and in need of correction.

---

[26] Ad for Orchid, *BioSciences*, September 2001, 31(3), 571.

[27] Ad for Thermo Hybaid US, *BioTechniques*, September 2001, 31(3), 575.

[28] Invitrogen ad, Elevate long PCR cloning to new heights, *BioTechniques*, February 2001, 30(2), 230.

[29] Gene Therapy Systems International ad, *BioTechniques*, January 2000, 28(1), 65.

[30] Fisher Scientific *BioTechniques*, April 2000, 28(4), 665.

Disease states, too, are portrayed as being associated with particular social groups. Current images of the body in advertising, do not create typologies along racial lines (necessarily) but do in terms of age, class, fitness level, or sex. Advertising for pharmaceuticals, for example, often features a "typical" patient for the drug concerned – an elderly male, or a middle-aged female (usually white). In these cases, the image reinforces prevailing ideas about the illness involved – men appear in ads focused on cluster headaches, for example, and women in ads about migraines, though both men and women suffer from both kinds of headaches (Kemper, 2004). Targeting audiences for advertisers is a state-of-the art promotional tool itself. As the media fragments its audiences to get more of the market share, let no group escape its psychological grip.

5.10 Pictorial Signs of Social Rank

In this section, we review some of the historical relationships between pictorial devices and the ways in which they encode signs of the body with regard to social rank. We revisit some of the history of racial science, and the manner in which images and measurements were deployed as scientific proof of ethnic and gender hierarchies. These pictorial propositions employed by both the scientific community and the public's growing engagement with popular culture embody manifold ideological assumptions with regard to race, class and women in society.

Before the mid-18th century, there was no clear boundary between empirical science and visual art. Artists and natural philosophers often collaborated, inspired by a mutual interest in direct, sensory knowledge of the human body. Enlightenment notions of the nature of perception and the process of learning, however, began to facilitate a change that was fully realized only in the 19th century. Scientists began to emphasize unadorned reality, a nature shorn of metaphysical and emblematic meanings, while artists increasingly moved away from naturalistic representations of the human body (Kemp and Wallace, 2000). As anatomy became defined as an empirical, experimental science, new technologies facilitated the shift. Photographic techniques replaced the work of draftsmen and artists, and new illustrations presented the deeper, invisible, microscopic structures of the body, rather than its visually accessible surface. Scientists increasingly sought to convey the ideal form of the body through the use of measurement and statistical calculations, rather than artistic interpretation (Shea, 2000).

Ironically, this increasing empiricism and reliance on objective means of producing images coexisted with the rise of a particularly virulent race science, and racial differences became one of the most important arenas in which the new techniques of visualizing and assessing the body were applied. The economic and social importance of the notion of race, across many centuries and in various places, engaged the attention of leading figures in the history of science, and the racial typologies that emerged in the 19th century had origins in European reactions to distant populations in the age of discovery.

The Linnaean classification system, as described in his 1758 *System Natura*, classified humanity into four kinds, groups that would roughly correlate with race. These were white Europeans, red Americans, yellow Asians, and black Africans. Linnaeus proposed that you could tell the difference between races at least partly by considering what they wore. Europeans for example wore tight clothing, and Africans wore grease. The races in his categorization also differed in temperament, Asians being melancholy and stern, and American Indians irascible and impassive (Koerner, 1999). Linnaeus thus constructed race as a diffuse phenomenon expressed in every aspect of the body and mind.

Physiognomy, a late 18th century science, was a field built around the study of external form as a guide to internal qualities. Those promoting this new science, most conspicuously Lavater (1789), proposed that the face, skull, and physique were outward expressions of the inner self. By the 19th century, the sciences of physiognomy and phrenology were developing ever more elaborate systems for defining racial difference. Phrenologists measured the skull to assess personality, ability, or character, and concluded that certain bumps (on the head) revealed positive qualities and other bumps revealed negative traits. The phrenologists, most conspicuously Franz Joseph Gall, managed to map the brain in ways that are still somewhat persuasive, but their interpretations of the meanings of localized brain conformation were simplistic and unsupported by data. Different races, they proposed, had differently shaped heads that correlated, unsurprisingly, with their relative position in the scale of civilization. Other groups were also scientifically sorted based on bodily traits. Women, the poor, criminals and ethnic groups such as the Irish and the gypsies were frequently interpreted as biologically marked by inferiority.[31]

The Italian criminal anthropologist and army doctor Cesar Lombroso performed autopsies on criminals and found that their bodies were like the bodies of chimpanzees, in the shape of the head, its symmetry and its size. Lombroso's prison studies supported the notion of the "born criminal" as an atavistic regression to man's evolutionary past, and his 1876 *L'uomo delinquente*, was a bold challenge to ideas about free will and individual responsibility. The text was filled with images of those destined by biology to break the law.[32] Large ears, bushy eyebrows, thin necks, long arms and other bodily traits were associated with criminality, and Lombroso found that African bodies were commonly marked by these criminalistic morphologies.

In the same period, scientific debate about human evolution often focused on racial difference and on the possibility that the races were descended through different lines, perhaps constituting different species or subspecies, a doctrine called polygenism (Bowler, 1986, 553–558). Illustrations of pathological faces, skulls, skeletons and body types filled textbooks and scientific papers. Many of these

---

[31] On the equivalence of race and gender, see Stepan (1986).
[32] On Lombroso, see Nye (1976); on the equivalence of race and gender, see Stepan (1986).

images implicitly compared gorillas and chimpanzees to humans in ways that linked Africans to nonhuman primates. Buffon in the mid-18th-century declared that apes and Africans engaged in fornication and interbreeding. But it wasn't just Africans who could be compared to apes. As Jonathan Marks has noted, the Irish were called "white chimpanzees" in the 19th century, by those who sought to deny their basic humanity (Marks, 2002).

The body has long been, in effect, a text in which physicians and scientists could read social and political meanings. Biological reasoning and images of corporeality were deployed to justify slavery, colonialism, the oppression of women and the poor. Primate faces could be drawn to seem "like" faces of Africans and women's faces were similarly "like" the faces of children.

Visual representations – photography, fingerprinting and drawing – were empowered to identify criminal types and explain cultural differences. And in the genetic age, how will other pictorializations and computations be employed to explain biological and ethnic differences? Will similar issues resurface in the masquerading guise of the genetic sciences, reinforcing the power relations already in place? Or will these advances in bio-science be applied to dispel social myths? Visualizations continue to be an effective underlying and even global force employed to mediate ideology. Although the genetic sciences have overtly stated that there is no such thing as race, how will assumptions regarding this and other issues attached to the life sciences be addressed?

In this next chapter we will look at interventions into living matter itself, and interventions that do in fact cross species lines. As combinations of living matter such as transgenic animals and plants enter the public consciousness, how will their existence be interpreted? In what follows, we explore the manipulation of life forms by visual artists and the role these visual practitioners play in being "outsiders" in scientific discourse.

5.11 The Transgenic Body: From Micro Venus to Alba

Scientist Vilém Flusser, in his 1988 *Artforum* column, "Curie's Children," suggested that the Walt Disneys of the future might be molecular biologists, who "may soon be handling skin color more or less as painters handle oils and acrylics" (Flusser, 1988). Posing the rhetorical question, "Why can't art inform nature?" he responded that, "When we ask why dogs can't be blue with red spots, we're really asking about art's role in the immediate future." Flusser recognized the potential of art to extend beyond representations of nature and science, to become integrated with both in genetically modified organisms. Yet, his vision of that intervention was only skin-deep: applying modernist principles of art and design to create "an enormous color symphony." Artists have used the materials and concepts of biotechnology in a rich panoply of ways. But only very recently has it been possible for them (usually with the help of scientific collaborators) to design genetically altered specimens by manipulating DNA. Responding to experiments in genetic engineering, artists have used biomaterials as their medium to create works that express concerns about transgenic research, cloning, and the commercialization of the body and its parts. By making the shift from *representation* to actual *embodiment*, such meta-critical reflections on these issues offer concrete examples of the state of the art of biotechnology and speculative models of its future.

Flusser was probably unaware that in 1986, artist Joe Davis and Harvard geneticist Dana Boyd began collaborating on an artwork, *Microvenus*, using DNA as the medium.[33] In protest of the sexually neutered representations of human female genitalia sent into space with the Pioneer voyage in the early 1970s, the Microvenus icon looked like the letters Y and I superimposed. In 1990, it was coded into a string of DNA nucleotides and transformed into *E. Coli* bacteria. Billions of the bacteria were produced, though they, like the icon they carried in them as part of their genetic makeup, were invisible to the naked eye. Galleries in the US were unwilling to risk displaying genetically engineered bacteria, so *Microvenus* was described in Scientific American as "the most highly reproduced graphic that almost no one had ever seen." Finally, in 2000, the work was shown in a pressurized containment facility at Ars Electronica in Linz, Austria. Although the work was conceived within the context of the search for extraterrestrial intelligence, no provision was made for transporting *Microvenus* bacteria into outer-space, which might contaminate extra-terrestrial environments.

In this ironic work, Davis satirizes the unnatural modification and misrepresentation of women's bodies by NASA scientists (Fig. 5.12). He not only corrects the design error but also constitutes the more accurate icon of female genitalia in the actual genetic material of a living organism. As a highly resilient life-form capable of withstanding the harsh environment of deep space, and one that quickly produces billions of copies of itself, *E. Coli* is a potentially viable medium for the dispersal of messages into space. Despite the tongue-in-cheek quality of this work, this substrate offered practical advantages to the materials used by NASA. For had the GMOs been transported into space and dispersed, the likelihood that the image would be discovered by extra-terrestrials arguably was much greater than the isolated images aboard the Pioneer.

5.12 Genesis as Living Text

In 1999, artist Eduardo Kac first exhibited *Genesis* (Fig. 5.13), also at Ars Electronica. In this artwork, bacteria were genetically modified to contain the verse from the biblical *Book of Genesis*, "Let man have dominion over the fish in the sea, and over the fowl of the air, and over every living thing that moves upon the earth." Kac chose this verse for its implications about "the dubious notion – divinely

---

[33] The work ultimately was realized by Davis and Boyd at Jon Beckwith's laboratory at Harvard Medical School and at Hatch Echol's laboratory at University of California, Berkeley.

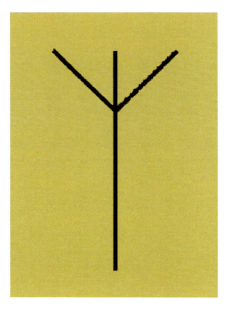

Fig. 5.12 *Microvenus*, 2000, by Joe Davis; printed with permission from Joe Davis

sanctioned – of humanity's supremacy over nature" (Kac, website). The English text was translated into Morse code, which then was converted into DNA base pairs, in order to synthesize an artificial gene that was spliced into the bacteria's genome. Participants, both locally and remotely over the Internet, could turn on an ultraviolet light, causing mutations in the bacteria's genetic code, which in turn caused alterations in the biblical verse when the mutant code was translated back into English. For Kac, the ability to alter the verse represents a refusal to "accept its meaning in the form we inherited it," and an insistence that, "new meanings emerge as we seek to change it." *Genesis* raises questions about the shared responsibility of individuals and communities to: (1) respect and protect nature, not just dominate it; (2) engage in dialogues about ethics and religion in order to reinterpret and give new meaning to traditional values; and (3) reflect on social and cultural implications of and policies regarding biotechnology.

It is important to note that these works offer public audiences an unusual opportunity to see and interact with living GMOs directly. The manner of their installation and presentation influences the nature of their reception, interpretation, and meaning. The pressurized containment facility in which *Microvenus* was exhibited implied a sense of danger – a need to quarantine the genetically modified *E. Coli* in order to protect the natural environment from this artificially produced genetic icon. The ambience of Kac's installation simultaneously evoked a sense of sublimity and clinical sterility. Original DNA-synthesized music based on Kac's Genesis gene and composed by Peter Gena was generated live in the gallery, contributing to the otherworldly and

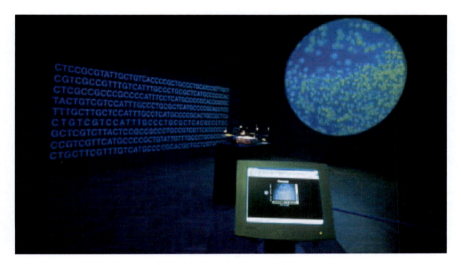

Fig. 5.13 *Genesis*, 1999, by Eduardo Kac, transgenic work with artist-created bacteria, ultraviolet light, internet, video (detail), edition of 2, dimensions variable. Collection Instituto Valenciano de ARte Moderno (IVAM), Valencia, Spain (1/2). Printed with permission from Eduardo Kac

ominous feeling. The main sculptural element consisted of a black pedestal on which a petri dish containing the genetically modified bacteria was sporadically awash in short-wave ultraviolet (UV) light. A greatly magnified real-time video of the bacteria, which glowed a greenish yellow against a blue background, was projected on one wall. The DNA sequence for the Genesis gene, inscribed in white, glowed under long-wave UV on an adjacent wall.[34] Similarly, visitors were illuminated by long-wave UV, suggesting a parallel between genetically modified bacterial code and wild-type human phylogeny as interrelated parts of a system. A computer terminal enabled local and remote visitors to access the website and actively participate in the installation by choosing either to irradiate the bacteria or shield them from the UV rays that cause mutations. By physically engaging visitors in these ways, *Genesis* attempted to create a context of empathy between its audience and GMOs. Moreover, the work insisted that the public be engaged in biotechnology and evolutionary processes, while refusing to permit its audience to remain untouched and external to the installation as outside observers.

Davis's piece refers to Greek mythology and the figure of Venus. A favorite traditional subject for artists, the goddess of love and beauty played a seminal role at the nexus of art and life as the animator of Galatea, the marble statue carved by Pygmalion in Greek myth. There appear to be no religious overtones in Davis's piece, which takes as its inspiration the design challenge of accurately and simply, though not idealistically, representing human genitalia in a robust medium; and the

---

[34] The audience was shielded from the harmful short-wave UV rays by a filter. Long-wave UV, typically known as "black light," does not pose the same risk.

search, not for a godhead, but for extraterrestrial forms of life. Kac's work, on the other hand, summons the force of the Old Testament, the authoritative word of God, which endows humanity with dominion over nature. Kac's work suggests that this dominion also implies responsibility. By failing to protect nature, by subjecting it to harmful radiation, humans cause irreversible mutations at the basic genetic level. In the context of the installation, such mutations result in irreversible changes in the text of Genesis – the word of God – encoded in the bacteria.[35] Although Kac embraces this malleability as a positive openness to the creation of new forms and the negotiation of new meanings, one might also be fearful that the misdirected alteration of genetic codes – and their religious and ethical corollaries – may result in undesirable physical aberrations that could sweep through the gene pool, while crippling the values that order life and give it meaning. Indeed, Kac's work must be interpreted as presenting both the positive and negative aspects of biotechnology. Rather than take a simple "pro or con" stand on a highly politicized and polemical issue, the artist removes transgenic species from the rarefied context of science and places them in the more public arena of art. The audiences of *Genesis* can have first-hand experiences of seeing and interacting with GMOs, participate in discourses about them, and form their own opinions.

5.13 GFP K9 Project: The Green Bunny

As part of the Ars Electronica symposium in 1999 on the theme of Life Science, Kac lectured on his *GFP K9* project, first proposed in 1998, and announced plans to produce his artwork, *GFP Bunny* (Fig. 5.14), which included creating a genetically modified albino rabbit that glows green when exposed to blue light because it has been engineered with "an enhanced version of the wild-type gene for green fluorescent protein (GFP) found in the jellyfish Aequorea Victoria" (Kac, "GFP Bunny"). While many people both inside and outside the art-world have embraced *GFP Bunny* as an important work, the response that afternoon in Linz anticipated the extremely negative reception of the work by many others. Just as one member of that audience exclaimed that she thought Kac must be a "terrible father," so five years later a visitor to the exhibition, *Gene(sis): Contemporary Art Explores Human Genomics*,[36] wrote in the comment book Kac furnished that s/he hoped the artist "has a child with Down's Syndrome."

---

[35] Scientists interested in extraterrestrial life are looking for what they term a "second genesis" as a reaffirmation that life on earth is not a single phenomenon. According to CNN's telecast, finding a second source of life would unlock mysteries into the way in which life began on earth. In addition it would impact theological explanations and religious practice. See CNN (2004).

[36] Organized by the Henry Art Gallery at the University of Washington, Seattle, the exhibition traveled to three other university art galleries, including the Block Museum at Northwestern University, where the quoted comment was entered in November 2004. See http://www.gene-sis.net/splash.html.

5 Technogenesis: Aesthetic Dimensions of Art and Biotechnology

Fig. 5.14 *GFP Bunny*, 2000, by Eduardo Kac, transgenic artwork. Alba, the fluorescent rabbit. Printed with permission from Eduardo Kac

Such personal attacks demand that one ask why the use of transgenic technology as an artistic medium incites people to character assassination and transgenerational curses. Kac believes that such responses result from the anger people experience when forced against their will to become aware of a polemical issue. They are no longer able to carry on in blissful ignorance but must respond (Kac, 2004). Rather than direct their anger towards the scientific laboratories that routinely create GMOs or the industries, such as agriculture, that commonly employ them, in these cases audience members take aim at the messenger, Kac, who lacks the authority of science and industry, and is, in any case, a much easier target. The artist is, to be sure, not simply the messenger but, from this perspective, must be considered an accomplice, for he claims that Alba was created solely for artistic purposes.

Despite some of the sharply negative statements in the *Gene(sis)* guestbook, (which included many supportive comments as well), the artist observed a shift in audience response to his work using bio-materials between 1999 and 2005. "In the beginning people were quite worked up about it; there was a greater concern and fear than exists now," he noted. "The discussion now is more philosophical; the sense that some impending doom is about to happen has completely vanished.

A level of discourse has developed that is much more complex than the original polarization" (Kac, 2005). If, in fact, audience responses have changed over time to become more subtle and sophisticated, then perhaps artistic research involving GMOs has a valuable contribution to make as a context for public discourse on the social and cultural implications of biotechnology. More research on the reception of Kac's work, and that of other artists working in this domain, may provide insight into the ethical and religious triggers pertaining to the public response to molecular biology and to the efficacy of art as a forum for dialogue and debate.

Contemporary art may be as elusive to scientists as molecular biology is to nonscientists. Just as one could not expect to comprehend genetic engineering without first understanding genetics, in order to comprehend the significance of *GFP Bunny*, one must understand its underpinnings in art history and aesthetic theory. To this end, the following discussion shall focus on the aesthetic foundations and implications of this work. Although on the surface Kac's *GFP Bunny* appears to have realized Flusser's fantasy of the artistic use of the techniques of molecular biology to create a "color symphony," its particular formal aspects are of less artistic significance than its theoretical propositions and ethical provocations. Since the 1970s, scientists routinely engineer transgenic animals for research. In science labs across the world, there are hundreds if not thousands of such mammals that, like Alba, express GFP and fluoresce when exposed to certain wavelengths. In this sense, Alba may be likened to an *objet trouvé*, or found object, employed by Marcel Duchamp beginning in the 1910s to create his "readymade" artworks. At the same time, *GFP Bunny* does not fit the definition of a readymade. Unlike a mass-produced *object* scavenged or acquired via retail markets, Kac emphasizes that Alba is a unique living *subject*, custom-engineered and bred in a government laboratory to the artist's specifications. Keeping this important distinction in mind, comparison with the Duchampian aesthetic strategy offers insight into how Kac's work is embedded in the history of art while expanding the field of artistic practice.

5.14 Laboratories of Art and Science

In response to the famous censorship of his artwork *Fountain* from an un-juried exhibition in 1917, Duchamp argued for its status as art. He claimed that by selecting a particular object (a common porcelain urinal), by giving it a title (and signing it), and by inserting it in an art exhibition, he gave that object a new meaning. Duchamp's *Fountain*, and his rationale for it, proposed that the meaning of an art object is not contained exclusively within the work and that an object's status as art cannot be determined solely by morphological characteristics. Rather, he insisted that contexts of reception and corresponding audience expectations are substantial factors in the production of meaning and value.

Similarly, Kac's selection of a transgenic mammal and his intention to recontextualize it within an artistic framework would have given new meaning to GMOs, the likes of which previously had been seen only in scientific laboratories. Like

Fountain, Kac claims *GFP Bunny* was censored; in this case by the Institut National de la Recherche Agronomique (INRA at Jouy-en-Josas, France). INRA created the rabbit with the knowledge that it was to be used for artistic purposes, then later refused to release it to the artist. Public encounters with the original *Fountain* and with *GFP Bunny* are possible only through photographs, taken serendipitously prior to their censorship. And, as Didier Ottinger has observed, "Following the example of its sanitary forerunner, the rabbit's 'prestige' grows in proportion to its invisibility… [though] never exhibited in the public space for which it was conceived …. its photograph did make the front page of the world's most important newspapers" (Ottinger, 2004). Ironically, by censoring Alba, INRA called attention to its collaboration on non-scientific projects, thereby generating, rather than stemming, negative publicity for itself, while at the same time helping to mythologize Kac's *GFP Bunny*.

Ottinger considers whether or not Alba might be considered an "assisted readymade" (a class of objects defined by Duchamp to include found objects that have been modified by the artist) but concludes that, "the absence of the rabbit's *déjà-là* prevents it from being strictly defined as a readymade." This point raises a dispute regarding Alba's uniqueness: Is she a unique, transgenic rabbit created solely for artistic purposes or is she one of many GFP rabbits generated by the lab? The Boston Globe article that broke the story in the US in September 2000 stated that Louis-Marie Houdebine (the INRA genetic researcher who bred the rabbit) was "intrigued by Kac's desire to involve the public, and had 'never considered' whether an entire animal would glow in the dark" (Cook, 2000). Two years later, however, after the controversy over INRA's refusal to grant Kac ownership of Alba, the French lab's story appeared to shift and the rabbit met an "untimely death," according to Wired News. The August 2002 article states that, according to Houdebine, Alba was "one of many GFP rabbits generated almost five years ago…" and quotes the geneticist's recollection that, "When E. Kac visited us, we examined three or four GFP rabbits…. He decided that one of them was his bunny, because it seemed a peaceful animal" (Philipkoski, 2002).

Although existing GMOs have been employed by artists such as Catherine Chalmers (Fig. 5.15), to create compelling work, the use of a readymade transgenic mammal would not have been of interest to Kac (2004). To do so would have been decidedly uncharacteristic for the artist, whose practice for over 20 years has focused on the creation of new art forms, not on the reuse of existing objects, of things "déjà-là." Moreover, it was of crucial importance to the *GFP Bunny* project that Alba embody characteristics that visually identify her as an icon of transgenics. So, for purely artistic purposes Kac specified that Alba must fluoresce green all over her body in order to manifest sufficient iconic and symbolic resonance.[37]

---

[37] Kac attributes the image to photographer Chrystelle Fontaine, whose name ironically bears an uncanny resemblance to Duchamp's *Fountain*. The photograph was taken with a digital camera through a special yellow filter designed to work in concert with the particular strain of GFP used in Alba (Kac, 2004). See http://www.artexetra.com/Kac.html.

Fig. 5.15 *Rhino*, 2000, by Catherine Chalmers; printed with permission from Catherine Chalmers

For scientific purposes, fluorescent markers, like GFP, have diagnostic utility only when attached to specific genes. Therefore, a rabbit that expresses GFP throughout its phenotype may have little or no use for scientists.[38] If the widely reproduced photograph of Alba fluorescing green throughout her phenotype is authentic – and that, too has been disputed – then it is likely that the rabbit was, as Kac maintains, bred specifically for his *GFP Bunny* project and would have been two-and-a-half years old and not nearly five years old at the time her death was reported.

Regardless of the controversy surrounding Alba's uniqueness and its bearing on the interpretation of *GFP Bunny* as a readymade, assisted readymade, or unique creation, Kac's attempt to place a GMO within an artistic framework draws on but supercedes the Duchampian strategy of recontextualization. Duchamp rejects the artist's and audience's traditional roles as the creator and beholder of beautiful images, respectively. The artist becomes, rather, the creator of enigmas that reveal and provoke debate over the very discursive conditions that make art possible. *Fountain* creates controversy by placing an object (urinal), loaded with abject meaning, within a fine art context. Similarly, *GFP Bunny* implies a rejection of the artist's and audience's traditional roles as the passive consumers of science and unwitting subjects of technocracy. The artist collaborates with the scientist to create new hybrid forms of life that provoke debate over the boundaries between art and science, between species, and between GMOs and wild-type organisms. Duchamp recontextualizes a pre-existing object to give it a new meaning and reveal the discursive conditions of art. Kac creates a unique and unprecedented form of subject

---

[38] Scientists from National Taiwan University produced pigs that they say "are the only ones that are green from the inside out." See Chris Hogg, "Taiwan breeds green-glowing pigs," BBC News, 12 January 2006. Also see Sue Broom, "Green-tinged farm points the way," BBC News, 28 April 2004. The author comments on the use of light to detect animals carrying fluorescent genes in this study. She states "Both chickens and pigs carrying the gene can be detected in normal light by their slight greenish tinge, but when viewed in blue light, all areas not covered with hair or feathers are seen to glow torch-light bright."

that opens up a new context for the negotiation of meaning and value with respect to both art and genetic science.

In upping the Duchampian ante, Kac's *GFP Bunny* recontextualized a problematic entity that, like *Fountain*, can be seen as abject. Its chimerical deviance might be considered abhorrent from certain religious and ethical views on nature. Alba is a living, breathing mammal, and some ethical positions are opposed to the exploitation of animals for human ends, regardless of their utility. Whereas the social utility of science, medicine, and agriculture provide an ethical rationale that many religious and ethical traditions accept as sufficient justification for sanctioning research on and application of transgenic technology, those traditions may not accept the use of the same techniques to serve artistic ends, the utility of which is more difficult to rationalize. Finally, for individuals and groups whose understanding of art is predicated on traditional aesthetic values of natural beauty and order, *GFP Bunny*, like *Fountain*, will not be considered art at all, much less good art. Such a position is as naïve as rejecting Watson and Crick (and everything built on their contributions to genetics) in favor of Lamarck.

It is important, moreover, to remember that the Duchampian strategy seeks to reveal and provoke debate over the very discursive conditions that make art possible. In other words, producing cognitive dissonance by appealing to the abject is intrinsic to the strategy and an integral part of the work itself. *GFP Bunny* extends this approach beyond the domain of aesthetics: it uses the context of art to *reveal and provoke debate over the discursive conditions in which genetic science operates*. Kac takes science out of the insular confines of the lab and metaphorically mounts it on the gallery wall, where it becomes the subject of social inquiry. Whereas *Fountain* gave an object a new meaning and, in the process, expanded the field of art, *GFP Bunny* not only gave a live, transgenic mammal a new meaning and expanded the field of art, but it contributed to broadening the discursive domain of molecular biology to include public debate over its social and cultural implications.

This line of reasoning would be easily understood by contemporary art historians. Indeed, Duchamp's strategy of recontextualization is as widely accepted in the field as the basic tenets of natural selection are accepted by evolutionary biologists. However, given the highly specialized nature of disciplinary knowledge and the abject quality of *GFP Bunny*, even public forums designed to foster communication over the ethical implications of biotechnology can reveal views that are polarized and closed to considering the aesthetic applications of genetic science. Such was the case at the symposium, "Art, Ethics, and Genetic Engineering: The Transgenic Art of Eduardo Kac," at Duke University on November 6, 2000.[39]

---

[39] Panelists (titles and affiliations pertain to their positions at the time of the panel) included Kalman P. Bland, Professor of Religion, Director of Judaic Studies Program, Duke University; Elizabeth Kiss, Director of Duke's Kenan Institute for Ethics; Associate Professor of the Practice of Political Science and Philosophy; Joseph Nevins, James B. Duke Professor of Genetics and Chair of the Department of Genetics; and Jeremy Sugarman, Director of the Center for the Study of Medical Ethics and Humanities, Associate Professor of Medicine; Associate Professor of Philosophy. See http://artexetra.com/Kac.html.

Taking *GFP Bunny* as the center of discussion, the event brought together a variety of intellectual perspectives, methods, and disciplines to exchange ideas and propositions about the social, cultural, and ethical ramifications of genetic engineering. The introductory remarks by the convener, Edward A. Shanken, shared with the panelists in advance, rhetorically asked why genetic engineering performed in the name of art would be considered more or less acceptable than the same processes carried out in the name of science. He noted that science and technology have made formidable advances towards the prediction and control of phenomena, including the remarkable ability to control illness and save human lives. But he also noted that the advances made by the arts towards interrogating the limits of knowledge and consciousness, and towards plumbing the depths and reconfiguring the conditions of human existence, are arguably of no less social significance. These considerations led to the question, "Is it inconsistent to argue on an ethical basis that animal research may be valid in science but not in the arts?"

Kac demonstrated to the audience how his artistic research for over a decade consistently addressed cultural and species hybridity by using technological media to create contexts for dialogical exchange. Panelist Joseph Nevins, a geneticist, was incapable of grasping what made *GFP Bunny* a work of art, much less a good one, and argued that it was a waste of a valuable scientific resource, the only useful function of which was for laboratory research. Neither he, nor the other panelists, including a religious scholar and two ethicists, seemed to appreciate how the socialization of a transgenic mammal as part of the artist's family, and the production of critical discourse about biotechnology and the conditions of art, were significant artistic statements. Transcending, on the one hand, the barrier between wild-type and genetically modified mammals and challenging, on the other, the closed circle of scientific discourse by opening up debate on biotechnology within an aesthetic context, did not register as valid aims for art, much less as useful contributions to culture. This misunderstanding was not a matter of specialized scientific or artistic nomenclature but consisted of an epistemological disagreement over what constituted valid methods of creating and disseminating knowledge, especially in regard to sentient creatures. Although opinions varied, the panel generally agreed that the creation of a transgenic rabbit was acceptable in the pursuit of scientific and medical knowledge, but not in the pursuit of art.

Despite the lessons of Duchamp's *Fountain* and over eight decades of artistic practice and art historical research building on and interpreting it, general audiences still have trouble accepting that artists undertake research that has value outside of the traditional aesthetic domain of beauty; that draws on, participates in, and challenges discourses in broad fields, including science and technology. Yet that is precisely what artists like Davis and Kac (and many others) succeed in doing. Despite the intransigence, if not inconsistency, of the Duke panelists' views, the fact that geneticists and ethicists appeared on the same stage with Kac to discuss *GFP Bunny* validates the work's success in opening up interdisciplinary debate. And indeed, the controversy surrounding *GFP Bunny* and the broad, international attention it received by the media have provided remarkable opportunities for Kac to participate in and interrogate the discourses of genetic science. By creating an unprecedented art subject, Alba, the artist established a context for dialogical exchange that has given molecular biology a new meaning.

GFP Bunny might be described as an artistic icon of the Age of Transgenic Reproduction (Shanken, 2004). Indeed, her image has been reprinted countless times and in diverse contexts, from international newspapers to biology textbooks. Like traditional icons, it signifies a larger concept. But Alba is also a living manifestation of that concept. As a result, there is more at stake, for Alba makes concrete the reality of living with the current state of biotechnology. *GFP Bunny* confounds the disengaged spectatorship that Walter Benjamin attributes to the popular consumption of mass culture, and demands a personal response. In contrast to the loss of aura that the German theorist claims to befall mechanically reproduced works of art, Alba possesses a fully present and hybrid aura – one that simultaneously serves the ritualistic value of art and the dialogical value of social discourse. *GFP Bunny* both *signifies* and *is* the actual embodiment of the possibility of communication and communion between animal kingdoms, between art and science, and between experts and the public.

5.15 Tissue Engineering as Culture

Other artists have also utilized the materials and techniques of microbiology and tissue engineering as artistic media to modify living organisms without any alteration in their DNA. As in art work of Kac and Davis, these works are not representations of biotechnology but actual, living embodiments of it. For example, in the Tissue Culture and Art (TC&A) project, initiated by artist Oron Catts in 1996, living tissues and non-living materials are conjoined and manipulated to create objects/beings that are "semi-living." Catts and collaborator Ionat Zurr were inspired in part by research conducted by tissue engineering pioneer Joseph P. Vacanti, who employed a living mouse as the biological substrate on which to "grow" an ear-shaped scaffold seeded with human cells. This mid-1990s icon of tissue engineering offered a glimpse of seemingly limitless possibilities of procedures in which living tissue is employed as a reconstructive sculptural technique. For the artist, this "semi-living new media" was adapted to aesthetic and philosophical ends.

In *Pigs Wings*, 2001 (Fig. 5.16) TC&A grew pig-bone tissue to mimic the shape of three different types of wings that enable flight in vertebrates, those of birds, bats, and pterosaurs. In *Fish and Chips*, 2001, the SymbioticA Research Group grew fish neurons over silicon chips connected to video and audio output devices, creating a cyborgian confluence of wetware, hardware, and software. This semi-living entity was endowed with the ability to make sound and images – to make art – begging questions about the future of human interaction with cyborgs whose behavior may be unpredictable, if not creative.

MEART (2004), which SymbioticA Research Group refers to as a "semi-living artist," asks similar questions. *MEART* is a bio-robotic drawing system which aspires,

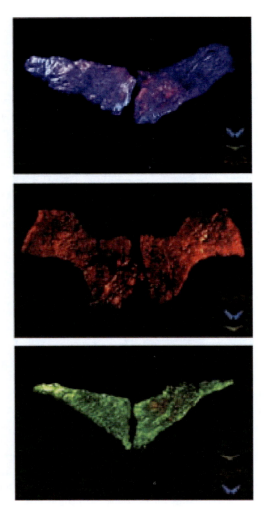

Fig. 5.16 *Pig Wings*, 2000-2001, by The Tissue Culture & Art Project. Medium: Pig mesenchymal cells (bone marrow stem cells) and biodegradable/bioabsorbable polymers (PGA, P4HB), 4cm × 2cm × 0.5cm each. Reprinted by permission of the artists

at least metaphorically, to learn how to draw portraits of gallery visitors in real time. Its "body" consists of a video camera and a robotic drawing arm, designed by artist Phil Gamblen, that can be installed on location at an exhibition. Its remote "brain" is comprised of thousands of mouse brain cells grown in a Multi-Electrode Array (MEA + ART = MEART) at Georgia Institute of Technology's Laboratory for NeuroEngineering, under the direction of Dr. Steve M. Potter. It also possesses a "nervous system" that enables the brain and the body to communicate via the Internet using software designed by Iain Sweetman at the University of Western Australia. Information about the difference between the subject's image and the current state of the portrait is fed to the neurons, stimulating them to control the robotic drawing arm in order to reduce the difference and hopefully learn through the process. As Potter noted, "I hope that we can look at the drawings it makes and see some evidence of learning. Then, we can scrutinize the cultured network under the microscope to help understand the learning process at the cellular level." Potter also hopes that this sort of interdisciplinary work will persuade scientists to start "thinking about what is art and what is the minimum needed to make a creative entity" (Symbiotica, 2004).

Davis, Kac, and TC & A take genetic engineering and tissue culturing out of the laboratory and use it as an artistic medium. In so doing, their work transforms the abstract complexities of biotechnology into an approachable form that simultaneously problematizes science and makes it more human. Few non-scientists feel they possess sufficient knowledge to judge science, but the public has few qualms about judging art. It is generally accepted that science is rightly incomprehensible to laypeople, but when the work of an artist challenges preconceptions of what art is, that breach of aesthetic expectation is greeted with charges of elitism, immorality, or frivolity. By using biotechnology to create art, Kac and Davis demystify genetic engineering, thus enabling related social and ethical issues to be studied in a broader cultural frame. At the same time, their work pushes the political, cultural, and ethical boundaries of aesthetics, and interrogates the relationship between art and science in a social context. Neither spokespeople for the benefits of genetic engineering, nor doomsayers of its cataclysmic effects, they ask questions and promote dialogue concerning the mounting acceleration of biotechnological practices. These artists insist on the potential, if not obligation, of art to play a role in representing, provoking, and complicating ethical issues, in disturbing ethical certainties, and even a role in ethical deliberation.

Perhaps the freedom of artistic license comes at the price of cultural authority, resulting in demonization or scapegoating of artists who use bio-materials to raise ethical issues. *GFP Bunny* made Kac the subject of belligerent criticism and hatred, particularly among religious conservatives. For many people uninitiated into the rarified interdisciplinary discourses at the nexus of art, science, and technology, he became an example of what is wrong with contemporary art and artists that not only stray beyond the bounds of traditional media but claim to have the right to use the materials and techniques of biotechnology to challenge the authority of science and, moreover, to force people to confront and take responsibility for chimera about which they would prefer to remain ignorant. Such dealings can, in addition, expose an artist to legal and political challenges, as in the widely reported case of Critical Art Ensemble (CAE) member Steve Kurtz.

CAE has consistently challenged the authority of science, the rhetoric of which it has likened to a Christian religious sect, as in their ironic project *Cult of the New Eve* (2000). CAE utilized bio-materials in works including *Contestational Biology* (2001) and *Free Range Grain* (2003) with the stated goal of raising public awareness of the relationship between politics, industry, and ideology, particularly with respect to genetically modified foods, which have been legislated against in Europe but are widespread and unmarked in the US. To this end, the group employs tactics including political theater, performance, and installation, typically weaving ironic and farcical elements into their politically-charged work. In May 2004, on suspicion of bioterrorism, Kurz's home was raided by agents from the FBI and the Joint Terrorism Task Force, which discovered only harmless bacteria cultures for CAE's art project, *Marching Plague*. According to the artist, this work was intended to interrogate and demystify issues surrounding germ warfare. Charges of bioterrorism were dropped but as of July 2006 Kurtz remained on trial for mail and wire fraud and faced a maximum sentence of 20 years in prison in connection with the manner by which he obtained $256 innocuous samples of *serratia marcescens* and *bacillus atrophaeus* from his collaborator, University of Pittsburgh scientist Robert Ferrell.[40] The arts community rallied in support of Kurtz, creating the Critical Art Ensemble Defense Fund, which has raised funds to help defray the artist's legal expenses.[41]

The use of biotechnology as an artistic medium brings to the fore philosophical, ethical and religious implications pertaining to the practice of biotechnology itself. For artworks that are made using bio-materials are not simply representations of speculative visions of the future, but are actual physical embodiments that give artistic shape and form to the bio-technological present. Whereas the artist and artwork that represent the implications of biotechnology may remain one step removed from the technical and conceptual problems associated with the handling of moist-media or wet-ware, those that get their hands dirty, so to speak, by employing bio-materials are more fully implicated in the turmoil surrounding biotechnology. Needless to say, the stakes are much higher. Compared to an artist who creates representations of GMOs, an artist whose practice actually creates and incorporates them has much greater responsibility for the care and keeping of his/her artwork and much greater social responsibility for ensuring that the work does not affect the ecosystem adversely.

5.16 Altering Nature: Laboratories of Knowledge in Art and Science

Highly sophisticated visualization tools and techniques have become an integral part of the scientific laboratory and its attendant culture. State of the art images utilize Photoshop filters, multidirectional lighting effects, re-calibrated color contrasts and even post-production simulations. Perhaps more than any other contemporary technical method, the digital image has become the *lingua franca* of communicating

---

[40] Dr. Donald A. Henderson, Professor of Medicine and Public Health at University of Pittsburg, claimed that these bacteria are "totally innocuous organisms" and applauded Kurtz's efforts to raise awareness of the "risks and threats of biological weapons" (Coyne, 2004).

[41] See the website at http://caedefensefund.org for more details on the case and the Defense Fund's fundraising activities. Also see Beard (2004).

systems, including digital mammograms and in-situ sonograms. Add to the mix real-time transport via the World Wide Web, and images traverse the networks at incredible speed. Moving into the field of visual culture, scientific images, like popular culture icons before them, now inhabit the public landscape. This migratory practice of the visual has, for scholar W.J.T. Mitchell, created a "social field" of images underscoring a "pictorial turn across disciplines" (Mitchell, 2006). As artists engage with scientific iconography within the terms of critical aesthetic inquiry, scientists employ visual images as diagrams of understanding. While visual art often relies on art historical resemblance for its expressive models, picturing in scientific practice is more explicatively causal or mechanistically bounded. In short, coded images in full color regalia have become part of the corporate culture of science, new media installations and special effects in Hollywood film, to name a few.

Developments in "picture science" are also attracting scholars to this area. The interdisciplinary research group entitled "The World As Image" at the Berlin-Brandenburg Academy of Sciences and Humanities studies "visual representations of world concepts and the analysis of scientific representations and models" (see http://www/bbaw.de). The following topics are currently being analyzed: (1) The world as icon: the globalization of visual memory. (2) The world as model: the diagrammatic representation of nature. (3) The world as artifact: the visual arts and the life sciences. (4) The world as number: algorithmic representation between 0 or 1. The research group includes author Ingeborg Reichle, Steffen Siegel and Achim Spelten.

On the material plane, the laboratory too has been radically transformed into what Karin Knorr Cetina calls production sites:

[T]he search for entities consisting of barely more than genes, and the strict regimes of breeding, growing, maintaining and documenting point to a deeper transformation; the change no longer concerns the transition between nature and the laboratory, or between fuzzy holistic practices and strict, standardized routines, but the transformation of organisms into production sites and into molecular machines (Cetina, 2003, 138–158).

Working with cells, molecules, microorganisms and bioinformatics, the laboratory itself has become a computational site equipped with microscopes, measuring devices, and computers. Animal models such as mice, fish or higher mammals reside in separate headquarters known as "animal house." The sensual connections to odor, sound, and visualized behavior, so prominently removed from the researcher's lab, are palpable reminders that sentient life itself is a thermodynamic system moving within its own intrinsic processes of equilibrium. Whereas, the researcher's lab is an antiseptic stainless steel and glass beakered cell, animal house reeks of squeals and food products in both digested and non-digested forms.

Various scientific institutions, likewise, are initiating artists-in-residence programs. AIL, artistsinlabs (Kunstschaffende in Laboratorien) in Zurich has hosted many international programs in various laboratories around the world. Under the auspices of Jill Scott and Irene Hediger, artists spend nine months or so working alongside scientists (see www.artists-in-labs.ch). The University of Leiden, as well, is offering bio-residencies for artists.

Inquiry into the subject of altering nature through biotechnology calls attention to the work of German philosopher Nicole C. Karafyllis.[42] Coining the term *biofakte* in 2001 to address the ontological status of organisms that have been fabricated in the laboratory, this neologism fuses the meanings of artifact and living entities. What ethical concerns abound when life forms are produced and reproduced through laboratory techniques? To what natural order do these life forms belong? What separates these living entities from mere "things?" As sentient utilitarian animals, their habitat and life is limited the lab. From Onco mouse, to Rhino mouse, to goats that produce pharmaceuticals or silk, to a featherless chicken (Fig. 5.17) how do these altered species affect the ecology at large?

These ethical questions get especially dicey when human genes are inserted into animal hosts. Science journalist Rick Weiss asks "How human must a chimera be before stringent research rules kick in?" "Would it be unethical for a human embryo to begin its development in an animal's womb?" (Weiss, 2004). As transgenic ani-

Fig. 5.17 *Unititled (the Featherless Chicken)*, by Adi Nes, 2002; printed with permission from the Jack Shainman Gallery, New York

---

[42] For further discussion and descriptions (in German) of the *biofakte* see Reichle (2005) and Karafyllis (2003).

mals continue to be fabricated as "living test-tubes" they set into motion unique, and even controversial research projects. Rick Weiss reports on research scientist Evan Balaban's work:

> Balaban took small sections of brain from developing quails and transplanted them into the developing brains of chickens. The resulting chickens exhibited vocal trills and head bobs unique to quails, proving that the transplanted parts of the brain contained the neuronal circuitry for quail calls. It offered astonishing proof that complex behaviors could be transferred across species (Weiss, 2004).

In conclusion, we may ask what social implications does this form of knowledge precipitate? How will the new biotechnologies change the ways in which we live? As animals and plants continue to be fashioned, mixed and matched from disparate molecular data, what kinds of alternative conceptions of evolution and natural history become ethical or moral futuristic narratives? From ideologies of the "miracles of science" to the hyperbolic fears of "science out of control" our understanding of the nature of experimental systems and their influence on the social order continues to expand. Symbolic models of the real continue to exude and reframe the profound philosophical implications of altering nature in the 21st century.

References

Alexander, Brian (2003). *Rapture: How Biotechnology Became the New Religion*. New York: Basic Books.
Anker, Suzanne (1996). "Cellular Archeology," *Art Journal* (Spring).
Anker, Suzanne (1997). "From Genesis to Gene," paper delivered at ARC: The Society for the Arts, Religion and Contemporary Culture, NYC at the conference *The Artist in/and Community: Millennial Visions*.
Anker, Suzanne (2004). Excerpt from "Picture Perfect: From Golden Rules to Golden Boys"; keynote address, "The Image in Science", sponsored by the Freie University in Berlin at the Hamburger Bahnof, December 13.
Anker, Suzanne, and Dorothy Nelkin (2004). *The Molecular Gaze: Art in the Genetic Age*. New York: Cold Spring Harbor Laboratory Press.
Appelyard, Brian (1998). *Brave New Worlds*. New York: Viking.
Armenini, Giovam Battista (1989 [1587]). *De'veri precetti della pictura*, in David Freedberg (ed.), *The Power of Images: Studies in the History and Theory of Response*. Chicago, IL/London: The University of Chicago Press.
BBC News (2001). "Gallery Puts DNA in the Frame," September 19.
Beard, Mark (2004). "Twisted Tale of Art, Death, DNA," *Wired* (June).
Bowler, Peter J. (1986). *Theories of Human Evolution: A Century of Debate, 1844–1944*. Baltimore, MD/London: John Hopkins University Press.
Celent, Germano (2000). *Marc Quinn*. Catalogue published on occasion of the exhibition Marc Quinn at the Fondazione Prada in Milan, 5 May–10 June.
CNN (2004). "Is Anybody Out There?," Sunday, August 8, 8:00PM.
Cook, Gareth (2000). "Cross Hare: Hop and Glow," *The Boston Globe*, September 17, A01.
Coyne, Brendan (2004). "Anti-biotech Artist Indicted for Possessing 'Harmless Bacteria'," *The New Scientist*, July 6. http://newstandardnews.net (Accessed July 3, 2006).
de Duve, Christian (1984). *A Guided Tour of the Living Cell*. New York: Scientific American Books.

Eco, Umberto (1984). *Semiotics and the Philosophy of Language*. Bloomington, IN: Indiana University Press.
Edgerton, Samuel (1975). *The Renaissance Rediscovery of Linear Perspective*. New York: Basic Books.
Flusser, Villem (1988). "Curie's Children", *Art Forum* XXXVI, Nr.7, March 1988, p. 15; XXVI, Nr.10, Summer 1988, p. 18; and XXVII, Nr.2, October 1988, p. 2.
Frankel, Felice (2002). *Envisioning Science: The Design and Craft of the Science Image*. Cambridge, MA: MIT Press.
Freedberg, David (1989). *The Power of Images: Studies in the History and Theory of Response*. Chicago, IL/London: The University of Chicago Press.
Fry, Northrup (1982). *The Great Code*. New York: Harcourt.
Kac, Eduardo (website). "Genesis," http://www.ekac.org/geninfo.html.
Kac, Eduardo (website). "GFP Bunny," online publication: http://www.ekac.org/gfpbunny.html.
Kac, Eduardo (2004). Interview with Edward A. Shanken, November 26, Oak Park, IL.
Kac, Eduardo (2005). Telephone interview with Edward A. Shanken, February 27.
Karafyllis, Nicole C. (ed.) (2003). *Biofakte:Versuch uber Menschen zwischen Artefakt und Lebewesen*. Paderborn, Germany: Mentis.
Kay, Lily (2000). *Who Wrote The Book of Life: A History of the Genetic Code*. California: Stanford University Press.
Kemp, Martin (2000). *Visualizations: The Nature Book of Art and Science*. Berkeley/Los Angeles, CA: The University of California Press.
Kemp, Martin, and Marina Wallace (2000). *Spectacular Bodies: The Art and Science of the Human Body from Leonardo da Vinci to Now*. Berkeley, CA: University of California Press.
Kemper, Joanna (2004). "What a Headache." Ph.D. dissertation, University of Pennsylvania, Pennsylvania.
Kevles, Bettyann (1998). *Naked to the Bone: Medical Imaging in the Twentieth Century*. Reading, MA: Perseus Books Group.
Kevles, Daniel, and Leroy Hood (eds.) (1992). *The Code of Codes: Scientific and Social Issues in the Human Genome Project*. Cambridge: Harvard University Press.
Koerner, Lisbet (1999). *Linnaeus*. Cambridge, MA: Harvard University Press.
Lewontin, Richard (1992). "The Dream of the Human Genome," *New York Review of Books*, 39(10), May 28.
Marks, Jonathan (2002). *What It Means to be 98% Chimpanzee: Apes, People and their Genes*. Berkeley, CA: University of California Press.
Mitchell, Syne (2002). *Technogenesis*. London: Roc.
Mitchell, William J.T. (2006). *What Do Pictures Want? The Lives and Loves of Images*. Chicago, IL: University of Chicago Press.
Nelkin, Dorothy, and Susan Lindee (2004). *The DNA Mystique: The Gene as a Cultural Icon*. Maryland: University of Maryland Press (originally published in 1996.).
Noble, David F. (1997). *The Religion of Technology*. New York: Penguin Books.
Nye, Robert A. (1976). "Heredity or Milieu: The Foundations of Modern European Criminological Theory," *Isis* 67(3).
Ottinger, Didier (2004). "Eduardo Kac in Wonderland," in Stephen Berg (trans.), Eduardo Kac (ed.), *Rabbit Remix* (exhibition catalog). Rio de Janeiro, Brazil: Laura Marsiaj Arte Contemporânea.
Petrullo, Lynn A. (2000). "The Church of DNA." Paper delivered at CAA Conference, February 25.
Philipkoski, Kristen (2002). "RIP: Alba, the Glowing Bunny," *Wired News*, August 12. Available online: http://www.wired.com/news/medtech/0,1286,54399,00.html.
Princenthal, Nancy (2000). Review of "codeX: genome", exhibition at Universal Concepts Unlimited, NYC, September.
Reichle, Ingeborg (2001). "Kunst und Genetik." Zur Rezeption der Gentechnik in der zeitgenössischen Kunst. *Die Philosophin. Forum für feministische Theorie und Philosophie*, Heft 23, Jg. 12, Tübingen, S. 28–42.

Reichle, Ingeborg (2005). *Kunst Aus Dem Labor: Zum Verhaltnis von Kunst und Wissenschaft im Zeitalter der Technoscience*. New York/Wien: Springer.Reodica, Julia (2004). "Un/Clean: Visualizing Im/Purity in Art and Science," *Art and Biotechnologies*. Montreal, Canada: Presses de l'Universite.

Rojek, Chris (2001). *Celebrity*. London: Reaktion Books.

Serres, Michael (2004). *Orlan: Carnal Art*. Paris: Editions Flammerions.

Shanken, Edward (2004). "Art, Ethics, and Genetic Engineering: The Transgenic Art of Eduardo Kac," in Dorothy Nelkin and Suzanne Anker (eds.), *The Molecular Gaze: Art in the Genetic Age*. Cold Spring Harbor, NY: Cold Spring Harbor Laboratory Press.

Shea, William R. (ed.) (2000). *Science and the Visual Image in the Enlightenment*. Canton, MA: Science History Publications.

Shlain, Leonard (1999). *The Alphabet Versus the Goddess: The Conflict Between Word and Image*. New York: Penguin.

Stafford, Barbara Maria, and Frances Terpak (2001). *Devices of Wonder: From the World in A Box to Images on a Screen*. Los Angeles, CA: The Getty Research Institute.

Staniszewski, Mary Anne (1995). *Believing Is Seeing: Creating the Culture of Art*. New York: Penguin Books.

Stepan, Nancy (1986). "Race and Gender: The Role of Analogy in Science" *Isis* 77(2).

Sulston, John, and Georgia Ferry (2002). *The Common Thread: A Story of Science, Politics, Ethics and the Human Genome*. London: Bantam; National Portrait Gallery Press Release, 2001, "Mark Quinn and John Sulston Unveil Genomic Portrait".

Symbiotica (2004). "Semi-Living Artist Performs in Bilbao, Spain." April (Symbiotica press release). See http://www.symbiotica.uwa.edu.au/.

Van Dijck, Jose (1998). *Imagenation: Popular Images of Genetics*. New York: New York University Press.

Walby, Catherine (2000). *The Visible Human Project: Informic Bodies and Posthuman Medicine*. London/New York: Routledge.

Weiss, Jeffrey, and Carol Mancusi-Ungaro (2000). *Mark Rothko*. New Haven, CT: Yale University Press.

Index

A
Abrahamic traditions, 33
   Christianity, 40–49
   Islam, 49–58
   Judaism, 34–40
Adam and Eve, 147
Aesthetics, 107, 111, 295, 311, 315
African myths, for creation of natural world, 18
African religious traditions, issues in, 18
Age-retardation, 175, 176, 187
Aggadah, 34
Aging
   anti-aging regimens, 187
   Bacon's attempts to improvement, 179
   conceptions of, 137, 175
   Godwin, end of improvement of, 183
   hygienics as supplement to, 185
   relationship with disease, 186
   in western civilization, 176
Agricultural production, 150
Ahimsa, 28, 30
Alba, 307–311, 313
Altruism, 110, 202
American Academy of Anti-Aging Medicine, 186
American intellectual property laws, 225
Ananda, 23
Anarabdha-karma, 23
Anatma, 29
Androcentric social biology, 98
Animal-human relationship, 17
Animals as food, purity of, 147
Anitya, 29
Anti-aging regimens, 175, 186, 187
Antireductionism, 89, 90
Antiretroviral (ARV) medication, for HIV/AIDS, 239
Anti-teleological vitalism, 85
Apologism, 176, 180

Art and science laboratories, 308–313
   knowledge, for altering nature, 316–319
Artist's cyrogenic/aesthetic technique, 293
Assisted reproduction technologies (ART), 139
Association of Southeast Asian Nations (ASEAN), 259
Atman, 23, 28, 29
Attitudes and practices of societies, 14
Autopsies on criminals, 301
Autoradiograph, exemplifying Peirce's notion of indexical sign, 288–289
Autosomal recessive disorder, 231
Avidya, 22, 27
Aye (earth), 18, 19, 255
Ayurveda, 22, 27

B
Babylonian Talmud, 35
Baconian impulse in modern technology, 27
Bal tashchit, 39
Basing moral value, 86
Bayh-Dole Act, 209, 217, 218, 229
Behavioral endocrinology, 96
Bhagavad Gita, 24, 25, 83
Biodiversity, 31, 259, 261, 263
Bioethics, 46, 87, 89, 138, 200, 228, 239
   appeals to nature and the natural as normative in, 138–142
   biological reductionism and challenge of, 86–88
Bioethics commission in United States, 139
Biogerontology, 187
Biomedical sciences, 79
Biophilia, 107, 110, 124
Bio-portrait, 291

323

Biotechnology
  as artistic medium, 316
  developments, 214
    ethical judgments and warrants for evaluation, 122
    implications of Hindu beliefs for altering nature and, 28
  ethical dilemmas inherent in, 276
  innovation and, 214
  intervention
    innovative forms of, 276
    into living forms, 278
    philosophical, ethical and religious implications associated with, 316
    social and ethical issues related to, 315
  patenting and importance, 203, 229
  policy's ethical assessment, implications of appeals to nature in, 187
    altering nature, historical difficulties in, 192–193
    constructing historically informed appeals, 189–190
    enthusiasm and caution in attempts to alter nature, 191–192
    historical conceptions of nature, importance of, 192
    implications of policy, 193–194
    nature alteration and evolution, 190–191
    nature as normative, 188–189
    scope of nature, 187–188
  religious perspectives on, 243
Blasphemy, 35
Bodhisattva, 30, 31
Body (sarira), 26
Born criminal, 301
Brahman, 22–25
Brahmananda, 22
British philosophy of science and medicine, 164
British technological superiority, 160
Buddha's teachings, 28
Buddhist culture, of India and Tibet, 32
Buddhist ethics, 28
Buddhist philosophical systems, 32
Buffon's *Epoques*, 83

C

Canvas of Genesis 1, 36
Carak Samhita, 22
Cartesian cleavage, of mind and body, 23
Catholic faith and fundamental, 43
Catholicism, 48

Catholic moral theology, 41–43
  implications of, 44
Catholic sacramentalism, 45
Caveats, need of, 146
Cessation (*nirodha*), 29, 30
Chemical fertilizers and pesticides, 150
Chemical-physical system, 87
Chinese confucians, 102
Christian dogmatism, 180
Christian idea of redemption, 281
Christianity, 28
  Eastern orthodoxy, 47–49
  Protestantism, 44–47
  Roman Catholic moral theology, 40–44
Christ's death and resurrection, 282
Chromosome
  as heredity's ancient bio-marker, 285
  images in works of art and transferred to, 286–287
Classical genetics, 94
Classical medicine teaching, in early modern Europe, 144
Climatological theory of disease, 160, 161
Clinical Laboratories Improvement Act, 207
Cohen-Boyer patent, on recombinant DNA, 231
Colonialism, 302
Columbia River, 20
Colville peoples, 20
Commodification, patent-eligibility for, 227
Communalism, 19
Compassion (*karuna*), 28, 29, 31, 53, 230, 266
Computer modeling and mathematics, developments in, 96
Conjugal intimacy, 44
Conservation of force (*Kraft*), 87
Conservative Judaism, 35
Cooperative Research and Development Agreement (CRADA), 232
Cosmic reality, 23
Covenant love (*hesed*), 45
Creation of man, 36
Critical Art Ensemble (CAE), 315
Cross-species fertilization, 191
Cultural developments, 201
Cultural disposition, 18
Cultural lineage, 293
Cyrogenic/aesthetic technique, 293

D

Daoism, 106–107, 124
Darwinian discourse and theoretical framework, nineteenth century, 146

Darwinian evolution, 110
Darwinism, 83
Death of nature, 65
Decaying system, 81
Deep ecology, 106
Deforestation, 19, 159, 255
Deification (*theosis*), 48
Deism, 165, 167, 173, 174, 189
Deontology, 102, 103, 109, 111–114, 125
Designer Gene, 289
Despiritualization, 76
Dharma, 22, 28–30
Dichotomy
    between discovery and isolation, 212
    fact and value, 66, 115–122
    nature and, 100
    no longer provides any clear guidance, 186
    politically valenced conventional, 98
    replacing old nature-nurture dichotomy with interactive model, 99
Diet, integral part of preventive medicine, 144
Divided Purusha, 25
Divine Command Theory. *See* Deontology
Divine transcendence, 75
DNA Garden, 292
DNA molecule
    Cohen-Boyer patent, 231
    controversy over patenting of, 208, 211
    fingerprint technique, 289
    in generating images and sculptures, 291
    patenting of, 209, 219, 227
    sequences, 210, 228
    social, cultural, and aesthetic dimensions, 284
    special status of, 278–279
Dogmatism, 165
Down's syndrome, 306
Dualism, 69, 150, 281
Dynamic nature, conceptions of aging, 175. *See also* Science and medicine, appeals to nature
    ancient Greece and Rome, 176–178
    contemporary, 185–186
    enlightenment, 182–183
    medieval and early modern Europe, 178–180
    nineteenth and early twentieth centuries, 183–185
    scientific method origins, 180–182

E
Earth's fragile ecosystems, maintenance of, 21
East Asian notions of ethics for nature, 66
East Asian perspectives, 101. *See also* Modern moral philosophies
Eastern orthodoxy, 47
    apophatic and kataphatic theology, 49
    approaches dominating orthodox ethical reflection, 48
    human likeness to God, 49
Ecofeminism, 99
E. Coli, 303, 304
Ecosystem peoples, 14
Elitism, 315
Empiricism, 82, 88, 89, 91, 92, 94, 118, 300
Enlightenment, 23, 29, 67, 201, 202
Environment
    colonial changes, 159
    as natural and improved, 148–149
Environmental origins of food, 147
Epistemes notion, incommensurable, 67
Epistemological indeterminacy, 72
Epoques, 80
Eso (caution), 18
Ethical irrationality, 64
Ethical theory. *See also* Ethics
    in conceptions of human nature and, 112
    conceptions of nature in development of, 102
    consistent with established scientific views, 126
    intuitionism and, 102
    Lovelock's Gaia hypothesis., 111
    mutation with epoch of enlightenment, 67
    pluralistic approach and diversity of, 125
    types and role of nature in, 101
Ethics, 15, 107
    abyss between science and, 64
    ambiguity of evolutionary, 83
    autonomy of, 119
    of biotechnology and, 77
    Buddhist, 28, 29
    domains of applied, 102
    ecocentric environmental and evolutionary, 66
    for guide human beings to, 113
    Jewish, 35, 40
    Kantian-style, 104
    medical, 173
    modern, 66
    nature and, 141
    Protestant theological, 40
    realm of values and, 115
    religious and secular traditions of, 266
    science from, 122
    sexual, 42
    virtue, 65, 102, 105–110, 112–114

European agriculture, 160
European Group on Ethics, 212
European Patent Convention (EPC), 207, 211, 213
Evolution of species, by natural selection, 84
Exhibitions, 277, 284, 297, 306, 308, 314
Experimental philosophy and science, 154
Expressed sequence tags (ESTs), 209

F
Fabricated hymen, as soft sculpture, 294
Factory food conditions, 148
Farmers, 150
    in conservation and development of plant genetic resources, 265
    dailie diet into seasons, 153
    legal opposition against the patent, 262
    rights of local, 261
    value placed on, 151
Farming calendar, 152
Farming scientization, 150
Fatir, 52
Fatwas, 50
Featherless chicken, 318
Federal food, drug, and cosmetic act, 207
Feminism, 66, 99, 125
Fiqh, 52
Fitrat allah, 52
Food, pure or impure judgment, 144
Food purity, in early modern period, 144
Forests and woods, 158
Fossil earth, 82
Frankenstein factor, 139
Fresh-water fish and environment, 146
Fundamentalism, 97, 124. *See also* Reductionism
Fundamental laws of nature, 97

G
Gaia hypothesis, role in science, 122
Garden of Eden, 147
Gemara, 35
Gender, role in religion, scientific practice and, 293
Gene patenting, 230
Genesis, 303–306
Gene splicing and nature, 139
Genetically engineered bacteria, 303
Genetically modified (GM)
    crops, 143, 145, 146, 150, 158, 161–163
    farming, 149
    foods, 145, 148

Genetically modified (GM) technology
    alteration of environment, 159
    and colonization, 161
    latest step in scientization of farming, 150
    in lessen dependence on seasons and, 153
    in nineteenth century, 162
    and older cultural image of farming, 151
    as technical innovation, 152
Genetic Code Copyright Certificates, 279
Genetic codes, 306
Genetic engineering, 139, 248, 302, 308, 311, 312, 315
    for use as artistic medium, 315
Genetic essentialism, 280, 291
Genetic heritage, 293
Genetic manipulation, 109, 146, 148, 155–157, 162, 248, 294
Genetic portrait, 290
Genetic reductionism, 93
Genetic sciences, 278
Genetics to agriculture, 154
Genome's environment, 99
Gerontology, 186
GFP Bunny, 307, 309–313
GFP K9 project, 306–308
Globalization, 153, 161, 263, 317
Glyphs, 288
God and humanity, 24
    covenant between, 38
God and universe, 36
God-given natural law violation, 139
God's covenant with Noah, 37
God's revelation, 37
God's sovereignty, 39
Green fluorescent protein (GFP), 306, 308

H
Hadith, 50
Halakhic materials, 35
Halakhic prohibition, 40
Healing, 39, 40, 141, 170, 172, 256, 257, 262
Health and disease, value-laden concepts, 172
Health and fitness, marketing metaphors, 298–300
    comical cultural allusions in marketing campaigns, 299
    commercial images of body, 299
    imbuing technology with human characteristics, 298
    pharmaceuticals advertising and disease states portray, 300
Health and healing, 28
Healthy food, 147

Hebrew Bible, 34, 36, 243
Heteronomy, 119
Heterosexuality, 138
Hindu dualism, 23
Hinduism, schools of thought, 22
Hindu *mythos,* 27
Hindu system of medical practice, 27
Historicism, 42
Human aging, 176
Human behavior, 76, 110, 202
Human beings, first appearance of, 80
Human civilization, 80
Human cloning, 139
Human creation, 17
Human endeavors, aggressive, 73
Human genetic disorders, 231
Human genome project, 99
Humanity, kinship relationship, 18
Human-machine hybrids, 191
Human nature, 22, 28, 41, 54, 104, 107, 112, 126, 141, 169, 202
  outraged, 246
  and relation to law, 202
  shared, 266
Human-selected and created organisms, 146
Human stem cell research, 139
Human T-cell lymphotrophic viruses (HTLV), 262
Humean normativity of nature, 173
Hume's Law, 117, 119. *See also* Dichotomy
Hygiene, 177, 179, 182, 184, 247

I

Iatrogenic risks, 163
Idolatry, 35, 40
Ifarabale (composure), 18
Ignorance (avidya), 27
Illusion (*Maya*), 29
Images marketing, 296
Immortality, 8, 25, 175, 178, 179, 183–185, 279, 293
Immutable laws of seasonality, 153
Imo (knowledge), 18
Indian corn, 158, 161
Indian tradition, 20
Indian tribal religions, 17
Indigenous communities, 14
Indigenous peoples, 15
  implications of biotechnology for, 21
  major elements, 14
Indigenous religions, 13, 200. *See also* Indigenous peoples
  attitudes and approaches of, 252, 256–258

issues of definition, 252–256
manifestations of beliefs, 258–264
and spiritualities, 14
Indigenous rights and values, legal recognition of, 258–259
Industrialization, 281
Industrial revolution, 151, 152, 160
Inorganic nature, 92
Instrumentalism, 89
Intellectualization, 64
Intellectual property, 203, 211, 213, 222, 225
  commercialization of, 229
International labor organization (ILO), 259
Intuitionism, 102, 105, 109, 110, 112
Islam, for science and nature, 28
  deals with animals and plant life, prophetic traditions, 53
  Hadith, elements of faith and practice, 50
  juridical-theology and Mu'tazila, 52
  modernity and historicist interpretations, 55
  muslim ethical teachings and act as stewards (*khalifa*), 53
  notion of *taskhir,* 54
  objectives of *shari'a,* 51
  politics of science and impact on religious communities, 54
  to reconcile modern science with Muslim moral tradition, 57
  responses to understanding nature, 58
  rhetorical resistance to scientism, 56
  testing ground for humans, nature, 55
  ulama moral discourse, 56
  umma and diversity of Muslim viewpoints on nature, 51
  view on invasive surgeries for transplantation, 57
Islamic hegemony, 281

J

Jewish ethical method, 35
Jewish legal system, 34
Jewish thought, aspects of, 39–40
Jicarilla Apache, 16
Judaeo-Christian tradition of utilizing nature, 154
Judaism, 28, 34
  and health care, 40

K

Kac's genesis gene, 304
Kalam, 52

Kantianism, 114
Kantian-style ethics, 104
Karma, 23, 24, 28
Karmic effect, 28, 31
Karyotype, 286, 288
Kuhn and post-positivist philosophy, 94–95. *See also* Philosophy of science

L

Laboratory-based bacteriological revolution, 161
Laboratory-based products, 148
Laboratory-produced crops, 161
Lamarck, theory for nature, 83
Land ethic, 105, 106
Laws
  of karma, 23
  of nature, 203
  of special sciences, 97
  of Torah, 40
Lebensphilosophie, 114
Life, biophysical conceptions, 87
Life extension, 175, 176, 185, 187, 293
Life sciences and nature, 77, 79
Lockean labor theory, 222
Logical empiricism, 88, 91, 92, 94, 95, 114
Logical positivism, 66, 90, 91, 93, 94, 124

M

Magical-religious ritual, 280
Magisterium, role in catholic church, 43, 44
Mahayana, 30
Male social supremacy, 99
Mammalian reproduction, 140
Maori, native peoples of New Zealand, 20
Mathematical regularities, 70–71
MEART bio-robotic drawing system, 313–315
Mechanization and seventeenth century departure, 68
  contemporary reception, 72–75
  mechanization significance, 68–72
Medical vitalism, 85
Medicine, aims to correct nature's deficiencies, 168
Meliorism, 176, 178, 180, 181
Metaphysical reductionism, 96
Metaphysics, banishing of, 94
Methodological reductionism, 96
Microscopic visualizations, 278
Microscopy advancement, 91
Microvenus, 303
Midrash, 34
Mishnah, 34, 35

Mishneh Torah, codification, 35
Model law (ML), 260
Modern biotechnology, 87
Modern life science, 77
Modern moral philosophies, 101
  criticism, 108–111
  ethical theories, 103–108
Modern neo-Darwinism, 86
Modern West, scientific-technological juggernaut, 65
Modoc world creation, 17
Moksha, 23, 24
Molecular biology
  conflicting interpretations of, 280
  development of, 283–284
  as modernist science, 278
  social and cultural implications of, 311
  use in artistic media to modify living organisms, 313
Molecular genetics, 94
Molecular structure of DNA, 97
Molecule of immortality. *See* DNA molecule
Moral revulsion, 139
Mount Shasta, 17
Multi-electrode array, 314

N

Nana Onyame's (God's) relations, 18
National bioethics advisory commission (NBAC), 139
Native American conception of nature, 16
Natural environment, 148
  shaping, 157–161
Naturalism of moral reasoning, 140
Naturalistic fallacy, 86, 102, 115
Natural law
  concepts, 202, 203, 222
  theories, 123
Natural science
  accounts, 103
  and ethical prescriptions, 107
Natural selection theory, 84
Natural world, 33
Natura naturans, 82
Natura naturata, 82
Nature
  approaches to bring back, 122–125
  default characterization of, 94
  as dialectically-developing system, 82
  as intermediary creative agency, 79
  negatively valorized, 138
  as normative, 138, 188–189
  ways, for improvement, 188

Index

reborn, 77
   biological reductionism and challenge of bioethics, 86–88
   evolutionary ethics, 83–86
   nature vitalization, 81–83
   rebirth, 78–81
religious perspectives, 243
valorization, 175
   reason, for ethically normative, 190
vitalization, 80
Nature deficiencies, medical intervention, 163. *See also* Science and medicine, appeals to nature
   accounts of normativity, implications for, 171–174
   Baconian experience-based method, 164–167
   Gregory on medicine's role in, 167–169
   Hume on observable normative, 169–171
Naturphilosophie, 81, 82, 86, 124
Neovitalism, 88, 114
New science and Baconian program, 153–157
Newtonian microforces, 80
Newtonian ontology, 90
Nirvana, 28, 29
Nonviolent attitude, 30
Non-Western cultures, 73
Normative reward theory, 223
Normative theory of monopoly, 222
Noumena, 32
Nucleotide sequence onto slot machine, 290

O

Obatala and *Oduduwa* divinities, 18
Official catholic teaching, 44
Ogbon (wisdom), 18
Olodumare, Supreme God, 18
OncoMouse and Humouse, 233
Ontological reductionism, 96
Ontology and epistemology of mechanical philosophy, 71
Optics, 281
   role in Anker's sculptural installation, 287
Organicism, 91
Organic nature, 92, 188, 193
Origin of Species, 84
Orthodox Church, 47
Orthodox ethical reflection, in biotechnology, 48
Orthodox Judaism, 35
Orun (heaven), 19
Owo (respect), 18
Oye (understanding), 18

P

Pancamahabhutika, 22
Pancasila, 30
Pantheism, 124
Pan-vitalism, 85
Paradigms and revolutions notion, 67
Parental and transgenerational relationships, 140
Patent and Trademark Act Amendments, 217
Patent Misuse Reform Act, 204
Patents. *See also* Protestantism; US patent system
   under Article 53(a), 214
   case studies, 233
      breast cancer, 238
      canavan disease, 235–237
      HIV/AIDS, 238–243
      OncoMouse and Humouse, 233–235
   for clinical HFE testing, 231
   for the coding sequence of erythropoietin, 213
   controversy over DNA patenting, 208–214
   Diamond vs. Chakrabarty, 205
   disputes over validity of, 206
   economic considerations
      and biotechnology patents, 214–219
      and DNA patenting, 219–220
   for higher organisms, 228
   incentive and reward features of, 223
   infringement suits, 204
   legal and economic understandings, 220
   legal standards for, 203
   liability for infringement, 232
   Madey vs. Duke University, 207
   for methods of creating transgenic animals, 214, 234
   nature, property and, 220–227
      contemporary critics of DNA patenting, 227
   protection and rate of innovation, 216
   for recombinant DNA molecule and tPA and cDNA vectors, 213
   scope of claims and adequacy of disclosure, 206
   standards related to issuance and infringement of, 205
Perspectiva, 282
Pharmaceutical companies, of nineteenth and twentieth centuries, 156
Philosophical reflections on nature and contemporary biotechnology, 77
Philosophy of Enlightenment, 202
Philosophy of science
   concepts of nature in twentieth-century, 89

Philosophy of science (*cont.*)
   derivation, 78
   feminist, 98–100
   hospitability to multiple conceptions of nature's complexity, 124
   and medicine, 165
   new directions in, 94
   series of challengers within twentieth-century, 66
Physical earth creation, 16–17
Physical science contribution, 94
Physiognomy, 301
Pictorializations, 285, 289, 296, 298, 302
Pictorial signs, of social rank, 300–302
Picturing DNA, avatar of 20th century molecules, 284
Pig as food, 145
Pig wings project, 314
Plastic surgeries, 294
Pluralism, 44, 45, 47, 66, 97, 124
Policy science, 201
Politically valenced conventional dichotomy, 98
Political movement, for women's liberation, 98
Portrait of Jeff Koons, 290
Positivism, 65
Postage stamps for popular culture, 277
Pragmatic holism, 120
Pragmatism, 118, 125
Prajna (wisdom), 28, 30
Prarabdha-karma, 23
Pratityasamutpada, 29
Pre-invasion and pre-colonial societies, 14
Premodern alchemy, 74
Preventive medicine, 144, 147
Primitive peoples, 15
Prolongevitism, 183
   culmination of eighteenth-century, 182
   in Western civilization, 178
Property rights, 221
Protein complexes, for gene activation, 96
Protestantism, 41, 43–45, 47, 243
   creation and redemption, doctrine of, 46
   ethical reflection and bioethics, 46
   orders of nature, 45
   and patenting, 248–252
   positive Christian responsibility under rubric of co-creation, 47
   and property, 243–248
   reasons for Protestant pluralism, 45, 46
   reflection about morality, 44
   unevangelical legalism and sacramentalism, 45

Protestant pluralism, 45
Public health service act, 207
Pure consciousness *(chit),* 23
Purism, 119
Putrefaction, 144, 145, 184

Q
Quinine, 161
Quinn, Marc, 291
Qur'an, 50

R
Rabbinic Judaism, 37
Racial superiority, 160
Ramanuja, 26, 27
Rang bzhin, 32
Rationalization, 64
Rational science and technology, 81
Realism, 89
Reciprocal altruism, 111
Reductionism, 86, 87, 89, 90
   associated with positivists, 95
   conventional notions of, 280
   genetic, 93
   important to avoid, 250
   metaphysical and ontological, 96, 98
Religion and science
   construct and repair the host organism, cell cycle, 282–283
   intersections and comparisons, 282
Religions, of Asian origin, 21
   Buddhism, 28–33
      four noble truths, 29
   Hinduism, 22–28
Religious and spiritual practice, 280
   role of gender in, 293
Responsa, 35, 40
Rhino, 310
Rig Veda, 22, 25
Roman Catholicism, 43
Roman Catholic teaching on reproductive technologies, 44
Roman stoicism, 79
Romanticism, 79, 82, 88, 122
Rural Advancement Foundation International (RAFI), 262

S
Sabbath, 39
Sakyamuni Buddha, 28
Samadhi (concentration), 29

Samsara, 23, 32
Sanatana dharma, 22
Sanctita, 23
Sangha, 30
Sanskrit, 32
Sarnath, 28
Sattelzeit notion, 67
Scholasticism, 41, 78, 79
Science. *See also* Science and medicine, appeals to nature
 advertising images, 296
 communication, language of mathematics, 65
 with the crafts, union of, 153–157
 and ethics, abyss between, 64
 limits of, 124
Science and medicine, appeals to nature, 142
 conception of aging, dynamic nature, 175–187
 food, animals, plants and environment in early modern period, 143–163
 medical intervention, 163–174
Scientific and medical advertising, 299
Scientific iconography, 276
Scientific realism, 89
Scientific reductionism, 27
Scientific revolution, 64–67, 73, 77, 102, 180, 188
Scientism, 56, 65, 117, 119
Scottish-enlightenment, 167, 173
Scripturalism, 52
Scripture in moral reflection, 43
Secularization, 55, 281
Semi-living artist, 313
Senescence, 175, 177, 187
Seventeenth-century chemists, 156
Sila (morality), 28, 30
Sixteenth and seventeenth centuries monstrosities, 146
Sixteenth-century English agriculture, changes, improvement, and stability, 151–153
Sixteenth-century farming, 149–151
Skin color and concepts, of biotechnology, 302–303
Slime mold, role of environment and diffusion gradients, 99
Small Reliquaries, 294–295
Social and cultural implications of biotechnology, GMOs contribution, 308
Social rank, pictorial signs of, 300–302
 empirical science *vs.* visual art, 300
 identification criminal types and explain cultural differences, 302
 Linnaean classification system, 301
 Physiognomy and autopsies on criminals, 301
Soul-body relationship, 26
Soul (saririn), 23, 26, 27, 73, 177, 257
Speciesism, 103
Spirituality, 14, 38, 47
Spiritual power, 16, 254
Spiritual science, 27
Stevenson-Wydler technology innovation act, 217
Subjectivism, 142
sub-Saharan Africa, deforestation and depletion of resources, 19
Suffering (*dukkha*), 29, 33, 169, 187, 212, 251, 255, 266
Sunna, 50
Supernatural, 22, 25, 41, 202
 power, religion as, 278
Supernature, 193
Sustainable development, of natural resources, 261
Suuru (patience), 18
Sympathy, 164, 167–171, 174, 245, 266

T

Taittiriya Upanishad, 25
Talmud, 35
Tay-Sachs disease, 238
Technical skill and labor, in sixteenth-century agriculture, 151
Technogenesis, 275
Teleology, 92, 140, 189
 value-laden teleologies of human nature, 141
Temiar people of Malaysia, 19
Theory of justice, 202
Theravada Buddhism, 30, 31
Thirst (*taha*), 29
Thousand-headed Purusha, 25
Thunderbolt, to penetrate universe, 18
Tibetan Buddhist tradition, 32
Tikkun olam, 40
Tissue Culture and Art (TCandA) project, 313
Tissue culture technologies, 293, 315
Tissue engineering, 294
 use in artistic media, 313, 315
Tobacco, 158, 161, 185
Trade related aspects of intellectual property rights (TRIPS), 211, 241, 259
Traditional cultural products, 150

Transcendent creation, 69
Transformation, laboratory iconography into color field painting, 289
Treatment action campaign (TAC), 240
Tribal peoples, 15
Tribal religion, task of, 17
Tutelage of nature, 76
Twentieth-century philosophy of science, for nature, 89
   logical positivism, 90–94
   new directions, 94–100
Tzedakah, 40

U
Unevangelical legalism, 45
Universal vitalism of Leibniz, 173
Unprecedented cultural investment, 277
Unseen Divine, 25
Upanishad, 22, 24–26
US patent and trademark office (USPTO), 262
US patent law, 226
US patent system, 203, 224
Utilitarianism, 102, 103, 108–110, 112–114, 125

V
Vajrayana, 30, 31
Variegated and copious assortment of flowers, 292–293
Vartamana-karma, 23
Vatican council, 43
Vatican II, natural law, 42
Vedas, 22, 24
Vestiges of the Natural History of Creation, 83
Vinaya Pitaka, 30
Vishistadvaita, 27
   God and humans in, 26

Visualization
   accessible through probes, arrays and scans, 299
   as global force employed to mediate ideology, 302
   in journals, textbooks or from the world-wide-web, 284
   role of, 278
   scientific apparatus for sale, 296
   for the scientist, 285
   secular visualization of divine, 278
   signification processes inherent in symbolic order, 280
   tools and techniques as integral part of, 316
Vitalism, 85, 89, 91, 101, 123, 173

W
Wanton destruction, prohibition of, 39
Western cultures, 65, 73, 117, 281
White River Sioux creation, 17
Women and nature, 73
Women in cultural settings, 42
World conservation union, 21
World religions, 13
World trade organization (WTO), 259

Y
Yanomami peoples, survival of, 20
Yekuana people of Venezuela, 20
Yoruba society, 18
Yuk factor, for biotechnological changes, 123

Z
Zoosemiotics, 287